ELECTROMAGNETIC
WAVE
THEORY

Electromagnetic Wave Theory

JAMES R. WAIT
The University of Arizona

1817

HARPER & ROW, PUBLISHERS, New York
Cambridge, Philadelphia, San Francisco,
London, Mexico City, São Paulo, Singapore, Sydney

Sponsoring Editor: John Willig
Project Editor: Eleanor Castellano
Cover Design: Michel Craig
Cover Illustration: Dana Ventura
Text Art: Vantage Art, Inc.
Production: Delia Tedoff
Compositor: Progressive Typographers, Inc.
Printer and Binder: Maple Press

ELECTROMAGNETIC WAVE THEORY

Library of Congress Cataloging in Publication Data

Wait, James R.
  Electromagnetic wave theory.

  Includes bibliographical references and index.
    1. Electromagnetic theory. I. Title.
QC670.W24 1985   530.1′41   84-6531
ISBN 0-06-046877-7

86 87 9 8 7 6 5 4 3

Contents

Preface

Electromagnetic Wave Theory is designed for an advanced course in electromagnetic waves at the senior undergraduate or first-year graduate level in an electrical engineering curriculum. It is assumed that an individual reading this book will have had at least one course in electromagnetics with knowledge of static fields, as well as an introduction to Maxwell's equations. Additionally, this book should appeal to practicing engineers as a useful reference. An attempt is made to present the subject in a self-contained format. The essential mathematics, such as vector analysis and special functions, are reviewed in a concise manner.

A central theme of the book is the impedance concept applied to the description of wave phenomena. This point of view has its origin in the profound ideas of Sergei Schelkunoff and Henry Booker in the 1940s. This approach is very fruitful in systematizing the solutions of a wide class of boundary value problems. We do not try to be comprehensive, but a representative selection of topics that yields to this approach is considered. In particular, problems of current relevance to radio wave propagation and antennas are emphasized. Some of these are drawn from the author's own research, but an attempt has been made to put the results in a general context.

Taking a somewhat traditional viewpoint, the book begins with a review of stationary fields. The concept of potential is important if one is to really appreciate the meaning of static-like fields. It is not sufficient just to let the frequency tend to zero in dynamic formulations. Also, the boundary value problems for stationary fields can be best illustrated when the fields can be derived from a scalar potential. This sets the scene for later vector formulations under truly dynamic conditions.

A somewhat novel departure from many texts on EM theory is that we develop many of the essential ideas in wave propagation by treating purely scalar problems in one, two, and three dimensions. The loss in generality is later recovered when we address some of the same problems when vector solutions of Maxwell's equations are required.

Not only is the impedance concept expounded, but transmission-line theory is used consistently to formulate boundary value problems even for fairly complicated geometries. Also, many of the wave theory results are interpreted by equivalent transmission line circuits. We feel this helps the reader see the essential unity of the subject.

Exercises for the reader are located at various spots in the text. These are phrased in such a manner that the desired results of the derivation are given. In some cases these exercises allow the reader the opportunity to test his understanding by working out a closely related problem.

A few bits of information should stand as guides for easier reading. The rationalized meter-kilogram-second (MKS) system of units is used consistently. This is now referred to as the System Internationale (S.I.). Equation numbers are in parentheses and cited as such in each chapter. Thus, for example, a reference to (9) in Chapter II refers to equation number 9 in this chapter. In a few cases where references are made in Chapter II to equations in other chapters, an explicit statement is made accordingly. References to the literature are listed at the end of each chapter. They are cited in the text by a number in square brackets. Thus, for example, [10] in Chapter II is the tenth reference cited. This is essentially the IEEE system. Appendixes to individual chapters are identified by lower case letters and cited as such in the foregoing chapter.

I am extremely grateful to Professor E. Bahar, Professor C. Balanis, and the other reviewers for their useful comments and critical remarks concerning the presentation of the derivations. Also, I wish to thank my colleagues at the University of Arizona and the University of Colorado for their support and encouragement over the years. In particular, I would like to mention Professors R. H. Mattson, D. G. Dudley, T. Triffet, and A. Q. Howard in Tucson and Professors D. C. Chang and S. W. Maley in Boulder. Some of the chapters were typed by Joann Main and Robin Voustas in Tucson, for which I am very appreciative.

JAMES R. WAIT

ELECTROMAGNETIC
WAVE
THEORY

Chapter 1
Electrostatics and
Magnetostatics

1.1 COULOMB'S LAW

We begin with a review of electrostatic field theory. This review serves as a useful background for treating more realistic and useful time-varying electromagnetic fields.

Electrostatics is founded on the basis that certain fundamental relations exist or characterize the field behavior. These relations have their origin primarily in experimental observations. For example, *Coulomb's law* is a statement that a force f exists between two point charges q_1 and q_2 that are separated by a distance r. This force is found experimentally to obey an inverse square law according to

$$f = \text{constant} \times \frac{q_1 q_2}{r^2} \tag{1.1}$$

In the rationalized MKS (meter-kilogram-second) system of units this constant is $(4\pi\epsilon)^{-1}$, where ϵ is the permittivity of the dielectric that contains these charges. In this same system of units, which has been accepted internationally, q_1 and q_2 are expressed in coulombs, f is in newtons, and ϵ is in farads per meter. Then for free space $\epsilon = \epsilon_0 \simeq 8.86 \times 10^{-12} \simeq 1/36\pi \times 10^{-9}$ F/m.

To be explicit, Coulomb's law in MKS units is

$$F = \frac{q_1 q_2}{4\pi\epsilon r^2} \tag{1.2}$$

where the direction of the force is along the line joining the two charges. As indicated, f is an attractive force when q_1 and q_2 are of opposite sign (for example, an electron and a proton), but the force is repulsive when q_1 and q_2 have the same sign.

If we think of q_1 as being a source point charge, then the force acting on test charge q_2 is

$$f = Eq_2 \tag{1.3}$$

where

$$E = \frac{q_1}{4\pi\epsilon r^2} \tag{1.4}$$

is, by definition, the electric field strength at the point charge q_2. This value of E does not depend on the test charge, and, in fact, we may allow q_2 to approach zero in the limit.

To indicate the vector nature of the electric field strength $\mathbf{E}$, we are led to write

$$\mathbf{E} = \frac{q_1}{4\pi\epsilon r^2} \mathbf{r} \tag{1.5}$$

where r is the unit vector in the r or radial direction from the charge q_1 to the point where the field is to be observed. In what follows we drop the subscript 1 on the charge q_1; this procedure should not cause any confusion.

It is evident that $\mathbf{E}$ associated with a charge q depends on the permittivity ϵ of the medium. Clearly, in the present context we can define a vector flux density $\mathbf{D}$ associated with a charge q according to

$$\mathbf{D} = \frac{q}{4\pi r^2} \mathbf{r} \tag{1.6}$$

where we may note that $4\pi r^2$ is the area of the enclosed spherical surface of radius r. Here the vector flux density or vector displacement has units of coulombs per square meter. Furthermore, $\mathbf{D}$ does not depend on the permittivity ϵ of the surrounding medium.

By comparing Equations (1.5) and (1.6), we observe that

$$\mathbf{D} = \epsilon\mathbf{E} \tag{1.7}$$

Actually, this result holds for all isotropic media even when the permittivity ϵ is dependent on the coordinates (that is, the medium is inhomogeneous).

When the medium is anisotropic, we generalize (1.7) to the form

$$\mathbf{D} = [\epsilon]\mathbf{E} \tag{1.8}$$

which, in fact, is a statement that

$$D_x = \epsilon_{11}E_x + \epsilon_{12}E_y + \epsilon_{13}E_z \tag{1.9}$$
$$D_y = \epsilon_{21}E_x + \epsilon_{22}E_y + \epsilon_{23}E_z \tag{1.10}$$
$$D_z = \epsilon_{31}E_x + \epsilon_{32}E_y + \epsilon_{33}E_z \tag{1.11}$$

In the limiting case where

$$\epsilon_{11} = \epsilon_{22} = \epsilon_{33} = \epsilon$$

and

$$\epsilon_{ij} = 0 \qquad \text{for } i \neq j$$

the medium becomes isotropic. We defer here any further reference to anisotropic media.

1.2 GAUSS' LAW

We now deal with *Gauss' law* by referring to the situation shown in Figure 1.1. As we see, the displacement or flux density vector $\mathbf{D}$ emanates radially from the charge q, and it subtends a local angle θ with the normal unit vector $\mathbf{n}$ to a closed surface S. The flux through the element da in the surface is

$$d\Psi = D \, da \, \cos \theta$$

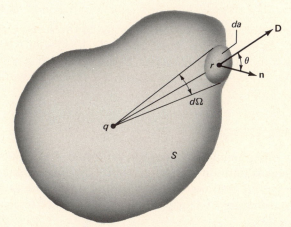

Figure 1.1   Point charge q enclosed by closed surface S and elemental area da of surface.

But by definition, the element of solid angle subtended by da at q is

$$d\Omega = \frac{da \cos \theta}{r^2} \tag{1.12}$$

Thus

$$d\Psi = Dr^2 \, d\Omega \tag{1.13}$$

But

$$D = \frac{q}{4\pi r^2} \tag{1.14}$$

so that we have the incredibly simple result that

$$d\Psi = q \, \frac{d\Omega}{4\pi} \tag{1.15}$$

If we now integrate over all solid angles, noting that

$$\oint d\Omega = 4\pi \tag{1.16}$$

we deduce that

$$\Psi = q \tag{1.17}$$

This equation is a statement of Gauss' law, which tells us that the total displacement or flux through any closed surface is equal to the amount of charge enclosed.

A simple extension of Equation (1.17), written in vector form, is

$$\oint_S \mathbf{D} \cdot \mathbf{da} = \int_V \rho \, dV \tag{1.18}$$

where $\mathbf{da}$ is a vector aligned with the normal to the surface with area da (note that $\mathbf{D} \cdot \mathbf{da} = D \, da \cos \theta$); on the right-hand side ρ is the charge density in coulombs per cubic meter. Equation (1.18) is a statement of Gauss' law. An alternative form is simply $\nabla \cdot \mathbf{D} = \rho$. Or in the case of a homogeneous medium, $\nabla \cdot \mathbf{E} = \rho/\epsilon$.

1.3 POTENTIAL CONCEPT

We may now introduce the concept of potential V at a point resulting from a charge q in coulombs. The potential V can be defined as the work done to move a unit test charge from infinity up to a point a distance r from the source charge q. Clearly, the work is obtained from the relation

$$\text{Work} = -\int_\infty^r E_r \, dr \tag{1.19}$$

where E_r is the force in the radial direction, given by

$$E_r = \frac{q}{4\pi\epsilon r^2} \tag{1.20}$$

Thus on integrating (1.20), we see that

$$V = \frac{q}{4\pi\epsilon r} \tag{1.21}$$

It is the so-called conservative property of static fields that the scalar quantity V is the same for any path drawn from infinity up to the field point at a radial distance r from q.

The relationship of electric field and potential follows quite easily. Here we might consider two points separated by a vector distance **ds** where the electric field **E** is measured. Now the work done dV in moving a unit charge through this infinitesimal distance is clearly

$$dV = -\mathbf{E} \cdot \mathbf{ds} \tag{1.22}$$

But we can also write

$$dV = \frac{\partial V}{\partial x} dx + \frac{\partial V}{\partial y} dy + \frac{\partial V}{\partial z} dz = \nabla V \cdot \mathbf{ds} \tag{1.23}$$

where

$$\nabla V = \mathbf{i}_x \frac{\partial V}{\partial x} + \mathbf{i}_y \frac{\partial V}{\partial y} + \mathbf{i}_z \frac{\partial V}{\partial z} \tag{1.24}$$

and

$$\mathbf{ds} = \mathbf{i}_x \, dx + \mathbf{i}_y \, dy + \mathbf{i}_z \, dz \tag{1.25}$$

Here, of course, $\mathbf{i}_x$, $\mathbf{i}_y$, and $\mathbf{i}_z$ are unit vectors in the x, y, and z directions, respectively. Now Equations (1.22)–(1.25) tell us that for conservative or static fields

$$\mathbf{E} = -\nabla V \tag{1.26}$$

where ∇ is the gradient operator defined by (1.24). An equivalent statement, preferred by this writer, is

$$\mathbf{E} = -\text{grad } V \tag{1.27}$$

where grad is the abbreviation for gradient. Of course, (1.26) and (1.27) hold in any orthogonal coordinate system.

1.4 DIPOLE CONCEPT

Another important concept is the dipole. To illustrate, we first consider two charges $+q$ and $-q$ separated by a distance ℓ, as indicated in Figure

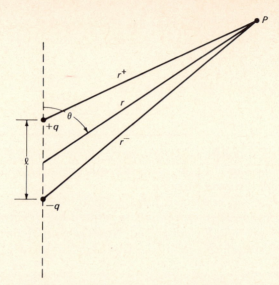

Figure 1.2 Geometry for calculating potential of two charges of equal and opposite sign.

1.2. The distance from $+q$ to the field point P is r^+, while the distance from $-q$ to the field point is r^-. The *resultant* potential at P is given by

$$V = \frac{1}{4\pi\epsilon}\left(\frac{q}{r^+} - \frac{q}{r^-}\right) \tag{1.28}$$

where ϵ is the permittivity of the homogeneous host medium. Now we will consider the case where $r \gg \ell$, whence

$$r^+ \simeq r - \left(\frac{\ell}{2}\right)\cos\theta \tag{1.29}$$

and

$$r^- \simeq r + \left(\frac{\ell}{2}\right)\cos\theta \tag{1.30}$$

where θ is the angle subtended by the line drawn from the center point of the charges to the field point and the vertical axis through $+q$ and $-q$.

Exercise: Derive an explicit expression for V, valid if r and ℓ are arbitrary, in terms of r and θ.

From (1.29) and (1.30) it follows that (1.28) simplifies in the manner

$$V \simeq \frac{q}{4\pi\epsilon}\left[\frac{1}{r - (\ell/2)\cos\theta} - \frac{1}{r + (\ell/2)\cos\theta}\right]$$

$$\simeq \frac{q\ell}{4\pi\epsilon r^2}\cos\theta \tag{1.31}$$

Now bearing in mind that symmetry about the polar axis prevails, it follows that in spherical coordinates (r, θ, ϕ)

$$E_r = -\frac{\partial V}{\partial r} = \frac{q\ell}{2\pi\epsilon r^3}\cos\theta \tag{1.32}$$

$$E_\Theta = -\frac{1}{r}\frac{\partial V}{\partial \theta} = \frac{q\ell}{4\pi\epsilon r^3}\sin\theta \tag{1.33}$$

$$E_\phi = 0$$

As perceptive readers will note, the potential expression in (1.31) and the field expressions in (1.32) and (1.33) are only valid if $\ell \ll r$. But they might also assert that these expressions are valid for all nonzero values of r if ℓ is infinitesimal or effectively so. In this case we should replace ℓ by, say, the differential ds; this replacement will be done in any further discussion of dipole fields. Incidentally, the term *dipole* should be reserved to describe the infinitesimal element and not be used in the context of linear antennas of finite length.

1.5 CHARGED LINE SOURCE

Another useful model is a uniform line charge. Such a configuration has an obvious cylindrical symmetry. Thus with reference to the cylindrical coordinates (ρ, ϕ, z), we locate the line charge along the z axis, and it extends from $z = -\infty$ to $+\infty$ with a uniform charge $\hat{q}$ C/m. Now it is not difficult to see that the flux density vector has only a radial ρ component, and it is given by

$$D_\rho = \frac{\hat{q}}{2\pi\rho} \quad \text{C/m}^2 \tag{1.34}$$

The corresponding electric field component is then

$$E_\rho = \frac{D_\rho}{\epsilon} = \frac{\hat{q}}{2\pi\epsilon\rho} \quad \text{V/m} \tag{1.35}$$

Now the potential V at the radial distance ρ is related to E_ρ by

$$-\frac{\partial V}{\partial \rho} = E_\rho \tag{1.36}$$

This equation tells us that

$$V = -\int^\rho E_\rho \, d\rho \tag{1.37}$$

where we have an indefinite integral on the right. On using (1.35), we see that

$$V = -\frac{\hat{q}}{2\pi\epsilon}\ln\rho + \text{constant} \tag{1.38}$$

where, again, it is understood that the surrounding medium of permittivity ϵ is homogeneous and isotropic.

A simple application of the preceding results tells us that the difference of potential ΔV, between the surfaces $\rho = a_1$ and $\rho = a_2$ for the line charge $\hat{q}$ at $\rho = 0$, is given by

$$\Delta V = \frac{\hat{q}}{2\pi\epsilon} \ln \frac{a_2}{a_1} \tag{1.39}$$

A slightly different formulation of the above problem would be to regard the surface $\rho = a$ as a perfectly conducting equipotential surface. Again we can let $\hat{q}$ be the total charge per unit length. Then (1.34) and (1.35) still apply for the region $\rho > a_1$. Now we also regard $\rho = a_2$ as a perfectly conducting surface. The difference of potential between these two concentric equipotential surfaces is given precisely by (1.39). The latter can be written in the equivalent form

$$\Delta V = \frac{\hat{q}}{C} \tag{1.40}$$

where

$$C = \frac{2\pi\epsilon}{\ln(a_2/a_1)} \tag{1.41}$$

by definition, is the capacitance per unit length. In other words, if we "apply" a voltage ΔV between these surfaces that are initially uncharged, we find that the inner surface has a total charge $\hat{q}$ per unit length given by

$$\hat{q} = C \, \Delta V$$

Then because total charge must be conserved, the charge on the outer concentric cylindrical is $-\hat{q}$ C/m.

1.6 PARALLEL CYLINDRICAL CONDUCTORS

Another cylindrical problem is to determine the capacity between two cylindrical conductors that are not concentric. One simple way to deal with this situation is to begin with the potential expression for two parallel line charges of strengths $\hat{q}$ and $-\hat{q}$, respectively. These charges are illustrated in Figure 1.3, where the line charges are located at points $(+)$ and $(-)$ separated by a distance d. The resultant potential at P with rectangular coordinates (x, y, z) is clearly

$$V = \frac{\hat{q}}{2\pi\epsilon} \ln \frac{r^-}{r^+} \tag{1.42}$$

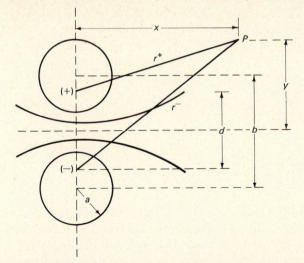

Figure 1.3 Geometry for calculating capacity of two parallel cylindrical conductors.

where

$$r^+ = \left[x^2 + \left(y - \frac{d}{2} \right)^2 \right]^{1/2}$$

and

$$r^- = \left[x^2 + \left(y + \frac{d}{2} \right)^2 \right]^{1/2}$$

Thus a family of equipotential surfaces are generated by setting

$$\frac{r^-}{r^+} = \text{constant} = K$$

This equation is equivalent to

$$K^2 \left[x^2 + \left(y - \frac{d}{2} \right)^2 \right] - \left[x^2 + \left(y + \frac{d}{2} \right)^2 \right] = 0 \qquad (1.43)$$

which, in turn, can be rewritten in the form

$$x^2 + \left(y - \frac{\alpha d}{2} \right)^2 - (\alpha^2 - 1) \left(\frac{d}{2} \right)^2 = 0 \qquad (1.44)$$

where

$$\alpha = \frac{K^2 + 1}{K^2 - 1} \qquad (1.45)$$

Of course, (1.44) is the same as

$$x^2 + \left(y - \frac{b}{2}\right)^2 - a^2 = 0 \tag{1.46}$$

where

$$b = \frac{K^2 + 1}{K^2 - 1} d \tag{1.47}$$

and

$$a = \frac{K}{K^2 - 1} d \tag{1.48}$$

Thus (1.44) or (1.46) describe a family of equipotential surfaces that have constant radii of curvature a, centered at $y = b/2$. We can now modify our problem and say that we are dealing with two parallel cylindrical conductors of radius a and separated by a distance b between centers. The total charge on the top conductor is $+\hat{q}$ C/m, while on the bottom conductor it is $-\hat{q}$ C/m. From (1.47) and (1.48) we deduce that

$$K = \frac{b + (b^2 - 4a^2)^{1/2}}{2a} \tag{1.49}$$

Now the potential V on the surface of the upper cylinder is

$$V = \frac{\hat{q}}{2\pi\epsilon} \ln K \tag{1.50}$$

which is relative to the equipotential plane $y = 0$. Clearly, the potential on the lower cylinder, also of radius a located at $y = -b/2$, is equal and opposite to (1.50). Thus the capacity between the two cylinders is obtained from

$$C = \frac{\hat{q}}{2V} = \frac{\pi\epsilon}{\ln\{[b + (b^2 - 4a^2)^{1/2}]/2a\}} \qquad \text{F/m} \tag{1.51}$$

Now if $b \gg a$, we have simply

$$C \simeq \frac{\pi\epsilon}{\ln(b/a)} \tag{1.52}$$

which is the value we would deduce from (1.38) on the assumption that the charges $+\hat{q}$ and $-\hat{q}$ are concentrated along the two axes at $y = +b/2$ and $-b/2$.

1.7 IMAGE CONCEPT

In many problems of electromagnetism the fields of sources in the presence of disturbing bodies can be estimated by introducing images of

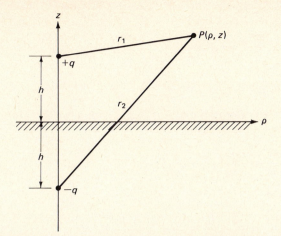

Figure 1.4   Negative image of point charge located over plane conductor.

the sources. This useful concept will be introduced by several simple examples.

In the first example we locate a point charge q over a plane surface that exhibits the property of being a perfect conductor. As indicated in Figure 1.4, we choose cylindrical coordinates (ρ, ϕ, z) with the perfect conductor located at $z \leq 0$ and the point charge q at $\rho = 0$ and $z = h$. Now we wish to deduce the potential function $V(\rho, \phi, z)$ at the point $P(\rho, \phi, z)$ everywhere for $z > 0$, which is a region of constant permittivity ϵ. First of all, we note that the potential must be symmetrical about the axis $\rho = 0$ so that there is no dependence on ϕ; thus we are dealing with a two-dimensional problem. Therefore we drop ϕ as an independent variable.

A perfect conductor in the context of electrostatics is at a constant potential. Thus the difference of potential between any two points on the surface must be zero. An equivalent statement is that the tangential electric field is zero.

The key observation that we now make is to note that the "primary" electric field from q is radially oriented. It would have a nonzero component E_ρ along the interface $z = 0$. Obviously, this result violates the required condition that the tangential electric field at a perfect conductor be zero. To correct the situation, we need to introduce a "secondary" electric field that produces an E_ρ at $z = 0$ with the opposite sign. Clearly, this condition is achieved by locating a charge $-q$ at $\rho = 0$ and $z = -h$ insofar as the observer at $P(\rho, z)$, for $z > 0$, is concerned. This situation is also illustrated in Figure 1.4. We now deduce the potential at $P(\rho, z)$ by superposition in the manner

$$V(\rho, z) = \frac{q}{4\pi\epsilon}\left(\frac{1}{r_1} - \frac{1}{r_2}\right) \tag{1.53}$$

where

$$r_1 = [\rho^2 + (z - h)^2]^{1/2}$$

and

$$r_2 = [\rho^2 + (z + h)^2]^{1/2}$$

We can now deduce explicit expressions for the electric field components at $P(\rho, z)$ as follows:

$$E_\rho = -\frac{\partial V}{\partial \rho} = \frac{q}{4\pi\epsilon}\left(\frac{\rho}{r_1^3} - \frac{\rho}{r_2^3}\right) \tag{1.54}$$

and

$$E_z = -\frac{\partial V}{\partial z} = \frac{q}{4\pi\epsilon}\left[\frac{(z - h)}{r_1^3} - \frac{(z + h)}{r_2^3}\right] \tag{1.55}$$

Clearly, we can verify that $E_\rho = 0$ at the interface $z = 0$. This result would also follow from the fact that $V(\rho, 0) = 0$.

We stress that Equations (1.53), (1.54), and (1.55) are only valid for the upper half space $z > 0$. We have eliminated the lower half space $z < 0$ from consideration by imposing the boundary condition $E_\rho = 0$ at $z = 0$ at the outset.

Exercise: In the above problem examine the limiting case where h tends to a vanishing small positive quantity. Show that the fields at $P(\rho, z)$ for $z > 0$ have the same character as that of a dipole source [that is, compared with (1.31)].

We now consider a slightly more complicated problem that shows the power of the image concept. Specifically, we consider a right-angled-corner reflector, as indicated in Figure 1.5. With respect to a rectangular coordinate system (x, y, z), the right-angled sector is defined by $x > 0$ and $y > 0$ with it's apex coincident with the z axis. A line charge $\hat{q}$ is located at (a, b, z), and it is of constant strength (that is, uniform) for $-\infty < z < +\infty$. The surfaces $x = 0$, $y > 0$ and $x > 0$, $y = 0$ are perfectly conducting.

Now it is again useful to talk about the primary potential V^p at $P(x, y)$. On the basis of (1.38) it would be given by

$$V^p = -\frac{\hat{q}}{2\pi\epsilon}\ln r_1 + \text{constant}$$

where $r_1 = [(x - a)^2 + (y - b)^2]^{1/2}$ is the linear distance from $\hat{q}$ to the observation point $P(x, y)$. To account for the presence of the right-angled reflector, we now construct a secondary field in terms of image charges. These charges are shown in Figure 1.5. Two of the images of strength $-\hat{q}$ are located at $(a, -b)$ and $(-a, b)$. In addition, we also need another

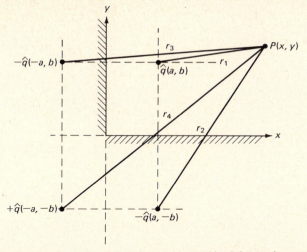

Figure 1.5 Image locations for point charge located in right-angle sector.

image at the antisymmetry point $(-a, -b)$ with strength $+\hat{q}$. The resultant potential for this system is

$$V(x, y) = -\frac{\hat{q}}{2\pi\epsilon}(\ln r_1 - \ln r_2 - \ln r_3 + \ln r_4)$$

$$= \frac{\hat{q}}{2\pi\epsilon}\left(\ln \frac{r_2 r_3}{r_1 r_4}\right) \qquad (1.56)$$

where

$$r_1 = [(x - a)^2 + (y - b)^2]^{1/2}$$
$$r_2 = [(x - a)^2 + (y + b)^2]^{1/2}$$
$$r_3 = [(x + a)^2 + (y - b)^2]^{1/2}$$
$$r_4 = [(x + a)^2 + (y + b)^2]^{1/2}$$

We have eliminated the "constant" by insisting that $V(x, y)$ vanishes on the sector surfaces.

Exercise: Obtain explicit expressions for the electric fields in the sector region $(x > 0, y > 0)$ in Figure 1.5 and verify directly that the tangential field components are zero on the sector surfaces.

We consider yet another problem that admits to an image type of solution. The situation is illustrated in Figure 1.6. A positive point charge is located at an infinitesimal height $ds/2$ above a perfectly conducting plane, $z = 0$, in terms of cylindrical coordinates (ρ, ϕ, z). The immediate consequence is that an accompanying negative point charge $-q$ appears as an image at $z = -ds/2$. The pair of charges acts as a source

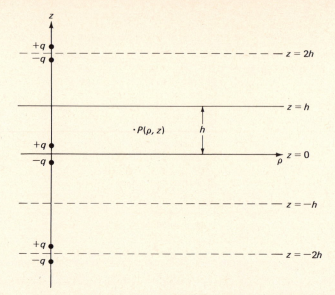

Figure 1.6 Image locations for point charge near bottom surface of parallel-plate region.

dipole aligned with the z axis and centered at the origin. At a height h above the ground plane we emplace another perfect conductor. The central problem is to determine the field at $P(\rho, z)$ within the homogeneous region $0 < z < h$ that has a constant permittivity ϵ.

Now the primary field is here defined as the field of the dipole at the origin in the absence of the upper conducting surface at $z = h$. The potential V^p is given by

$$V^p = \frac{q \, ds}{4\pi\epsilon} \frac{z}{(\rho^2 + z^2)^{3/2}} \tag{1.57}$$

Clearly, the associated electric field $E_\rho = -\partial V^p/\partial \rho$ vanishes at $z = 0$. To account for the influence of the reflector at $z = h$, we are led to emplace an image $z = 2h$ with the same polarity as the primary dipole. The combination of these two dipoles would produce a vanishing E_ρ field at $z = h$. But now we have to do something about the lower surface. Clearly, this situation requires an additional image dipole to be located at $z = -2h$. To balance its effect on the upper surface requires yet a further image dipole at $z = 4h$ (beyond the top of Figure 1.6). This dipole, in turn, needs another image at $z = -4h$. The process continues indefinitely, and we end with an infinite string of dipoles on the z axis.

From the image development described above, we can now write the expression for the total potential at $P(\rho, z)$:

$$V(\rho, z) = \frac{q \, ds}{4\pi\epsilon} \sum_{n=-\infty}^{+\infty} \frac{(z - 2nh)}{[\rho^2 + (z - 2nh)^2]^{3/2}} \tag{1.58}$$

Here the summation is over all integer values of n from $-\infty$ to $+\infty$, including $n = 0$. An explicit expression for the electric field component E_ρ is

$$E\rho = -\frac{\partial V}{\partial \rho} = \frac{q \, ds}{4\pi\epsilon} \sum_{n=-\infty}^{+\infty} \frac{3\rho(z - 2nh)}{[\rho^2 + (z - 2nh)^2]^{5/2}} \tag{1.59}$$

Now obviously E_ρ vanishes at $z = 0$ as it must. But it is not quite so obvious that it vanishes at $z = h$. This fact can be demonstrated by noting that the summation term evaluated at $z = h$ is given by

$$\sum_{n=-\infty}^{+\infty} \left\{ \frac{3\rho(1 - 2n)h}{[\rho^2 + (1 - 2n)^2 h^2]^{5/2}} \right\} = \sum_{n=1}^{\infty} (\cdots) + \sum_{n=0}^{-\infty} (\cdots)$$

$$= \sum_{n=1}^{\infty} \frac{3\rho}{[\rho^2 + (1 - 2n)^2 h^2]^{5/2}} [(1 - 2n)h$$
$$+ (-1 + 2n)h] = 0 \tag{1.60}$$

The image concept can be applied to various electrostatic problems involving boundaries between homogeneous dielectric regions. We will identify a few examples. The first case to consider is shown in Figure 1.7(a), where a positive point charge is located over a plane interface separating the two homogeneous semi-infinite regions. Cylindrical coordinates are chosen with the interface at $z = 0$ and the charge $+q$ located at $\rho = 0$ and $z = h$ in the dielectric region of permittivity ϵ_1. The lower-dielectric region (that is, for $z < 0$) has permittivity ϵ_2. The objective, of course, is to determine the fields everywhere, including the lower half space.

In considering the upper-half-space region (that is, $z > 0$), we take a hint from Figure 1.4 and locate an image at $z = -h$. We postulate its charge as being equal to Kq, where $K \to -1$ if the lower region becomes a conductor. Now we do not know K at this stage. But certainly K should tend to zero if the whole region became homogeneous (that is, if $\epsilon_2 \to \epsilon_1$).

In dealing with the potential and the resultant fields in the lower half space, we might surmise that the effective source strength was modified. Thus insofar as the region $z < 0$ is concerned, we have the picture shown in Figure 1.7(b), where the modified source is Tq.

Thus for $z > 0$, we write

$$V(\rho, z) = \frac{q}{4\pi\epsilon_1} \left(\frac{1}{r_1} + K \frac{1}{r_2} \right) \tag{1.61}$$

where

$$r_1 = [\rho^2 + (z - h)^2]^{1/2}$$

and

$$r_2 = [\rho^2 + (z + h)^2]^{1/2}$$

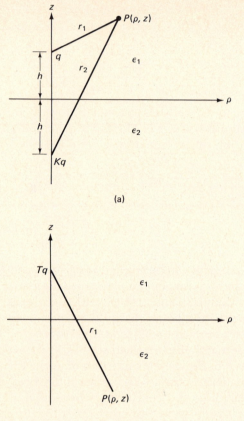

(a)

(b)

Figure 1.7(a)   Modified image of strength Kq for charge q located over interface between homogeneous dielectrics; (b) effective source of strength Tq when observing field on other side of interface.

For the case $z < 0$ we write

$$V(\rho, z) = \frac{Tq}{4\pi\epsilon_2} \frac{1}{r_1} \qquad (1.62)$$

We now apply boundary conditions at the plane interface $z = 0$. The two conditions are (1) $V(\rho, z)$ is continuous and (2) $D_z(\rho, z)$, the normal flux density, is continuous. Clearly, these conditions must be true if there is no surface charge. In the second condition we note that

$$D_z = -\epsilon_1 \frac{\partial V}{\partial z} \qquad \text{for } z > 0 \qquad (1.63)$$

while

$$D_z = -\epsilon_2 \frac{\partial V}{\partial z} \qquad \text{for } z < 0 \qquad (1.64)$$

Application of these two boundary conditions leads to the two algebraic equations

$$\frac{1}{\epsilon_1}(1 + K) = \frac{T}{\epsilon_2} \tag{1.65}$$

$$-1 + K = -T \tag{1.66}$$

Solving leads to

$$K = \frac{\epsilon_1 - \epsilon_2}{\epsilon_1 + \epsilon_2}. \tag{1.67}$$

and

$$T = \frac{2\epsilon_2}{\epsilon_1 + \epsilon_2} \tag{1.68}$$

Exercise: In the above problem derive explicit expressions for the electric field components E_ρ and E_z and show that E_ρ is continuous across the interface $z = 0$. Also, note that $\epsilon_1 E_z$ for $z \rightarrow +0$ is equal to $\epsilon_2 E_z$ for $z \rightarrow -0$.

Exercise: In place of the single point charge at $z = h$ in Figure 1.7(a), locate a vertical dipole (that is, a charge pair). Show, by using image theory, that the potential in the upper half space is given by

$$V(\rho, z) = \frac{q \, ds}{4\pi\epsilon_1} \left(\frac{z - h}{r_1^3} - K \frac{z + h}{r_2^3} \right) \tag{1.69}$$

where K is defined by (1.67).

Exercise: Locate a line charge $\hat{q}$ at $x = 0, y = b$, uniform in the z direction. The region $y > 0$ is a dielectric with permittivity ϵ_1, while the region $y < 0$ is a dielectric with permittivity ϵ_2. Show that the resultant potential has the form

$$V(x, y) = -\frac{\hat{q}}{2\pi\epsilon_1} (\ln r_1 + K \ln r_2) + \text{constant} \tag{1.70}$$

for $y > 0$, while

$$V(x, y) = -\frac{\hat{q}}{2\pi\epsilon_2} T \ln r_1 + \text{constant} \tag{1.71}$$

for $y < 0$. Verify that D_y, the vertical flux density, and E_x, the horizontal electric field, are continuous over the horizontal interface. In the above problem K and T are defined by (1.67) and (1.68), while $r_1^2 = x^2 + (y - b)^2$ and $r_2^2 = x^2 + (y + b)^2$.

1.8 METHODS BASED ON SOLVING LAPLACE'S EQUATION

The scope of the image method is somewhat limited. We quickly run out of doable problems by this method. A more systematic procedure is to deal with the solutions of the basic equations.

Now in any homogeneous dielectric region of permittivity ϵ, we can write

$$\mathbf{E} = -\mathrm{grad}\ V \tag{1.72}$$

where V is the potential. Furthermore,

$$\mathrm{div}\ \mathbf{E} = -\mathrm{div\ grad}\ V = 0$$

in homogeneous regions where there are no charges. But

$$\mathrm{div\ grad}\ V = \nabla^2 V \tag{1.73}$$

where ∇^2 is the laplacian operator. In rectangular coordinates (x, y, z)

$$\nabla^2 = \frac{\partial^2}{\partial x^2} + \frac{\partial^2}{\partial y^2} + \frac{\partial^2}{\partial z^2} \tag{1.74}$$

In cylindrical coordinates (ρ, ϕ, z)

$$\nabla^2 = \frac{1}{\rho} \frac{\partial}{\partial \rho} \left(\rho \frac{\partial}{\partial \rho} \right) + \frac{\partial^2}{\rho^2 \partial \phi^2} + \frac{\partial^2}{\partial z^2} \tag{1.75}$$

In spherical coordinates (r, θ, ϕ):

$$\nabla^2 = \frac{1}{r^2} \frac{\partial}{\partial r} \left(r^2 \frac{\partial}{\partial r} \right) + \frac{1}{r^2 \sin \theta} \frac{\partial}{\partial \theta} \left(\sin\theta \frac{\partial}{\partial \theta} \right)$$

$$+ \frac{1}{r^2 \sin^2 \theta} \frac{\partial^2}{\partial \phi^2} \tag{1.76}$$

We will now consider a selection of problems that are based on solving Laplace's equation,

$$\nabla^2 V = 0$$

subject to certain boundary conditions. The latter are dictated by the nature of the specific physical problem.

The approach is illustrated with reference to the situation shown in Figure 1.7(a). Except right at the point charge, the potential $V(\rho, z)$ satisfies Laplace's equation,

$$\frac{1}{\rho} \frac{\partial}{\partial \rho} \left(\rho \frac{\partial V}{\partial \rho} \right) + \frac{\partial^2 V}{\partial z^2} = 0 \tag{1.77}$$

bearing in mind that azimuthal symmetry prevails so that $\partial/\partial \phi = 0$. Now an appropriate solution of (1.77) that is finite at $\rho = 0$ is of the form

$$J_0(\lambda \rho) e^{\pm \lambda z}$$

where λ is arbitrary. Here $J_0(x)$ is the Bessel function* of order zero and argument x, and it satisfies

$$\frac{1}{x}\frac{d}{dx}\left[x\frac{dJ_0(x)}{dx}\right] + J_0(x) = 0 \tag{1.78}$$

The Bessel function $Y_0(\lambda\rho)$ would not be admissible because the solution must be bounded at $\rho = 0$.

We will now construct our solutions as superpositions of the elementary forms, where we allow λ to take all real values. In carrying out this step, we utilize the fact that the primary potential given by

$$V^p = \frac{q}{4\pi\epsilon_1 r_1} \tag{1.79}$$

where $r_1 = [\rho^2 + (z-h)^2]^{1/2}$, can be written in equivalent integral form

$$V^p = \frac{q}{4\pi\epsilon_1}\int_0^\infty J_0(\lambda\rho)e^{-\lambda|z-h|}\,d\lambda \tag{1.80}$$

This equation is a consequence of the Weber integral identity:

$$\int_0^\infty J_0(x)e^{-yx}\,dx = \frac{1}{(1+y^2)^{1/2}} \tag{1.81}$$

for $y > 0$ (see Chapter 3).

In view of the preceding remarks we are led to write the following integral representations for the potentials in the two regions:

$$V(\rho, z) = \frac{q}{4\pi\epsilon_1}\int_0^\infty (e^{-\lambda|z-h|} + Ke^{-\lambda(z+h)})J_0(\lambda\rho)\,d\lambda \tag{1.82}$$

for $z > 0$, while

$$V(\rho, z) = \frac{q}{4\pi\epsilon_2}\int_0^\infty Te^{\lambda(z-h)}J_0(\lambda\rho)\,d\lambda \tag{1.83}$$

for $z < 0$. At this stage we would not know the form K and T take except to say that they are independent of ρ and z.

The key point here is that we have constructed an integral representation for the potential in the upper region that has the right singularity at the source (that is, at $\rho \to 0$ and $z = h$), and it becomes a vanishing quantity as $z \to +\infty$. Similarly, the potential in the lower region vanishes as $z \to -\infty$. The various multiplying factors in (1.82) and (1.83) were introduced for convenience and could have been deduced. The boundary conditions as before are that $V(\rho, z)$ and $D_z(\rho, z)$ are continuous at

---

* Actually,

$$J_0(x) = \sum_{m=0}^\infty \frac{(-1)^m x^{2m}}{(m!)^2 2^{2m}} = 1 - \frac{x^2}{2^2} + \cdots$$

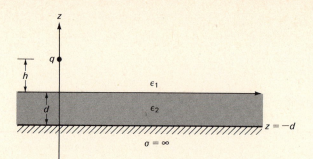

Figure 1.8   Charge q located over a grounded dielectric slab.

$z = 0$. Not surprisingly, these conditions lead to the results for K and T given by (1.67) and (1.68). Furthermore, from (1.81) the integrals given by (1.82) and (1.83) lead right back to (1.61) and (1.62).

To show how more complicated situations can be handled by solving Laplace's equation, let us consider the situation shown in Figure 1.8. The point charge is now located at a height h over the top surface of a dielectric slab of permittivity ϵ_2 and thickness d. The slab is resting on a perfect conductor located at $z = -d$.

We now write the appropriate form of the potential solutions, bearing in mind that the primary field, the region $z > 0$, will be the same as before. Thus

$$V(\rho, z) = \frac{q}{4\pi\epsilon_1} \int_0^\infty [e^{-\lambda|z-h|} + R(\lambda)e^{-\lambda(z+h)}] J_0(\lambda\rho)\, d\lambda \tag{1.84}$$

for $z > 0$, while

$$V(\rho, z) = \int_0^\infty [A(\lambda)e^{-\lambda z} + B(\lambda)e^{\lambda z}] J_0(\lambda\rho)\, d\lambda \tag{1.85}$$

for $0 > z > -d$. Here we do not know the specific forms for $R(\lambda), A(\lambda)$, or $B(\lambda)$.

At the perfectly conducting base at $z = -d$, we require that

$$E_\rho\Big|_{z=-d} = -\frac{\partial V}{\partial \rho}\Big|_{z=-d} = 0 \tag{1.86}$$

Applying this result to (1.85) tells us immediately that $A(\lambda) = -B(\lambda)e^{-2\lambda d}$, thus we can write

$$V(\rho, z) = \int_0^\infty A(\lambda)(e^{-\lambda z} - e^{\lambda(z+2d)}) J_0(\lambda\rho)\, d\lambda \tag{1.87}$$

Continuity of potential at $z = 0$ leads to

$$\frac{q}{4\pi\epsilon_1}(e^{-\lambda h} + Re^{-\lambda h}) = A(1 - e^{2\lambda d}) \tag{1.88}$$

Continuity of the normal flux density at $z = 0$ leads to

$$\frac{q}{4\pi\epsilon_1}(e^{-\lambda h} - Re^{-\lambda h}) = -A(1 + e^{2\lambda d})\frac{\epsilon_2}{\epsilon_1} \tag{1.89}$$

On dividing these equations, we get

$$\frac{1 - R}{1 + R} = \frac{\epsilon_2}{\epsilon_1}\left(\frac{1 + e^{-2\lambda d}}{1 - e^{-2\lambda d}}\right) \tag{1.90}$$

or

$$R = -\left[\frac{1 - (\epsilon_1/\epsilon_2)\tanh \lambda d}{1 + (\epsilon_1/\epsilon_2)\tanh \lambda d}\right] \tag{1.91}$$

Of course, if $d \to \infty$, we recover the expressions (1.69) and (1.67) for the point charge over the semi-infinite half space of permittivity ϵ_2. The other special case is when $d \to 0$, whence we have a point charge located over a perfectly conducting ground plane. The solution then reduces to that given by (1.53), where again we have utilized the Weber integral given by (1.81). For the general case here we need to deal with the integral formula given by (1.84), with the expression for $R(\lambda)$ specified by (1.91). While series expressions in terms of d, ϵ_1/ϵ_2, and so on can be developed, it is usually easier to work directly with the integral formula, using a variety of numerical methods.

Exercise: Show that in the case where $(\epsilon_1/\epsilon_2)d$ is sufficiently small, Equation (1.91) can be approximated by

$$R \simeq -\exp[-2(\epsilon_1/\epsilon_2)\lambda d] \tag{1.91'}$$

Then show that (1.84) is given by

$$V(\rho, z) \simeq \frac{q}{4\pi\epsilon_1}\left(\frac{1}{r} - \frac{1}{r_a}\right) \tag{1.84'}$$

where

$$r = [\rho^2 + (z - h)^2]^{1/2}$$

and

$$r_a = \left\{\rho^2 + \left[z + h + 2\left(\frac{\epsilon_1}{\epsilon_2}\right)d\right]^2\right\}^{1/2}$$

Another example of a problem that yields to a potential theory approach is shown in Figure 1.9. Specifically, we have a semi-infinite, parallel-plate region that consists of two perfectly conducting planar surfaces defined by $x > 0$, $y = 0$ and $x > 0$, $y = \ell$. The region in between is a pure homogeneous dielectric of permittivity ϵ. The end surface,

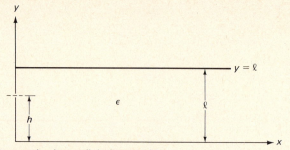

Figure 1.9   Gap-excited, parallel-plate region.

$x = 0$, $\ell > y > 0$, is perfectly conducting except for a narrow slit at height h. The slit is excited by a voltage that is constant for all values of z. The object is to determine the potential $V(x, y)$ everywhere in the region $x > 0$, $\ell > y > 0$.

First of all, we note that $V(x, y)$ satisfies Laplace's equations:

$$\left(\frac{\partial^2}{\partial x^2} + \frac{\partial^2}{\partial y^2} \right) V(x, y) = 0 \tag{1.92}$$

Again we use a product solution technique. For a start we would try something of the form

$$V = f(y)e^{-\lambda x} \tag{1.93}$$

where λ is a real positive constant suggested by the fact that V must be bounded as $x \to \infty$.

Thus

$$\left(\frac{\partial^2}{\partial y^2} + \lambda^2 \right) f(y) = 0 \tag{1.94}$$

Solutions are linear combinations of $\sin \lambda y$ and $\cos \lambda y$, but they must be chosen to satisfy the boundary conditions at $y = 0$ and ℓ that $E_x = 0$. This result suggests that $\sin(n\pi y / \ell)$, where n is an integer, is acceptable, so that $\lambda = \lambda_n = n\pi / \ell$. But, also, $f(y) = a + by$, where a and b are constants, is a possible solution. This result suggests we choose

$$V = a + by + \sum_{n=1}^{\infty} A_n \sin \frac{n\pi y}{\ell} e^{-\lambda_n x} \tag{1.95}$$

where a, b, and A_n are yet to be chosen.

Now the vertical electric field must have the form

$$E_y = -\frac{\partial V}{\partial y} = -b - \sum_{n=1}^{\infty} \frac{n\pi}{\ell} A_n \cos \frac{n\pi y}{\ell} e^{-\lambda_n x} \tag{1.96}$$

where $\lambda_n = n\pi/\ell$. Now at $x = 0$ our source condition is that

$$E_y = \hat{f}(y) = \begin{cases} 0 & \text{for } 0 < y < h - \Delta \\ \dfrac{v_0}{2\Delta} & \text{for } h - \Delta < y < h + \Delta \\ 0 & \text{for } h + \Delta < y < \ell \end{cases}$$

where 2Δ is the width of the slit.

To obtain A_n, we insist that

$$-b - \sum_{n=1}^{\infty} \frac{n\pi}{\ell} A_n \cos \frac{n\pi y}{\ell} = \hat{f}(y) \qquad (1.97)$$

This equation is solved by multiplying both sides by $\cos n'\pi y/\ell$ and integrating over y from 0 to ℓ. Here n' is an integer from 1 to ∞. That is,

$$-\sum_{n=1}^{\infty} \frac{n\pi}{\ell} A_n \int_0^{\ell} \cos \frac{n\pi y}{\ell} \cos \frac{n'\pi y}{\ell} \, dy = \int_0^{\ell} \hat{f}(y) \cos \frac{n'\pi y}{\ell} \, dy \qquad (1.98)$$

The integral on the left vanishes for $n' \neq n \neq 0$, as can be readily verified. But it is equal to $\ell/2$ when $n' = n \neq 0$. The integral on the right only has a contribution when $h - \Delta < y < h + \Delta$. Thus

$$\frac{n\pi A_n}{\ell} \frac{\ell}{2} = -\int_{h-\Delta}^{h+\Delta} \frac{v_0}{2\Delta} \cos \frac{n\pi y}{\ell} \, dy$$

$$\simeq -\frac{v_0}{2\Delta} \cos \frac{n\pi h}{\ell} 2\Delta \qquad (1.99)$$

When 2Δ is sufficiently small. Thus

$$A_n \simeq -\frac{2}{n\pi} v_0 \cos \frac{n\pi h}{\ell} \qquad (1.100)$$

Also, we note that

$$\int_0^{\ell} E_y \bigg|_{x=0} dy = \int_0^{\ell} \hat{f}(y) \, dy \qquad (1.101)$$

or

$$-b\ell = v_0$$

which specifies b.

Thus our solution for the potential is given by

$$V(x, y) = a - \frac{v_0}{\ell} y$$

$$-\frac{2}{\pi} v_0 \sum_{n=1}^{\infty} \frac{1}{n} \cos \frac{n\pi h}{\ell} \sin \frac{n\pi y}{\ell} \exp\left(-\frac{n\pi}{\ell} x\right) \qquad (1.102)$$

Now a is still undetermined, and it would remain so unless we specified the potential on the lower plate (that is, at $y = 0$). For example, if we said that $V(x, 0) = 0$, then clearly $a = 0$ and, as a consequence, $V(x, \ell) = -v_0$.

It is useful to note that

$$E_y = -\frac{v_0}{\ell} \sum_{n=0}^{\infty} \hat{\epsilon}_n \cos \frac{n\pi h}{\ell} \cos \frac{n\pi y}{\ell} \exp\left(-\frac{n\pi}{\ell} x\right) \tag{1.103}$$

where $\hat{\epsilon}_0 = 1$, $\hat{\epsilon}_n = 2$ $(n \neq 0)$, is an equivalent statement for the final solution.

1.9 DIELECTRIC CYLINDER IN UNIFORM APPLIED FIELD

A good example where Laplace's equation can be solved is when a dielectric cylinder of infinite length is immersed in a uniform applied field. The situation is illustrated in Figure 1.10, where the cylinder is defined by $\rho \leq a$ for $-\infty < z < \infty$ with respect to a cylindrical coordinate system (ρ, ϕ, z). The permittivity of the cylinder is ϵ, and the permittivity of the external medium is ϵ_0.

The primary field, by definition, is that existing in the absence of the cylinder. In terms of the cartesian coordinate system (x, y, z) the primary electric field $\mathbf{E}^p$ is specified to be

$$E_x^p = E_0$$

and $E_y^p = E_z^p = 0$. In other words, E_0 is directed in the direction transverse to the axis of the cylinder. In cylindrical coordinates we

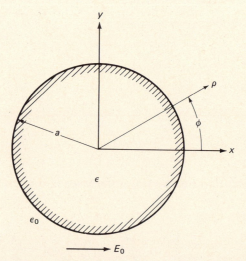

Figure 1.10   Homogeneous dielectric cylinder immersed in uniform, static electric field E_0 that is transverse to axis of cylinder.

would write

$$E_\rho^p = E_0 \cos \phi \tag{1.104}$$

$$E_\phi^p = -E_0 \sin \phi \tag{1.105}$$

and

$$E_z^p = 0$$

The "primary" potential function $V^p(\rho, z)$ that gives rise to the "primary" electric field $\mathbf{E}^p$ is related to the latter by

$$\mathbf{E}^p = -\operatorname{grad} V^p \tag{1.106}$$

or, to be specific,

$$E_\rho^p = -\frac{\partial V^p}{\partial \rho} \tag{1.107}$$

and

$$E_\phi^p = -\frac{1}{\rho} \frac{\partial V^p}{\partial \phi} \tag{1.108}$$

Clearly, if we choose

$$V^p = -E_0 \rho \cos \phi \tag{1.109}$$

we would regenerate the required primary electric field. We also confirm that

$$\nabla^2 V^p = 0 \tag{1.110}$$

where

$$\nabla^2 = \frac{1}{\rho} \frac{\partial}{\partial \rho} \left(\rho \frac{\partial}{\partial \rho} \right) + \frac{1}{\rho^2} \frac{\partial^2}{\partial \phi^2} \tag{1.111}$$

is the laplacian operator in cylindrical coordinates (for the case $\partial/\partial z = 0$).

Now we wish to augment the primary potential by a secondary potential that accounts for the presence of the cylinder. Thus for the region $\rho > a$, we write the resultant potential as

$$V = V^p + V^s \tag{1.112}$$

where V^s is the secondary potential. Now since both V and V^p satisfy Laplace's equation, we can also say that

$$\nabla^2 V^s = 0 \tag{1.113}$$

where ∇^2 in cylindrical coordinates is given by (1.111). Two solutions of (1.113) that have the required $\cos \phi$ dependence are $\rho \cos \phi$ and

$\rho^{-1} \cos \phi$. Of these two, only the latter is acceptable if V^s is to be bounded as $\rho \to \infty$.

Within the cylinder the resultant potential must have the form $\rho \cos \phi$ because $\rho^{-1} \cos \phi$ is infinite at the central axis.

In view of the above, we are led to write

$$V = -E_0 \rho \cos \phi + A\rho^{-1} \cos \phi \tag{1.114}$$

for $\rho > a$, while

$$V = B\rho \cos\phi \tag{1.115}$$

for $\rho < a$. Here A and B are constants yet to be determined.

The boundary conditions at $\rho = a$ are that the potential and normal flux densities are continuous. That is,

$$V(a - 0, \phi) = V(a + 0, \phi) \tag{1.116}$$

and

$$\epsilon \left. \frac{\partial V(\rho, \phi)}{\partial \rho} \right| = \epsilon_0 \left. \frac{\partial V(\rho, \phi)}{\partial \rho} \right|_{\rho=a-0} \tag{1.117}$$

These lead to the algebraic pair

$$Ba = -E_0 a + \frac{A}{a} \tag{1.118}$$

and

$$\epsilon B = -\epsilon_0 E_0 - \frac{\epsilon_0 A}{a^2} \tag{1.119}$$

Solving gives

$$A = \frac{\epsilon - \epsilon_0}{\epsilon + \epsilon_0} a^2 E_0 \tag{1.120}$$

and

$$B = -\frac{2\epsilon_0}{\epsilon + \epsilon_0} E_0 \tag{1.121}$$

In particular, we note that

$$V^s = \frac{\epsilon - \epsilon_0}{\epsilon + \epsilon_0} a^2 E_0 \frac{1}{\rho} \cos \phi \tag{1.122}$$

It is of interest to compare the above expression for V^s with that produced by an equivalent line-dipole source. To this end, we first examine the potential of two line charges $\hat{q}$ and $-\hat{q}$ C/m located at $x = \delta h/2$ and $-\delta h/2$ in the plane $x = 0$. The situation is illustrated in

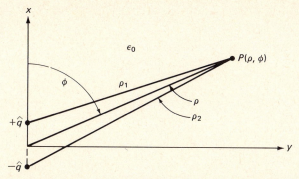

Figure 1.11 Geometry for deducing potential of two line charges of equal and opposite sign at separation δh.

Figure 1.11. In accordance with (1.38), we can write the expression for the potential $V(\rho, \phi)$ at P as the superposition at the two line charges:

$$V(\rho, \phi) = \frac{\hat{q}}{2\pi\epsilon_0} \ln \frac{\rho_2}{\rho_1} \tag{1.123}$$

where ϵ_0 is the permittivity of the ambient medium. Now $\rho_2 \approx \rho + (\delta h/2)\cos \phi$ and $\rho_1 \approx \rho - (\delta h/2)\cos \phi$ subject to $(\delta h)^2 \ll \rho^2$. Within the same approximation

$$\ln \frac{\rho_2}{\rho_1} \approx \ln\left(1 + \frac{\delta h}{\rho} \cos \phi\right) \approx \frac{\delta h}{\rho} \cos \phi \tag{1.124}$$

Thus we can write (1.122) in the form

$$V^s = \frac{\hat{q}}{2\pi\epsilon_0} \delta h \frac{\cos \phi}{\rho} \tag{1.125}$$

where the "dipole moment" is given by

$$\hat{q} \, \delta h = 2\pi\epsilon_0 \frac{\epsilon - \epsilon_0}{\epsilon + \epsilon_0} a^2 E_0 \tag{1.126}$$

As indicated, a homogeneous dielectric cylinder immersed in a uniform electric field produces a secondary field that is equivalent to a line electric dipole.

1.10 DIELECTRIC SPHERE IN UNIFORM APPLIED FIELD

A simple three-dimensional problem analogous to the preceding cylinder problem is when a dielectric sphere is immersed in a uniform electric field E_0. As before, the objective is to determine the resultant potentials and fields both inside and outside the spherical body. The situation is illustrated in Figure 1.12, where a spherical coordinate system (r, θ, ϕ) is chosen with the dielectric sphere of permittivity ϵ defined by $r < b$.

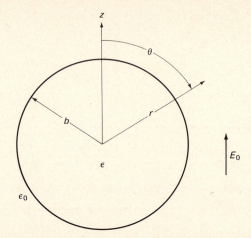

Figure 1.12 Homogeneous dielectric sphere immersed in uniform, static electric field E_0.

Without loss of generality the uniform electric field is taken to be parallel to the z axis (that is, $\theta = 0$ direction); as a consequence, the whole problem becomes azimuthally symmetric (that is, $\partial/\partial\phi) = 0$).

The primary potential Φ^p is now given by

$$\Phi^p = -E_0 r \cos\theta = -E_0 z \tag{1.127}$$

which, of course, leads back to

$$E_z^p = -\frac{\partial\Phi^p}{\partial z} = E_0 \tag{1.128}$$

The resultant potential Φ satisfies Laplace's equation, which we write, in spherical coordinates (for $\partial/\partial\phi = 0$), as follows:

$$\nabla^2\Phi = \frac{1}{r^2}\frac{\partial}{\partial r}\left(r^2\frac{\partial\Phi}{\partial r}\right) + \frac{1}{r^2\sin\theta}\frac{\partial}{\partial\theta}\left(\sin\theta\frac{\partial\Phi}{\partial\theta}\right) = 0 \tag{1.129}$$

As we can verify, Φ^p satisfies this equation. Thus Φ^s, the secondary potential in the region $r > b$, also must satisfy this equation. In fact, the only solution that has the same angular behavior $\cos\theta$ and decays to zero as $r \to \infty$ is found to be of the form $r^{-2}\cos\theta$. Thus we write

$$\Phi = -E_0 r \cos\theta + \alpha r^{-2}\cos\theta \tag{1.130}$$

for $r > b$, while

$$\Phi = \beta r \cos\theta \tag{1.131}$$

for $r < b$ where α and β are unknown coefficients.

The boundary conditions require that potential and normal displacement D_r be continuous at $r = b$. Specifically,

$$\Phi(b - 0, \theta) = \Phi(b + 0, \theta) \tag{1.132}$$

and

$$\epsilon \left. \frac{\delta\Phi(r, \theta)}{\delta r} \right|_{r=b-0} = \epsilon_0 \left. \frac{\delta\Phi(r, \theta)}{\delta r} \right|_{r=b+0} \tag{1.133}$$

These lead to the pair

$$\beta b = -E_0 b + \frac{\alpha}{b^2} \tag{1.134}$$

$$\epsilon\beta = -\epsilon_0 E_0 - \frac{2\epsilon_0 \alpha}{b^3} \tag{1.135}$$

Solving, we find

$$\alpha = \frac{\epsilon - \epsilon_0}{\epsilon + 2\epsilon_0} b^3 E_0 \tag{1.136}$$

and

$$\beta = -\frac{3\epsilon_0}{\epsilon + 2\epsilon_0} E_0 \tag{1.137}$$

In particular, we see that the secondary potential, for $r > b$, is given by

$$\Phi^s = \frac{\epsilon - \epsilon_0}{\epsilon + 2\epsilon_0} b^3 E_0 \frac{1}{r^2} \cos \theta \tag{1.138}$$

This equation has the form

$$\Phi^s = \frac{q\,\delta\ell}{4\pi\epsilon_0 r^2} \cos \theta \tag{1.139}$$

where

$$q\,\delta\ell = 4\pi\epsilon_0 b^3 \frac{\epsilon - \epsilon_0}{\epsilon + 2\epsilon_0} E_0 \tag{1.140}$$

is the effective moment of the equivalent electric dipole [that is, compare with (1.31)]. Thus the secondary fields, for $r > b$, can be derived from two equivalent point charges $+q$ and $-q$ located at $\delta\ell/2$ and $-\delta\ell/2$ on the z axis. Here $\delta\ell \ll r$, but otherwise the equivalence is exact.

1.11 STATIC MAGNETIC FIELDS

A (vector) magnetic flux density **B** is best defined in terms of the force $d\mathbf{F}$ produced on a current element $d\ell$ carrying a current I by the relation

$$d\mathbf{F} = I\,d\ell \times \mathbf{B} \tag{1.141}$$

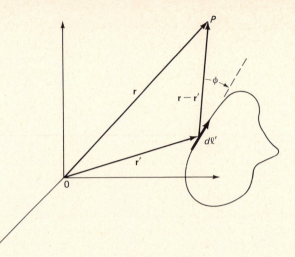

Figure 1.13 Geometry for deducing magnetic field at *P* due to current-carrying circuit.

In scalar form this vector relation is written

$$dF = I \, d\ell \, B \sin \theta \tag{1.142}$$

where θ is the angle subtended by $d\ell$ and **B**. The force of strength dF is directed perpendicularly from the plane containing $d\ell$ and **B**. The direction or sense of $d\mathbf{F}$ is in the direction of a right-handed screw if $d\ell$ is rotated into **B**. In our rationalized system of units dF is in newtons, where $1 \text{ N} = 1 \text{ kg} \cdot \text{m/s}^2 = 10^5$ dyn. Then B is in webers per square meter, or teslas by most recent decree.

A closely related quantity is the magnetic field vector **H**. It is defined by Ampere's law in terms of a current $\mathbf{I}'(\mathbf{r}')$ in an associated circuit. The situation is illustrated in Figure 1.13. The vector magnetic field $d\mathbf{H}(\mathbf{r})$ at *P* in a homogeneous isotropic medium due to a vector current element $\mathbf{I}'(\mathbf{r}') \, d\ell'$ is given by

$$\mathbf{H}(\mathbf{r}) = \frac{\mathbf{I}'(\mathbf{r}') \, d\ell' \times (\mathbf{r} - \mathbf{r}')}{4\pi |\mathbf{r} - \mathbf{r}'|^3} \tag{1.143}$$

As indicated, **r** is the vector coordinate from the arbitrary origin 0 to the field point *P*. On the other hand, **r**′ is the vector coordinate from 0 to the vector current element. Then, of course, **r** − **r**′ is the vector from the current element to the field point or observer at *P*.

The scalar counterpart of (1.143) is really what we mean by Ampere's law that had originally been established experimentally. Then (1.43) takes the form

$$dH(\mathbf{r}) = \frac{I'(\mathbf{r}') \, d\ell' \sin \phi}{4\pi |\mathbf{r} - \mathbf{r}'|^2} \tag{1.144}$$

where ϕ is the angle subtended by the direction of the current element $d\ell$ and the vector $\mathbf{r} - \mathbf{r'}$.

The total magnetic field $H(\mathbf{r})$ at P due to the current $I(\mathbf{r'})$ in the circuit shown in Figure 1.14 is found by integrating over $d\ell'$, to give

$$\mathbf{H}(\mathbf{r}) = \int \frac{I'(\mathbf{r'})\,d\boldsymbol{\ell'} \times (\mathbf{r} - \mathbf{r'})}{4\pi|\mathbf{r} - \mathbf{r'}|^3} \tag{1.145}$$

The circuit need not be closed, but the magnetic field expression is strictly only valid for purely static fields (that is, the circuit has a filamental direct current). We can note that **H** has dimensions of amperes per meter.

Now as we have seen, the flux density or magnetic induction vector **B** is related to the force acting on a current element. On the other hand, the magnetic field vector **H** is defined at a point in relation to the strength and direction of the current element. At a point in the medium these quantities are related by

$$\mathbf{B} = \mu\mathbf{H} \tag{1.146}$$

where μ is a scalar quantity for an isotropic region. In fact, μ is defined as the magnetic permeability, and it has dimensions of henrys per meter. In free space

$$\mu = \mu_0 = 4\pi \times 10^{-7} \text{ henries/m}$$

which is a value that should be committed to memory.

The reader will observe or already know that (1.6) is fully analogous to the equation $\mathbf{D} = \epsilon\mathbf{E}$ in electrostatics, where ϵ is the permittivity. As in electrostatics, the constitutive relation connecting the flux density and field is valid even for inhomogeneous media. Also, μ may be a tensor if the magnetic properties are anisotropic.

Some of the alternative forms of Ampere's law can be derived from

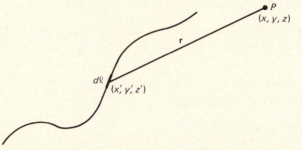

Figure 1.14 Simplified method for calculating magnetic field of current circuit of finite length.

(1.145), which is the primordial form. It can be simplified to

$$\mathbf{H}(x,\, y, z) = \int \frac{I'(x',\, y',\, z')\, d\ell' \times \mathbf{r}}{4\pi r^3} \tag{1.147}$$

where we move the origin of the coordinate system to be at the current element at each point in the integration. The observation point is (x, y, z) in a fixed cartesian frame. Also, $I'(x',\, y',\, z')$ is the current in the contributing element $d\ell'$, and $\mathbf{r}$ is the vector distance from $d\ell'$ to the observer point. To be explicit,

$$\mathbf{r} = \mathbf{i}_x(x - x') + \mathbf{i}_y(y - y') + \mathbf{i}_z(z - z') \tag{1.148}$$

and

$$r = |\mathbf{r}| = [(x - x')^2 + (y - y')^2 + (z - z')^2]^{1/2} \tag{1.149}$$

Now (1.147) is modified by first noting that

$$\frac{d\ell' \times \mathbf{r}}{r^3} = \nabla\left(\frac{1}{r}\right) \times d\ell' \tag{1.150}$$

Note that operator

$$\nabla = \mathbf{i}_x \frac{\partial}{\partial x} + \mathbf{i}_y \frac{\partial}{\partial y} + \mathbf{i}_z \frac{\partial}{\partial z} \tag{1.151}$$

Also, since in rectangular coordinates

$$\nabla \times (\psi \mathbf{F}) = \psi \nabla \times \mathbf{F} - \mathbf{F} \times \nabla \psi \tag{1.152}$$

we can deduce that

$$\nabla\left(\frac{1}{r}\right) \times d\ell' = \nabla \times \left(\frac{d\ell'}{r}\right) - \frac{1}{r} \nabla \times d\ell' \tag{1.153}$$

where the latter term on the right is zero. This equation enables us to write (1.147) in the form

$$\mathbf{B}(x, y, z) = \nabla \times \mathbf{A}(x, y, z) \tag{1.154}$$

where

$$\mathbf{A} = \mu \int \frac{I'(x',\, y',\, z')}{4\pi r} \, d\ell' \tag{1.155}$$

1.12 CONNECTIONS BETWEEN ELECTROSTATIC AND MAGNETOSTATIC PROBLEMS

We return to (1.18) and note that

$$\oint_s \mathbf{D} \cdot d\mathbf{a} = \int \text{div } \mathbf{D} \, dv \tag{1.156}$$

Thus

$$\operatorname{div} \mathbf{D} = \rho \tag{1.157}$$

where ρ is the charge density in the dielectric. But

$$\mathbf{D} = \epsilon \mathbf{E}$$

and

$$\mathbf{E} = -\nabla V \tag{1.158}$$

so that

$$\operatorname{div} \mathbf{E} = -\nabla \cdot \nabla V = -\nabla^2 V \tag{1.159}$$

or

$$\nabla^2 V = -\frac{\rho}{\epsilon}$$

which is Poisson's equation. The solution is

$$V = \frac{1}{4\pi\epsilon_0} \int_V \frac{\rho \, dv}{r} \tag{1.160}$$

which is a statement that the resultant potential is the linear superposition of all the elementary charges $\rho \, dv$.

There is an obvious similarity between the electrostatic formula (1.160) and the magnetostatic formula (1.155). To push the analogy even further, we can write (1.155) in component form. For example, choosing the x component of the vector $\mathbf{A}$, we see that

$$A_x = \mu \int \frac{I_x'(x', y', z') \, d\ell_x}{4\pi r} \tag{1.161}$$

$$= \mu \int_V \frac{J_x'(x', y', z')}{4\pi r} \, dv' \tag{1.162}$$

where $J_x'(x', y', z')$ is the x component of current density (in amperes per square meter) at the point (x', y', z') that is a distance r to the observer at (x, y, z). There is now a complete analogy between the scalar electrostatic equation (1.160) and the corresponding scalar magnetostatic equation (1.162).

A further analogy is to the Poisson equation,

$$\nabla^2 V = -\frac{\rho}{\epsilon} \tag{1.163}$$

This equation has a counterpart in

$$\nabla^2 A_x = -\mu J_x \tag{1.164}$$

where we can regard μJ_x as the source of the x component of the vector potential $\mathbf{A}$. Also, of course,

$$\nabla^2 A_y = -\mu J_y \tag{1.165}$$

and

$$\nabla^2 A_z = -\mu J_z \tag{1.166}$$

A vector form of (1.164), (1.165), and (1.166) is

$$\nabla^2 \mathbf{A} = -\mu \mathbf{J} \tag{1.167}$$

where the ∇^2 operating on a vector only has a strict meaning when we are dealing with the cartesian components.

Some other basic relations in magnetostatics for the homogeneous medium of permeability μ follow.

First of all, we assume that

$$\nabla \cdot \mathbf{A} = \text{div } \mathbf{A} = 0 \tag{1.168}$$

Then

$$\nabla \times \nabla \times \mathbf{A} = -\nabla^2 \mathbf{A} + \nabla(\nabla \cdot \mathbf{A}) \tag{1.169}$$

where the latter term on the right is zero. Thus we can write

$$\nabla \times \nabla \times \mathbf{A} = \mu \mathbf{J} \tag{1.170}$$

Then in view of (1.154)

$$\nabla \times \mathbf{H} = \mathbf{J} \tag{1.171}$$

Furthermore, it is easy to see from (1.154) that

$$\text{div } \mathbf{B} = 0 \tag{1.172}$$

because div curl is identically zero. Actually, both (1.171) and (1.172) apply for inhomogeneous media, but then

$$\text{div } \mathbf{H} = 0 \tag{1.173}$$

is valid only if μ is spatially constant.

Also, we note that since (1.171) applies at every point in the medium, both sides can be integrated over a surface; that is,

$$\int_s (\nabla \times \mathbf{H}) \cdot \mathbf{n} \, dS = \int \mathbf{J} \cdot \mathbf{n} \, dS \tag{1.174}$$

where $\mathbf{n}$ is the normal to this surface and dS is an element of area. Now Stokes' theorem is

$$\oint \mathbf{H} \cdot d\ell = \int (\nabla \times \mathbf{H}) \cdot \mathbf{n} \, dS \tag{1.175}$$

where the contour integral on the left is for the closed circuit around the boundary of the surface. On equating (1.174) and (1.175), we arrive at another basic form of Ampere's law:

$$\oint \mathbf{H} \cdot d\ell = \int \mathbf{J} \cdot \mathbf{n} \, dS \qquad (1.176)$$

In words, this important law says the line integral of H around a closed circuit is equal to the total current enclosed.

1.13 MAGNETIC FIELDS OF CURRENT DISTRIBUTIONS

There are many important practical situations where we need to estimate the magnetic fields of specified electric currents. We will consider a number of problems of this type where we will assume that the surrounding medium is either a vacuum or a region of constant magnetic permeability.

The first example deals with a linear wire of length L that carries a constant current I. The situation is indicated in Figure 1.15, where we have chosen appropriate coordinate systems (x, y, z) and (ρ, ϕ, z). To avoid confusion, we designate z' to indicate the variable point on the current-carrying wire. Here $-L/2 < z' < L/2$ is the specified range of z', while dz' is an infinitesimal element.

The particular form of Ampere's law that we will use is given by (1.147). It is sometimes called the Biot-Savart law when written in this

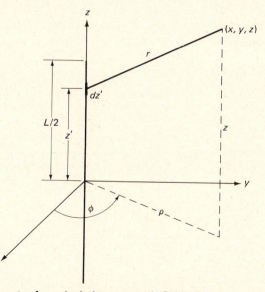

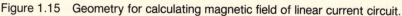

Figure 1.15 Geometry for calculating magnetic field of linear current circuit.

form. In any case for our problem we have

$$\mathbf{H} = \frac{I}{4\pi} \int_{-L/2}^{L/2} \frac{d\boldsymbol{\ell}' \times \mathbf{r}}{r^3} \qquad (1.177)$$

The vector $\mathbf{r}$ is given by

$$\mathbf{r} = \mathbf{i}_x x + \mathbf{i}_y y + \mathbf{i}_z(z - z')$$

and the vector $d\boldsymbol{\ell}'$ is simply

$$d\boldsymbol{\ell}' = \mathbf{i}_z \, dz'$$

Now

$$d\boldsymbol{\ell}' \times \mathbf{r} = \begin{bmatrix} \mathbf{i}_x & \mathbf{i}_y & \mathbf{i}_z \\ 0 & 0 & dz' \\ x & y & z - z' \end{bmatrix} \qquad (1.178)$$
$$= -\mathbf{i}_x y \, dz' + \mathbf{i}_y x \, dz'$$

Then we see that

$$H_x = -\frac{Iy}{4\pi} \int_{-L/2}^{L/2} \frac{dz'}{[x^2 + y^2 + (z - z')^2]^{3/2}} \qquad (1.179)$$

and

$$H_y = \frac{Ix}{4\pi} \int_{-L/2}^{L/2} \frac{dz'}{[x^2 + y^2 + (z - z')^2]^{3/2}} \qquad (1.180)$$

and, of course,

$$H_z = 0$$

Now in vector form our result reads

$$\mathbf{H} = H_x \mathbf{i}_x + H_y \mathbf{i}_y \qquad (1.181)$$

But note that the unit vector in the ϕ direction is

$$\mathbf{i}_\phi = -\mathbf{i}_x \sin \phi + \mathbf{i}_y \cos \phi = -\frac{y}{\rho} \mathbf{i}_x + \frac{x}{\rho} \mathbf{i}_z$$

Also,

$$H_\phi = -\frac{y}{\rho} H_x + \frac{x}{\rho} H_y \qquad (1.182)$$

Thus

$$H_\phi = \frac{I\rho}{4\pi} \int_{-L/2}^{L/2} \frac{dz'}{[x^2 + y^2 + (z - z')^2]^{3/2}} \qquad (1.183)$$

The integration can be carried out to give

$$H_\phi = \frac{I}{4\pi\rho} \left(\frac{z + (L/2)}{\{\rho^2 + [z + (L/2)]^2\}^{1/2}} \right.$$
$$\left. - \frac{z - (L/2)}{\{\rho^2 + [z - (L/2)]^2\}^{1/2}} \right) \tag{1.184}$$

Now for an infinitely long wire we obtain the limiting expression

$$H_\phi \bigg]_{L\to\infty} = \frac{I}{4\pi\rho}[+1 - (-1)] = \frac{I}{2\pi\rho} \tag{1.185}$$

This equation, of course, is consistent with Ampere's law, which says, in the present context, that the line integral of the magnetic field $2\pi\rho H_\phi$ around a closed circuit of radius ρ is equal to the current I.

The other limiting case of no less interest is when all dimensions such as ρ and z are large compared with L. Then, for example, we see that (1.184) is simplified to

$$H_\phi \simeq \frac{I\rho L}{4\pi R^3} \tag{1.186}$$

where

$$R = (x^2 + y^2 + z^2)^{1/2}$$

In this case the external field is that of an electric dipole of current moment IL.

Now it is instructive here to solve the preceding problem by use of the vector potential. Thus by reference to (1.154) and (1.155) it follows that

$$\mathbf{H}(r, \rho, \phi) = \frac{1}{\mu} \operatorname{curl} \mathbf{A} \tag{1.187}$$

where

$$\mathbf{A} = \mu \mathbf{i}_z \int_{-L/2}^{L/2} \frac{I}{4\pi r} \, dz' = \mathbf{i}_z A_z \tag{1.188}$$

where

$$r = [\rho^2 + (z - z')^2]^{1/2}$$

Because $\mathbf{A}$ only has a z component and noting that $\delta/\delta\phi = 0$, we immediately deduce that $\mathbf{H}$ only has a ϕ component given by

$$H_\phi = -\frac{1}{\mu} \frac{\partial A_z}{\partial \rho} \tag{1.189}$$

This equation immediately leads back to (1.183). In the electric dipole approximation (that is, ρ and $z \gg L$), we have the remarkably simple result

$$A_z = \frac{\mu IL}{4\pi R} \tag{1.190}$$

where

$$R = (x^2 + y^2 + z^2)^{1/2} = (\rho^2 + z^2)^{1/2}$$

A related problem of some interest is to deduce the field of a closed loop of radius a and carrying a uniform current I (see Figure 1.16). We calculate an expression for the vector potential at (x, y, z). It is deduced from

$$\mathbf{A} = \frac{\mu I}{4\pi} \oint \frac{d\ell}{r} \tag{1.191}$$

where

$$\begin{aligned}
d\ell &= \mathbf{i}_\phi a \, d\phi' \\
&= (-\mathbf{i}_x \sin \phi' + \mathbf{i}_y \cos \phi') a \, d\phi'
\end{aligned}$$

and

$$\begin{aligned}
r &= [(x - a \cos \phi')^2 + (y - a \sin \phi')^2 + z^2]^{1/2} \\
&= [x^2 + y^2 + z^2 + a^2 - 2ax \cos \phi' - 2ay \sin \phi']^{1/2}
\end{aligned}$$

The integration with respect to ϕ' over the range 0 to 2π is certainly possible, and the result can be expressed in terms of complete elliptic

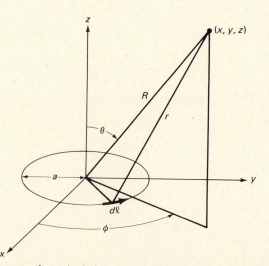

Figure 1.16 Geometry for calculating magnetic field of circular loop of current.

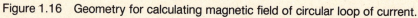

integrals. But for our purpose we will restrict attention to the case when $R \gg a$, where $R = (x^2 + y^2 + z^2)^{1/2}$. Then clearly,

$$\frac{1}{r} \simeq \frac{1}{R}\left(1 + \frac{ax}{R^2}\cos\phi' + \frac{ay}{R^2}\sin\phi'\right) \tag{1.192}$$

where neglected terms contain higher powers of ax/R^2 and ay/R^2.

We now need to deal with

$$\mathbf{A} = \frac{\mu Ia}{4\pi R}\int_0^{2\pi} (-\mathbf{i}_x\sin\phi' + \mathbf{i}_y\cos\phi')$$
$$\times\left(1 + \frac{ax}{R^2}\cos\phi' + \frac{ay}{R^2}\sin\phi'\right)d\phi' \tag{1.193}$$

The integration can be carried out to give

$$\mathbf{A} = \frac{\mu I\pi a^2}{4\pi R^3}(-\mathbf{i}_x y + \mathbf{i}_y x) \tag{1.194}$$

Now

$$\mathbf{i}_\phi = -\mathbf{i}_x\sin\phi + \mathbf{i}_y\cos\phi$$
$$= -\mathbf{i}_x\left(\frac{y}{\rho}\right) + \mathbf{i}_y\left(\frac{x}{\rho}\right)$$

Thus we find that

$$\mathbf{A} = \mathbf{i}_\phi A_\phi$$

where

$$A_\phi = \frac{\mu I(\pi a^2)}{4\pi R^2}\sin\theta \tag{1.195}$$

We can obtain the expressions for the magnetic field components by working from

$$\mathbf{H} = \frac{1}{\mu}\text{ curl }\mathbf{A} \tag{1.196}$$

Then we deduce that in spherical coordinates

$$H_R = \frac{(\pi a^2)I}{4\pi R^3}\,2\cos\theta \tag{1.197}$$

and

$$H_\theta = \frac{(\pi a^2)I}{4\pi R^3}\sin\theta \tag{1.198}$$

and

$$H_\phi = 0$$

We have here the magnetic field configuration associated with a magnetic dipole. Its strength or moment is defined as the area times the circulating current (that is, $\pi a^2 I$). Earlier, we had derived the corresponding field of the electric dipole with an electric moment $q\ell$ in a homogeneous dielectric of permittivity ϵ. For the same spherical coordinate system (R, θ, ϕ) *used here*, we deduced that

$$E_R = \frac{q\ell}{4\pi\epsilon R^3}\, 2\cos\theta \tag{1.199}$$

and

$$E_\theta = \frac{q\ell}{4\pi\epsilon R^3}\,\sin\theta \tag{1.200}$$

$$E_\phi = 0$$

As indicated, the two situations are completely analogous. This type of duality turns out to be very useful in the development of the subject.

Exercise: Derive the result stated by (1.198) but begin with the Biot-Savart law, which for the loop geometry in Figure 1.16 is given by

$$\mathbf{H} = \frac{I}{4\pi}\oint \frac{d\boldsymbol{\ell}' \times \mathbf{r}}{r^3} \tag{1.201}$$

1.14 IMAGE CONCEPT IN MAGNETOSTATICS

We can deal with the effects of disturbing bodies in magnetostatics in much the same manner as in electrostatics. First of all, we note that in a homogeneous material medium of magnetic permeability μ, we have

$$\operatorname{div}\mathbf{H} = 0 \tag{1.202}$$

in regions devoid of sources. Now in analogy to electrostatics we set

$$\mathbf{H} = -\operatorname{grad}\Phi \tag{1.203}$$

where Φ is a magnetic potential. Then clearly

$$\operatorname{div}\operatorname{grad}\Phi = \nabla^2\Phi = 0 \tag{1.204}$$

which of course is Laplace's equation.

It is useful to illustrate this potential concept to describe the fields of a magnetic dipole. For example, we will locate the small current-carrying loop (with area-current product $I\,dA$) at the origin of a cylindrical coordinate system (ρ, ϕ, z), and the loop axis is chosen with its axis pointing in the z direction. By converting the expressions given by

(1.197) and (1.198) in spherical coordinates to cylindrical coordinates, we readily deduce that

$$H_\rho = \frac{I \, dA \, 3\rho z}{4\pi R^5}$$

(1.205)

and

$$H_z = \frac{I \, dA}{4\pi}\left(\frac{3z^2}{R^5} - \frac{1}{R^3}\right)$$

(1.206)

where $R = (\rho^2 + z^2)^{1/2}$. But we confirm immediately that these results are obtained from

$$H_\rho = -\frac{\partial \Phi}{\partial \rho}$$

(1.207)

and

$$H_z = -\frac{\partial \Phi}{\partial z}$$

(1.208)

if

$$\Phi = \frac{I \, dA z}{4\pi R^3}$$

(1.209)

The latter expression for the potential of the magnetic dipole written in spherical coordinates is

$$\Phi = \frac{I \, dA \cos \theta}{4\pi R^2}$$

(1.210)

Then we readily confirm that (1.197) and (1.198) are retrieved from the gradient operations

$$H_R = -\frac{\partial \Phi}{\partial R}$$

(1.211)

and

$$H_\theta = -\frac{1}{R}\frac{\partial \Phi}{\partial \theta}$$

(1.212)

We said above that **H** can be derived from the gradient of a scalar potential at any point of a homogeneous region devoid of sources. But, of course, we can approach the source (for example, $R \rightarrow$ small as desired), where we note for a magnetic dipole H_ρ and H_z both vary as R^{-3} while the potential itself varies as R^{-2}.

It is also useful (with reference to later discussions) to rewrite

(1.210) for the magnetic potential of a magnetic dipole in the form

$$\Phi = -\frac{\partial \Pi^*}{\partial z} \qquad (1.213)$$

where

$$\Pi^* = \frac{I\,dA}{4\pi R} \qquad (1.214)$$

The scalar Π^* is here an auxiliary function that has perfect spherical symmetry for this problem. It can be called a magnetic Hertz potential that is a special form of the Hertz vector that we discuss later. There is an analogy between the magnetic Hertz potential for a magnetic dipole given by (1.214) and the electric potential of a point charge as given by (1.21).

We are now in the position to deal with some basic problems involving inhomogeneous magnetic media. We consider the specific configuration shown in Figure 1.17. With respect to a cylindrical coordinate system (ρ, ϕ, z) we have two homogeneous half spaces separated by a plane interface at $z = 0$. The upper region $z > 0$ has a permeability μ_1 and the lower region $z < 0$ has a permeability μ_2. We locate a magnetic dipole or small loop of area dA carrying a current I at $z = h > 0$ on the z axis. The dipole is oriented in the positive z direction (that is, the loop axis is vertical relative to the horizontal interface).

First of all, we write the expressions for the primary fields at $P(\rho, z)$,

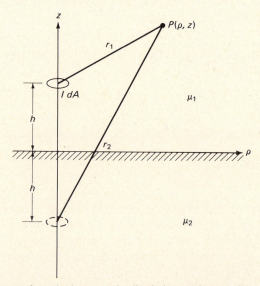

Figure 1.17   Image of vertical magnetic dipole located over plane interface between two homogeneous media of contrasting magnetic permeabilities.

which are independent of ϕ, for an infinite medium of permeability μ_1. These expressions are obtained in terms of the primary potential Φ^p via

$$\mathbf{H}^p = -\operatorname{grad} \Phi^p \tag{1.215}$$

where

$$\Phi^p = \frac{I \, dA(z - h)}{4\pi r_1^3} \tag{1.216}$$

where

$$r_1 = [\rho^2 + (z - h)^2]^{1/2} \tag{1.217}$$

This form for Φ^p follows directly from (1.209).

Now as in the analogous electrostatic problems, we need to augment the primary potential by a secondary potential that will properly account for the presence of the lower half space of permeability μ_2. In other words, for $z > 0$ we are saying that resultant fields are to be obtained from

$$\mathbf{H} = -\operatorname{grad} \Phi \tag{1.218}$$

where

$$\Phi = \Phi^p + \Phi^s \tag{1.219}$$

where Φ^s is the secondary potential.

We now take a hint from the corresponding electrostatic solutions and postulate that Φ^s results from an image magnetic dipole of relative strength K_m located at $z = -h$ just beneath the source magnetic dipole at $z = h$. This postulation leads us to write

$$\Phi^s = K_m \frac{I \, dA(z + h)}{4\pi r_2^3} \tag{1.220}$$

where

$$r_2 = [\rho^2 + (z + h)^2]^{1/2} \tag{1.221}$$

We now turn our attention to the case where the observer is in the lower half space. Then, again in analogy with the electrostatic problems [for example, Equation (1.62)], we express the potential in the region $z < 0$ in the form

$$\Phi = \Phi^\ell = T_m \frac{I \, dA(z - h)}{4\pi r_1^3} \tag{1.222}$$

where we imagine that T_m is a transmission coefficient that would be 1 in the case where $\mu_2 = \mu_1$.

An intermediate step now is to confirm that

$$\nabla^2 \Phi^p = \nabla^2 \Phi^s = 0 \tag{1.223}$$

everywhere in the region $z > 0$ except right at the source point $\rho = 0$, $z = h$. Also,

$$\nabla^2 \Phi^\ell = 0 \tag{1.224}$$

everywhere in the region $z < 0$ with no exceptions. Also, the postulated forms for the potentials vanish as $\rho \to \infty$ and $|z| \to \infty$.

The task remains to determine the coefficients K_m and T_m from the boundary conditions at $z = 0$. The latter, in words, can be stated as follows: (1) Normal magnetic flux density is continuous, and (2) tangential magnetic field is continuous. Thus

$$\mu_1 \frac{\partial}{\partial z}(\Phi^p + \Phi^s)\bigg|_{z=0} = \mu_2 \frac{\partial}{\partial z} \Phi^\ell \bigg|_{z=0} \tag{1.225}$$

and

$$(\Phi^p + \Phi^s)\bigg|_{z=0} = \Phi^\ell \bigg|_{z=0} \tag{1.226}$$

After carrying out the derivative operations before setting $z = 0$, we find that

$$\mu_1(1 + K_m) = \mu_2 T_m \tag{1.227}$$

and

$$-1 + K_m = -T_m \tag{1.228}$$

Solving this pair gives

$$K_m = \frac{\mu_2 - \mu_1}{\mu_2 + \mu_1} \tag{1.229}$$

and

$$T_m = \frac{2\mu_1}{\mu_2 + \mu_1} \tag{1.230}$$

Now that we have solved the problem of the magnetic dipole or small loop located *above* the interface, it is now instructive to let $h \to 0$. Then the resultant potentials in the two regions are seen to be

$$\Phi = \frac{I\,dA}{4\pi r^3} z(1 + K_m) = \frac{I\,dA}{4\pi r^3} \frac{2\mu_2}{\mu_2 + \mu_1} \tag{1.231}$$

for $z > 0$, where $r = (\rho^2 + z^2)^{1/2}$ and

$$\Phi = \frac{I\,dA}{4\pi r^3} \frac{2\mu_1}{\mu_2 + \mu_1} \tag{1.232}$$

for $z < 0$. In both cases the magnetic field has the configuration of a magnetic dipole, but for the upper region ($z > 0$) the moment is modified by $2\mu_2/(\mu_2 + \mu_1)$, while in the lower region the effective moment is modified by $2\mu_1/(\mu_2 + \mu_1)$.

Exercise: Consider the same problem as shown in Figure 1.17 but orient the source magnetic dipole in the horizontal direction. It is now convenient to convert to cartesian coordinates (x, y, z) with the dipole located at $z = h$ on the z axis and oriented in the x direction. Show that the resultant potential in the upper half space is

$$\Phi = \frac{I\,dA\,x}{4\pi}\left(\frac{1}{r_1^3} - \frac{K_m}{r_2^3}\right) \tag{1.233}$$

where K_m is defined by (1.229).

While the image representation can be gotten rather directly for the magnetic dipole located over a half space, there is a certain amount of intuition involved in the initial postulates. It is worthwhile to solve this problem by two additional methods that proceed in a more logical fashion. Such an approach is more complicated but it is also more general.

Alternative Method I

We refer to Figure 1.17 for the same configuration involving a vertical magnetic dipole located at height h over the interface. Now we write the primary potential for $z > 0$ in the form

$$\Phi^p = -\frac{I\,dA}{4\pi}\frac{\partial}{\partial z}\frac{1}{r_1} \tag{1.234}$$

which, of course, is identical to (1.216). Now we note the identity

$$\frac{1}{r_1} = \int_0^\infty J_0(\lambda\rho)e^{\pm\lambda(z-h)}\,d\lambda \tag{1.235}$$

where the minus sign in the exponent is used for $z > h$ and the plus sign for $z < h$. Thus the integral formula for the primary potential is

$$\Phi^p = \pm\frac{I\,dA}{4\pi}\int_0^\infty \lambda J_0(\lambda\rho)e^{\pm\lambda(z-h)}d\lambda \tag{1.236}$$

where the upper (lower) signs are to be used for $z > h(z < h)$.

Now rather than "guessing" that the secondary potential for $z > 0$ is due to an image, we now choose an integral form by taking a hint from (1.216). This action leads to the form

$$\Phi^s = \frac{I\,dA}{4\pi}\int_0^\infty R(\lambda)\lambda J_0(\lambda\rho)e^{-\lambda(z+h)}\,d\lambda \tag{1.237}$$

for $z > 0$, where $R(\lambda)$, in general, could be a function of λ. The other factors are inserted for convenience when we later apply boundary conditions at $z = 0$. We also confirm that

$$\nabla^2 \Phi^s = 0$$

and also observe that the form $e^{-\lambda z}$ (not $e^{+\lambda z}$) is needed in the integral if the solution is to be finite as $z \to \infty$.

To construct the integral form in the lower half space, we now use the basic solution type $J_0(\lambda \rho) e^{\lambda z}$ that remains finite as $z \to -\infty$. This solution suggests we write for $z < 0$ that

$$\Phi^\ell = -\frac{I\, dA}{4\pi} \int_0^\infty T(\lambda)\lambda J_0(\lambda \rho) e^{\lambda(z-h)}\, d\lambda \tag{1.238}$$

where the function $T(\lambda)$ is not known.

The boundary conditions for the present problem are specified by (1.225) and (1.226). These conditions can be applied directly to (1.236), (1.237), and (1.238), with the result that

$$\mu_1[1 + R(\lambda)] = \mu_2 T(\lambda) \tag{1.239}$$

and

$$-1 + R(\lambda) = -T(\lambda) \tag{1.240}$$

Thus

$$R(\lambda) = K_m \tag{1.241}$$

and

$$T(\lambda) = T_m \tag{1.242}$$

as given by (1.229) and (1.230).

Noting that (1.237) is equivalent to

$$\Phi^s = -\frac{I\, dA}{4\pi} K_m \frac{\partial}{\partial z} \frac{1}{r_2} \tag{1.243}$$

and (1.238) is equivalent to

$$\Phi^\ell = -\frac{I\, dA}{4\pi} T_m \frac{\partial}{\partial z} \frac{1}{r_1} \tag{1.244}$$

we recover (1.220) and (1.222) based on the image formulation. The relative simplicity of the present problem arises from the fact that $R(\lambda)$ and $T(\lambda)$ are independent of λ.

Alternative Method II

We now wish to tackle the problem without having to introduce an expression for the primary potential from prior knowledge. Instead we will just invoke Ampere's current law at the source. To facilitate the solution and at the same time provide a more general result, we will let the radius of the loop be unrestricted. The uniform current in the loop will still be I, and the loop itself is contained in the plane $z = h$. The situation is illustrated in Figure 1.18.

We now consider that we have three regions. The first, designated (1), is above the loop, (1') is between the loop plane and the interface, and (2) is below the interface. In each region the respective total potential satisfies Laplace's equation, and the boundary conditions at $z = 0$ must be satisfied. In addition, we must invoke a source condition at the loop plane.

In view of the above considerations we are led to write for region (1), $z > h$,

$$\Phi = \int_0^\infty f(\lambda) J_0(\lambda\rho) e^{-\lambda z} \, d\lambda \tag{1.245}$$

for region (1'), $0 < z < h$,

$$\Phi = \int_0^\infty [g(\lambda) e^{\lambda z} + p(\lambda) e^{-\lambda z}] J_0(\lambda\rho) \, d\lambda \tag{1.246}$$

for region (2), $z < 0$,

$$\Phi = \int_0^\infty q(\lambda) e^{\lambda z} J_0(\lambda\rho) \, d\lambda \tag{1.247}$$

Here f, g, p, and q, which may be functions of λ, are yet to be determined. Clearly, these expressions for the potential satisfy $\nabla^2 \Phi = 0$ everywhere with the exclusion of the surface $z = h$, although we may approach it from above or below.

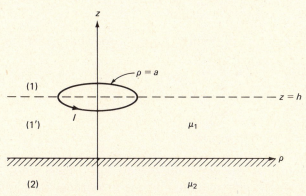

Figure 1.18 Finite circular loop of current over plane interface between homogeneous regions.

Now our source condition is

$$\lim_{\epsilon \to 0} [H_\rho(\rho, z)|_{z=h+\epsilon} - H_\rho(\rho, z)|_{z=h-\epsilon}] = j_\phi(\rho) \qquad (1.248)$$

where $j_\phi(\rho)$ is the surface current density, in amperes per meter, at the surface $z = h$. This equation is really a statement or consequence of Ampere's law being a special case of (1.176). In the present case the contour goes from $\rho = 0$ just above the loop at $z = h + \epsilon$ along a radial to $\rho = \infty$, then returns along the radial just below the loop at $z = h - \epsilon$. The normal current threading this narrow rectangle is $j_\phi(\rho)$ A/m. In our case

$$j_\phi(\rho) = I\,\delta(\rho - a) \qquad (1.249)$$

where $\delta(\rho - a)$ is the unit impulse or Dirac function. This function has the property that it is infinite at $\rho = a$ but of unit area. Thus, for example,

$$\lim_{\Delta x \to 0} \int_{x-\Delta x}^{x+\Delta x} \delta(x)\,dx = 1 \qquad (1.250)$$

In other words, we are saying that the source current in the plane $z = h$ results in a discontinuity of the horizontal component of the magnetic field. As stated, this equation is in accord with Ampere's current law.

Another source condition that applies to the present problem is that the vertical magnetic field is continuous at $z = h$. That is,

$$\lim_{\epsilon \to 0} [H_z(\rho, z)|_{z=h+\epsilon} - H_z(\rho, z)|_{z=h-\epsilon}] = 0 \qquad (1.251)$$

To implement the source and boundary conditions, we note that for region (1)

$$H_\rho(\rho, z) = -\frac{\partial \Phi}{\partial \rho} = \int_0^\infty f(\lambda) J_1(\lambda \rho) e^{-\lambda z} \lambda\, d\lambda \qquad (1.252)$$

where

$$J_1(x) = -\frac{d}{dx} J_0(x) \qquad (1.253)$$

is the Bessel function of the first type of order one. Also, for region (1)

$$H_z(\rho, z) = -\frac{\partial \Phi}{\partial z} = \int_0^\infty f(\lambda) J_0(\lambda \rho) e^{-\lambda z} \lambda\, d\lambda \qquad (1.254)$$

We now write

$$j_\phi(\rho) = \int_0^\infty A(\lambda) J_1(\lambda \rho) \lambda\, d\lambda \qquad (1.255)$$

which obviously is the form needed to make both sides of (1.248) compatible. Here we invoke the Fourier-Bessel transform to tell us that

$$A(\lambda) = \int_0^\infty j_\phi(\rho) J_1(\lambda \rho) \rho\, d\rho \qquad (1.256)$$

In the case of the circular loop we insert (1.249) into (1.256), yielding

$$A(\lambda) = IJ_1(\lambda a)a \tag{1.257}$$

The source condition given by (1.248) is now seen to be satisfied if

$$f(\lambda)e^{-\lambda h} - g(\lambda)e^{\lambda h} - p(\lambda)e^{-\lambda h} = A(\lambda) \tag{1.258}$$

The continuity condition at $z = h$ given by (1.251) is satisfied if

$$f(\lambda)e^{-\lambda h} + g(\lambda)e^{\lambda h} - p(\lambda)e^{-\lambda h} = 0 \tag{1.259}$$

From (1.258) and (1.259) we deduce that

$$g(\lambda) = -(\tfrac{1}{2})A(\lambda)e^{-\lambda h} \tag{1.260}$$

and

$$f(\lambda) = (\tfrac{1}{2})A(\lambda)e^{\lambda h} + p(\lambda) \tag{1.261}$$

Thus our forms for the potentials above the interface are for (1), $z > h$,

$$\Phi = \frac{1}{2}\int_0^\infty A(\lambda)e^{-\lambda(z-h)}J_0(\lambda\rho)\,d\lambda + \int_0^\infty p(\lambda)J_0(\lambda\rho)e^{-\lambda z}\,d\lambda \tag{1.262}$$

and for (1′), $0 < z < h$,

$$\Phi = -\frac{1}{2}\int_0^\infty A(\lambda)e^{\lambda(z-h)}J_0(\lambda\rho)\,d\lambda + \int_0^\infty p(\lambda)J_0(\lambda\rho)e^{-\lambda z}\,d\lambda \tag{1.263}$$

The corresponding form for the region (2), $z < 0$, is

$$\Phi = \int_0^\infty q(\lambda)J_0(\lambda\rho)e^{\lambda z}\,d\lambda \tag{1.264}$$

as we indicated before. To solve for $p(\lambda)$ and $q(\lambda)$, we apply the continuity conditions

$$\mu_1\frac{\partial\Phi}{\partial z}\bigg|_{z=+0} - \mu_2\frac{\partial\Phi}{\partial z}\bigg|_{z=-0} = 0 \tag{1.265}$$

and

$$\Phi\bigg|_{z=+0} - \Phi\bigg|_{z=-0} = 0 \tag{1.266}$$

Applying these conditions yields

$$p(\lambda) = \frac{1}{2}\frac{\mu_2 - \mu_1}{\mu_2 + \lambda_1}A(\lambda)e^{-\lambda h} \tag{1.267}$$

and

$$q(\lambda) = -\frac{1}{2}\frac{2\mu_1}{\mu_2 + \mu_1}A(\lambda)e^{-\lambda h} \tag{1.268}$$

When these expressions are inserted in (1.261), (1.262), and (1.263), we have explicit forms for the potentials in every region. In general, $A(\lambda)$ is the Fourier-Bessel transform of the specified current distribution in the surface $z = h$. As indicated by (1.257), the form for $A(\lambda)$ simplifies if we are dealing with a circular loop. If the loop radius is sufficiently small, we have the further simplification to

$$A(\lambda) = \frac{I\lambda a^2}{2} \qquad\qquad (1.269)$$

where we have made use of the Bessel function approximation

$$J_1(x) = \frac{x}{2}$$

which is valid when $x \ll 1$ (neglected terms in the ascending power series contain x^3, x^5, and so on).

Not surprisingly, the limit of the vanishing small loop is identical to the image solution obtained earlier. The present approach, while much more complicated, can be applied to more realistic models, including cases involving electromagnetic induction where the simple image concepts do not apply.

1.15 FURTHER WORKED EXAMPLES

Exercise: Deduce the dc magnetic field of an electric dipole given the electric fields produced by this dipole in a homogeneous conducting medium of conductivity σ.

SOLUTION: A spherical coordinate system is chosen with the electric dipole of current moment $I\,d\ell$ located at the origin and oriented in the z direction. The electric field components are

$$E_r = \frac{I\,d\ell\,\cos\theta}{2\pi\sigma r^3}$$

$$E_\theta = \frac{I\,d\ell\,\sin\theta}{4\pi\sigma r^3}$$

Now Ampere's law can be used to deduce H_ϕ, which is the only component of the magnetic field. Thus

$$2\pi r\,\sin\theta\,H_\phi = \int_0^{2\pi}\int_0^{\theta} J_r r^2\,\sin\theta\,d\phi\,d\theta$$

Now

$$J_r = \sigma E_r$$

Thus

$$r \sin \theta \, H_\phi = \frac{I \, d\ell}{2\pi r} \int_0^\theta \cos \theta \sin \theta \, d\theta$$

which easily leads to

$$H_\phi = \frac{I \, d\ell \sin \theta}{4\pi r^2}$$

(See Chapter 2 for further developments of this quasi-static approach.)

Exercise: Deduce the magnetic field from an insulated linear wire carrying a constant current I and grounded at the endpoints in a homogeneous conducting medium.

SOLUTION: A cylindrical coordinate system is chosen with endpoints of the wire at $z = \pm L/2$ on the z axis. The situation is illustrated in Figure 1.19. It is also convenient to choose spherical coordinates (r_1, θ_1) and (r_2, θ_2) about current electrodes C_1 and C_2, as indicated, bearing in mind that we have azimuthal symmetry (that is, $\delta/\delta\phi = 0$). Now the current density emanating from C_1 is purely radial and given by

$$J_{r1} = \frac{I}{4\pi r_1^2}$$

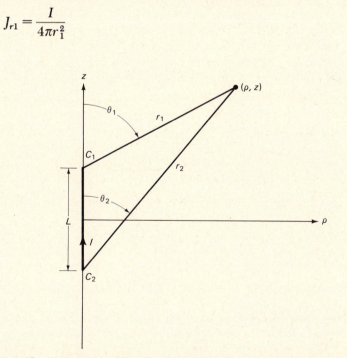

Figure 1.19   Linear wire of length L carrying uniform current I with grounded endpoints in homogeneous conducting medium.

while that from C_2 is

$$J_{r2} = \frac{-I}{4\pi r_2^2}$$

Now clearly the total resultant current density in the vertical (that is, z) direction is obtained from

$$J_z = J_{r1}\cos\theta_1 + J_{r2}\cos\theta_2 = \frac{I}{4\pi}\left[\frac{z-(L/2)}{r_1^3} - \frac{z+(L/2)}{r_2^3}\right]$$

We now invoke Ampere's law in its most basic form to deduce the magnetic field at (ρ, ϕ), which only has a component H_ϕ. Thus we write

$$2\pi\rho H_\phi = 2\pi \int_0^\rho J_z\rho\,d\rho$$

where the left-hand side is the line integral of H around the circuit of radius ρ, while the right-hand side is the total current threading the loop of radius ρ. Thus

$$H_\phi = \frac{I}{4\pi\rho}\left[\left(z-\frac{L}{2}\right)\int_0^\rho \frac{\rho}{r_1^3}\,d\rho - \left(z+\frac{L}{2}\right)\int_0^\rho \frac{\rho}{r_2^3}\,d\rho\right]$$

Noting that

$$\left(\frac{\partial}{\partial\rho}\right)r_1^{-1} = -\rho r_1^{-3}$$

and

$$\left(\frac{\partial}{\partial\rho}\right)r_2^{-1} = -\rho r_2^{-3}$$

we readily deduce that

$$H_\phi = \frac{I}{4\pi\rho}\left(\frac{z+(L/2)}{\{\rho^2+[z+(L/2)]^2\}^{1/2}} - \frac{z-(L/2)}{\{\rho^2+[z-(L/2)^2]\}^{1/2}}\right)$$

which, by no coincidence, is identical to (1.184) obtained by a different method.

References

1. J. C. Maxwell: *A Treatise on Electricity and Magnetism*, Clarendon Press, 1891.
2. J. A. Stratton: *Electromagnetic Theory*, McGraw-Hill, New York, 1941. Fundamental and authorative text written at a rather advanced level.
3. S. Ramo, J. Whinnery, and T. van Duzer: *Fields and Waves in Communication Electronics*, Wiley, New York, 1964. Excellent introduction to static electric and static magnetic fields with applications in electrotechnology.

4. R. Plonsey and R. E. Collin: *Principles and Applications of Electromagnetic Fields*, McGraw-Hill, New York, 1961. Designed for the serious student with many worked examples and good chapters on static fields.

5. S. A. Schelkunoff: *Electromagnetic Waves*, Van Nostrand, New York, 1943. Mathematically satisfying treatment of static field concepts and clearly presented sections on related circuit approaches.

6. E. C. Jordan and K. G. Balmain *Principles and Applications of Electromagnetic Fields*, McGraw-Hill, New York, 1968. An intermediate text written with exceptional clarity and physical insight with many applications.

7. W. R. Smythe: *Static and Dynamic Electricity*, 3d ed., McGraw-Hill, New York, 1968. Encyclopedic coverage of solutions.

8. J. R. Wait: *Electromagnetics and Plasmas*, Holt, Rinehart and Winston, New York, 1969.

Chapter 2
Quasi-Static Current Flow in Heterogeneous Conductors

2.1 INTRODUCTION

In dealing with electrostatics, we usually assign permittivities to the constituent media. The resulting electric fields are derived from electric scalar potentials, which themselves are solutions of Laplace's equation in piecewise homogeneous regions. While the stated condition on solving electrostatic problems is time invariance, the results are generally valid for slowly varying fields and for media that have finite conductivity.

It is our purpose here to demonstrate the extension of electrostatics to conditions of time-varying current flow in conducting media. All we really need to do is to recognize that Ohm's law for a medium of conductivity σ and permittivity ϵ relates the electric field $\mathbf{E}$ to the current density $\mathbf{J}$ by the exact formula

$$\mathbf{J} = (\sigma + i\epsilon\omega)\mathbf{E} \tag{2.1}$$

when the fields vary as $\exp(i\omega t)$. By convention, the actual physical quantity for the electric field vector $\mathbf{e}(t)$ as a function of time is

$$\mathbf{e}(t) = \operatorname{Re} \mathbf{E} \exp(i\omega t)$$

Equivalent forms of Equation (2.1) are

$$\mathbf{J} = i\hat{\epsilon}\omega\mathbf{E} \tag{2.2}$$

and

$$\mathbf{J} = \hat{\sigma}\mathbf{E} \tag{2.3}$$

where

$$\hat{\epsilon} = \epsilon - \frac{i\sigma}{\omega}$$

and

$$\hat{\sigma} = \sigma + i\epsilon\omega$$

Here $\hat{\epsilon}$ is the complex permittivity and $\hat{\sigma}$ is the complex conductivity in terms of the real parameters σ and ϵ.

As we have indicated, (2.1), (2.2), and (2.3) are equivalent statements. Once we have recognized this elegantly simple assertion, it is obvious that electrostatic solutions can be extended or adapted to current flow in conducting media by a simple change of variable. The only restriction or proviso is that the resulting electric field components are derivable from a scalar potential. In homogeneous regions the latter should satisfy Laplace's equation.

The concepts mentioned above provide a simplicity that is often overlooked. A good example is in electrical geophysics [1] where we often deal with time-varying current flow in conducting media. Some examples will be discussed here.

2.2 BASIC IDEAS

To be explicit, our basic assumption is that the electric field $\mathbf{E}$ can be derived from

$$\mathbf{E} = -\operatorname{grad}\psi \tag{2.4}$$

where ψ is a scalar potential. Further, except at source points,

$$\operatorname{div}\mathbf{J} = 0 \tag{2.5}$$

which simply states that the sum of the current into and out of any closed volume is zero. Obviously, this equation is not true if there is a current electrode in the volume!

By combining (2.1), (2.4), and (2.5) and assuming σ and ϵ are independent of the coordinates, we obtain Laplace's equation,

$$\nabla^2\psi = \operatorname{div}\operatorname{grad}\psi = 0 \tag{2.6}$$

This equation is valid for source-free homogeneous regions for dc fields or for fields that are sufficiently slowly varying. Without discussing such problems in the context of vector solutions of Maxwell's equations (see Chapter 4), we assert here that the potential theory results are valid when $|\gamma\ell| \ll 1$, where γ is the medium propagation constant and ℓ is a typical distance in the physical problem. Actually,

$$\gamma = [(i\mu_0\omega)(\sigma + i\epsilon\omega)]^{1/2} \tag{2.7}$$

where μ_0 is the magnetic permeability. In what follows here we will assume that the inequality above is met. On the other hand, we need make no assumption on the relative magnitudes of the conduction current σE and the displacement current $i\epsilon\omega E$ in the regions under consideration. This is what we mean by quasi-static theory. It encompasses a whole range of problems in electrical geophysics. Many related situations occur in the electrical studies of biological systems.

2.3 THE CONCENTRIC-SPHERE PROBLEM

We begin with a very basic problem that has a close counterpart in electrostatics. We consider a spherical body of outer radius b that is immersed in a homogeneous medium of complex conductivity $\hat{\sigma}$. The situation is illustrated in Figure 2.1, where we have chosen spherical coordinates (r, θ, ϕ) centered at the body. To make the problem somewhat interesting, we treat the body as a concentric system where the inner core of radius a has a complex conductivity $\hat{\sigma}_1$ and the outer shell of thickness $b - a$ has a complex conductivity $\hat{\sigma}_2$.

We are interested in calculating the fields produced by this spherical target when the exciting uniform electric field E_0 is taken to be parallel to the polar axis $\theta = 0$. At this point we can proceed rather quickly since the problem closely resembles the electrostatic formula-

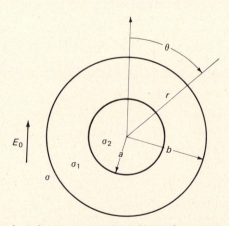

Figure 2.1 Concentric-sphere system placed in uniform electric field; regions are characterized by complex conductivities [Interchange σ_2 and σ_1].

tion of a dielectric sphere immersed in a uniform field when the external region is also a homogeneous dielectric.

For the problem as posed, the potential ψ in the external region satisfies Laplace's equation (2.6). In the present case we have azimuthal symmetry $\partial/\partial\phi = 0$. Thus for $r > b$

$$\frac{\partial}{\partial r}\left(r^2\frac{\partial\psi}{\partial r}\right) + \frac{1}{\sin\theta}\frac{\partial}{\partial\theta}\left(\sin\theta\frac{\partial\psi}{\partial\theta}\right) = 0 \qquad (2.8)$$

The two independent solutions that have the correct $\cos\theta$ angular dependence are $r\cos\theta$ and $r^{-2}\cos\theta$. Then the desired form for the external potential, for $r > b$, is

$$\psi = -E_0 r\cos\theta + \frac{a^3 E_0 \chi}{r^2}\cos\theta \qquad (2.9)$$

where the dimensionless coefficient χ is yet to be determined. We note that if $r \to \infty$, the expression for the potential $-E_0 r \times \cos\theta$ has the correct form.

Appropriate expressions for the potentials inside the spherical body are given as follows:

$$\psi_1 = -E_1 r\cos\theta \qquad (2.10)$$

for $0 < r < a$, while

$$\psi_2 = \alpha r\cos\theta + \beta r^{-2}\cos\theta \qquad (2.11)$$

for $a < r < b$. Here E_1, α, and β are coefficients yet to be determined.

Now it is useful to write expressions for the radial current density in each of the three regions. Thus

$$J_r = -\hat{\sigma}\frac{\partial\psi}{\partial r} = \hat{\sigma}E_0\left(\cos\theta + \frac{2\chi a^3}{r^3}\cos\theta\right) \qquad (2.12)$$

for $r > b$, while

$$J_{r,1} = \hat{\sigma}_1 E_1\cos\theta \qquad (2.13)$$

for $0 < r < a$, while

$$J_{r,2} = -\hat{\sigma}_2\left(\alpha\cos\theta - \frac{2\beta}{r^3}\cos\theta\right) \qquad (2.14)$$

for $a < r < b$. Corresponding forms for the tangential components J_θ, $J_{\theta,1}$, and $J_{\theta,2}$ can easily be written. For example,

$$J_\theta = -\hat{\sigma}\frac{1}{r}\frac{\partial\psi}{\partial\theta} \qquad (2.15)$$

for $r > b$. But we do not need these expressions for what follows.

Now the boundary conditions of the problem are that radial current density and potential are continuous at boundaries between the homogeneous regions. Thus

$$\left.\begin{array}{c} \psi_1 = \psi_2 \\ J_{r,1} = J_{r,2} \end{array}\right]_{r=a} \qquad (2.16)\\(2.17)$$

and

$$\left.\begin{array}{c} \psi_2 = \psi \\ J_{r,2} = J_r \end{array}\right]_{r=b} \qquad (2.18)\\(2.19)$$

are explicit statements of the boundary conditions. When these equations are applied to (2.9)–(2.14), we end up with four linear algebraic equations to solve for E_1, α, β, and χ. These equations are simplified to

$$-E_1 a = \alpha a + \frac{\beta}{a^2} \qquad (2.20)$$

$$-E_0 b + \left(\frac{a^3}{b^2}\right) E_0 \chi = \alpha b + \frac{\beta}{b^2} \qquad (2.21)$$

$$-\hat{\sigma}_1 E_1 = \hat{\sigma}_2 \alpha - \hat{\sigma}_2 \left[2\left(\frac{\beta}{a^3}\right)\right] \qquad (2.22)$$

$$-\hat{\sigma} E_0 - \hat{\sigma}\left(\frac{2a^3}{b^3}\right) E_0 \chi = \hat{\sigma}_2 \alpha - \hat{\sigma}_2\left(\frac{2\beta}{b^3}\right) \qquad (2.23)$$

Using simple algebra, we eliminate E_1 from (2.20) and (2.22) to yield

$$\frac{\beta}{\alpha} = -a^3 \left(1 - \frac{\hat{\sigma}_2}{\hat{\sigma}_1}\right)\left(1 + \frac{2\hat{\sigma}_2}{\hat{\sigma}_1}\right)^{-1} \qquad (2.24)$$

Then χ can be obtained from (2.21) and (2.23) to give

$$\chi = \left(1 - \frac{\hat{\sigma}}{\hat{\sigma}_2}\left\{\frac{1 - [(\hat{\sigma}_1 - \hat{\sigma}_2)/(\hat{\sigma}_1 + 2\hat{\sigma}_2)](a^3/b^3)}{1 + 2[(\hat{\sigma}_1 - \hat{\sigma}_2)/(\hat{\sigma}_1 + 2\hat{\sigma}_2)](a^3/b^3)}\right\}\right)\frac{b^3}{a^3}$$

$$\times \left(1 + 2\frac{\hat{\sigma}}{\hat{\sigma}_2}\left\{\frac{1 - [(\hat{\sigma}_1 - \hat{\sigma}_2)/(\hat{\sigma}_1 + 2\hat{\sigma}_2)](a^3/b^3)}{1 + 2[(\hat{\sigma}_1 - \hat{\sigma}_2)/(\hat{\sigma}_1 + 2\hat{\sigma}_2(a^3/b^3))]}\right\}\right)^{-1} \qquad (2.25)$$

Using this result for χ in (2.9), we then have an explicit expression for the electric potential in the external region. In the case where $\sigma_2 \to \infty$ (that is, perfectly conducting spherical body of radius b), we have $\chi = b^3/a^3$. In general, we can refer to χ as the normalized dipole moment.

Now as we have tried to indicate, the present formulation in terms of complex conductivities encompasses both pure dielectric media and pure conducting media. Thus, for example, if the conductivities σ, σ_1, and σ_2 are identically zero, we replace $\hat{\sigma}$, $\hat{\sigma}_1$, and $\hat{\sigma}_2$ in (2.25) by $i\epsilon\omega$, $i\epsilon_1\omega$, and $i\epsilon_2\omega$, respectively. In this case the $i\omega$'s cancel everywhere, so, in effect, the expression for χ contains only the real permittivities ϵ, ϵ_1, and

ϵ_2. The other limiting case is where the conduction currents greatly exceed the respective displacement currents in the various regions. Normally, this requirement is achieved if the frequency is sufficiently low. In this case we merely replace $\hat{\sigma}$, $\hat{\sigma}_1$, and $\hat{\sigma}_2$ by the real conductivities σ, σ_1, and σ_2.

2.4 HOMOGENEOUS SPHERE LIMIT

While (2.25), for the dipole moment, is not really that complicated even for all conductivities complex, it is useful to examine special limiting forms. The case of a homogeneous sphere can be obtained in two ways. In the first we let $a \to 0$, whence (2.9) goes over to

$$\psi = -E_0 r \cos\theta + \frac{b^3 E_0 \chi_0}{r^2} \cos\theta \qquad (2.26)$$

where

$$\chi_0 = \left(1 - \frac{\hat{\sigma}}{\hat{\sigma}_2}\right)\left(1 + 2\frac{\hat{\sigma}}{\hat{\sigma}_2}\right)^{-1} \qquad (2.27)$$

This expression for ψ corresponds to a homogeneous sphere of radius b with complex conductivity $\hat{\sigma}_2$ immersed in a homogeneous medium of complex conductivity $\hat{\sigma}$. In the second way we let $a \to b$, in which case (2.9) reduces to

$$\psi = -E_0 r \cos\theta + \frac{a^3 E_0 \chi_0}{r^2} \cos\theta \qquad (2.28)$$

where

$$\chi_0 = \left(1 - \frac{\hat{\sigma}}{\hat{\sigma}_1}\right)\left(1 + 2\frac{\hat{\sigma}}{\hat{\sigma}_1}\right)^{-1} \qquad (2.29)$$

This equation corresponds, of course, to a homogeneous sphere of radius a of complex conductivity $\hat{\sigma}_1$ placed in a medium of complex conductivity $\hat{\sigma}$.

2.5 CONDUCTING-SHELL LIMIT

The static problem of a homogeneous sphere placed in a uniform exciting field is solved in virtually every textbook. But the layering makes the problem more interesting. We will examine two special forms of the layered sphere that have relevance in both geophysics and biology. The first limiting form is to allow the outer shell of thickness $b - a$ to become very small compared with, say, the radius a. At the same time the ratio

$|\hat{\sigma}_2/\hat{\sigma}_1|$ becomes very large. Thus

$$\frac{\hat{\sigma}_1 - \hat{\sigma}_2}{\hat{\sigma}_1 + 2\hat{\sigma}_2} \frac{a^3}{b^3} = -\frac{1}{2}\left[\frac{1 - (\hat{\sigma}_1/\hat{\sigma}_2)}{1 + \frac{1}{2}(\hat{\sigma}_1/\hat{\sigma}_2)}\right]\left\{\frac{1}{1 + [(b-a)/a]}\right\}^3$$

$$\doteq \left(-\frac{1}{2} + \frac{3}{4}\frac{\hat{\sigma}_1}{\hat{\sigma}_2}\right)\left[1 - \frac{3(b-a)}{a}\right]$$

$$\doteq -\frac{1}{2} + \frac{3}{4}\frac{\hat{\sigma}_1}{\hat{\sigma}_2} + \frac{3}{2}\frac{b-a}{a} \tag{2.30}$$

As indicated, we have neglected second-order quantities.

Using (2.30) we find that (2.25) can be cast into the form

$$\chi = \frac{1 - \delta}{1 + 2\delta} \tag{2.31}$$

where

$$\delta = \frac{\hat{\sigma}}{\hat{\sigma}_1 + (2y/a)} \tag{2.32}$$

and

$$y = \hat{\sigma}_2(b - a) \tag{2.33}$$

Here y can be recognized as the admittance, in siemens, of the thin conducting shell.

In a mathematical sense (2.31) can be regarded as the limit of (2.25) when $|\hat{\sigma}_2| \to \infty$ and $(b-a) \to 0$ such that their product remains finite. Obviously, such a result is only an approximation to a highly conducting shell of nonzero thickness.

It is very instructive to examine an alternative boundary-value approach to the above thin-shell problem. In this method we bypass the need to solve Laplace's equation for the potential within the shell. But we must then develop the appropriate boundary conditions to connect up or relate the fields on the two sides of the shell.

To facilitate our discussion, we depict a small section of the vanishingly thin shell of angular width $\delta\theta$ in Figure 2.2. When this area is rotated about the polar axis $\theta = 0$, we have a volume element $2\pi a^2 \sin\theta \, \delta\theta \, \delta r$. Now because current flow is divergentless, the total current into this volume must equal the total current flowing out.

Now the net current flowing across the spherical surfaces into the element volume is

$$(J_{r,1} - J_r)2\pi a^2 \sin\theta \, \delta\theta$$

The current flowing out of the annular end sections is

$$\hat{\sigma}_2 \, \delta r \, 2\pi a[E_\theta(\theta + \delta\theta)\sin(\theta + \delta\theta) - E_\theta(\theta)\sin\theta]$$

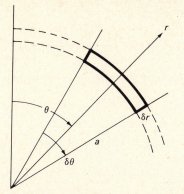

Figure 2.2   Annular section of spherical shell of thickness δr and width $a\,\delta\theta$.

or, in the limit of small $\delta\theta$,

$$\hat{\sigma}_2\,\delta r\,2\pi a\,\frac{\partial}{\partial\theta}\,(\sin\theta\,E_\theta)\,\delta\theta$$

Thus

$$(J_{r,1}-J_r)a = y\,\frac{1}{\sin\theta}\,\frac{\partial}{\partial\theta}\,(\sin\theta\,E_\theta) \tag{2.34}$$

where we note that

$$y = \hat{\sigma}_2\delta r = \hat{\sigma}_2(b-a)$$

is the admittance of the shell as previously defined. The boundary equation (2.34) still needs to be simplified. First of all, we note that the left-hand side of (2.34) is equivalent to

$$\left[-\left(\hat{\sigma}_1\,\frac{\partial\psi_1}{\partial r}\right)_{r=a-0}+\left(\hat{\sigma}\,\frac{\partial\psi}{\partial r}\right)_{r=a+0}\right]a$$

As indicated, these derivative operations are to be carried out just inside and just outside the thin shell, respectively. On the right-hand side of (2.34) we note first that E_θ is essentially continuous across the shell. Thus for convenience we write $E_\theta = -\partial\psi_1/r\,\partial\theta$, to yield the right-hand side of (2.34) in the form

$$\left[-y\,\frac{1}{\sin\theta}\,\frac{\partial}{\partial\theta}\left(\sin\theta\,\frac{1}{r}\,\frac{\partial\psi_1}{\partial\theta}\right)\right]_{r=a-0}$$

But since ψ_1 satisfies Laplace's equation, this expression is equal to

$$\frac{y}{a}\,\frac{\partial}{\partial r}\left(r^2\,\frac{\partial\psi_1}{\partial r}\right)\bigg]_{r=a-0}$$

Thus (2.34) is fully equivalent to

$$\left(\hat{\sigma}\,\frac{\partial\psi}{\partial r}\right)_{r=a+0} = \left[\hat{\sigma}_1\,\frac{\partial\psi_1}{\partial r} + \frac{y}{a^2}\,\frac{\partial}{\partial r}\left(r^2\,\frac{\partial\psi_1}{\partial r}\right)\right]_{r=a-0} \tag{2.35}$$

This thin-conducting-sheet boundary condition is to be supplemented by

$$(\psi)_{r=a+0} = (\psi_1)_{r=a-0} \tag{2.36}$$

which is simply a statement that the potential is continuous across the sheet.

Using the boundary equations (2.35) and (2.36) and the forms for ψ and ψ_1 given by (2.9) and (2.10), respectively, we can derive the result for χ given by (2.31). We have achieved a simplification in the algebra at the expense of a slightly more complicated boundary condition for the normal derivative of the potential. Also, by using this method, we have recognized at the outset that the significant parameter is the product of the (complex) conductivity and the thickness of the shell and not their individual values.

It is useful here to illustrate the equivalent electric circuit for the spherical body with the thin conducting shell. To this end, we first define an input impedance Z as follows:

$$Z = \left[-\frac{1}{r^2}\,\frac{\psi}{J_r}\right]_{r=a+0} \tag{2.37}$$

From (2.9) this impedance is given by

$$Z = \frac{1}{a\hat{\sigma}}\,\frac{1-\chi}{1+2\chi} \tag{2.38}$$

where, in general, χ is defined by (2.25). But in the case of the thin conducting shell of admittance y, the appropriate form of χ is given by (2.31), in which case

$$\frac{1}{Z} = 2y + \hat{\sigma}_1 a \tag{2.39}$$

The equivalent circuit is very simple, as indicated in Figure 2.3. The input impedance Z, in ohms, is merely the parallel combination of two impedances $1/2y$ and $1/\hat{\sigma}_1 a$.

2.6 RESISTIVE-SHELL LIMIT

A second thin-shell approximation of the general expression, for the induced dipole factor χ, is also useful. In this case we envisage the outer shell to be of small thickness, as before, but now the complex conductivity $\hat{\sigma}_2$ is small compared, say, with $\hat{\sigma}_1$. This limit can be illustrated by

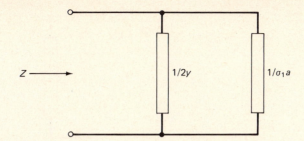

Figure 2.3   Equivalent circuit for conductive shell for homogeneous interior.

noting that

$$\frac{\hat{\sigma}_1 - \hat{\sigma}_2}{\hat{\sigma}_1 + 2\hat{\sigma}_2}\frac{a^3}{b^3} = \left(1 - \frac{\hat{\sigma}_2}{\hat{\sigma}_1}\right)\left(1 + 2\frac{\hat{\sigma}_2}{\hat{\sigma}_1}\right)^{-1}\left(1 + \frac{b-a}{a}\right)^{-3}$$

$$\cong 1 - 3\frac{\hat{\sigma}_2}{\hat{\sigma}_1} - 3\frac{b-a}{a} \tag{2.40}$$

when terms of second order and higher order are neglected.

Using (2.40), we now find that (2.25) can be approximated by

$$\chi = \frac{1 - \delta}{1 + 2\delta} \tag{2.41}$$

where now

$$\delta = \frac{\hat{\sigma}}{\hat{\sigma}_1} + \frac{z\hat{\sigma}}{a} \tag{2.42}$$

and where $z = (b - a)/\hat{\sigma}_2$ is the impedance of the shell. The corresponding mathematical statement of this limit is that $|\hat{\sigma}_2| \to 0$ and $(b - a) \to 0$ such that their ratio z remains finite.

The alternative boundary-value approach to the homogeneous sphere with a shell of low conductivity is now outlined briefly [4]. In this case we can talk about a voltage v or difference of potential across the shell. By Ohm's law $v = zJ_r$, where J_r is the radial current density that is continuous through the shell. Now for $r > b$, $J_r = -(\partial \psi/\partial r)\hat{\sigma}$; thus we have the following boundary condition:

$$\left[\psi - z\hat{\sigma}\frac{\partial \psi}{\partial r}\right]_{r=a+0} = [\psi_1]_{r=a-0} \tag{2.43}$$

As a consequence of the continuity of normal current density at the shell, we also have the boundary condition

$$\left[\hat{\sigma}\frac{\partial \psi}{\partial r}\right]_{r=a+0} = \left[\hat{\sigma}_1\frac{\partial \psi_1}{\partial r}\right]_{r=a-0} \tag{2.44}$$

From the general forms for ψ and ψ_1 given by (2.9) and (2.10), the

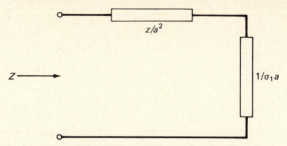

Figure 2.4   Equivalent circuit for shell with interface impedance z.

boundary equations (2.43) and (2.44) can now be invoked to yield directly the form for χ given by (2.41).

The equivalent circuit for the sphere encased by a low-conductivity shell now follows immediately. Using the impedance definition stated by (2.37), we see that

$$Z = \frac{z}{a^2} + \frac{1}{\hat{\sigma}_1 a} \qquad (2.45)$$

The corresponding circuit is shown in Figure 2.4.

2.7 DOUBLE-SHELL MODEL

More complicated spherical models can be handled easily by extending the number of concentric layers. The unknown coefficients are determined by the boundary conditions that require the continuity of the potential and normal current density at each of the concentric interfaces. If any one of the layers is sufficiently thin, there is a simplification if the boundary condition, appropriate for thin shells, is used. For a given system this technique reduces the number of equations to be solved. Furthermore, the recognition of the form of the equivalent circuit eliminates much of the repetitive algebra.

An illustrative example of a more complicated geometry is shown in Figure 2.5. The target or spherical body of radius a has a complex conductivity $\hat{\sigma}_1$ and is surrounded by two concentric thin shells. The innermost shell has an admittance y, the outer shell has an impedance z. The thicknesses of these shells are both much less than the radius a. Thus the designation $r = a - 0$ refers to the inside surface of the double shell, and $r = a + 0$ refers to the outside surface.

We can solve the above problem directly by first noting that the forms of the potential ψ and ψ_1 inside and outside the body are given by (2.9) and (2.10), respectively. Now the boundary conditions must incorporate the fact that both potential and normal current density are effectively discontinuous in going from $a - 0$ to $a + 0$. In a sense we are thus combining the boundary conditions from the two preceding problems.

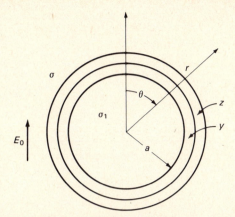

Figure 2.5   Double-shell model characterized by interface impedance z and surface admittance y.

For the present double-shell system the boundary conditions are found to be

$$\left(\hat{\sigma}\frac{\partial\psi}{\partial r}\right)_{r=a+0} = \left[\hat{\sigma}_1\frac{\partial\psi_1}{\partial r} + \frac{y}{a^2}\frac{\partial}{\partial r}\left(r^2\frac{\partial\psi_1}{\partial r}\right)\right]_{r=a-0} \qquad (2.46)$$

and

$$\left[\psi - z\hat{\sigma}\frac{\partial\psi}{\partial r}\right]_{r=a+0} = [\psi_1]_{r=a-0} \qquad (2.47)$$

When these conditions are applied to (2.9) and (2.10), we find that the induced dipole moment χ is given by

$$\chi = \frac{1-\delta}{1+2\delta} \qquad (2.48)$$

where now

$$\delta = \frac{\hat{\sigma}}{\hat{\sigma}_1 + (2y/a)} + \frac{z\hat{\sigma}}{a} \qquad (2.49)$$

The equivalent circuit for the equivalent Z as defined by (2.37) is now as shown in Figure 2.6. In this case the outer shell, representing a thin sheath of low conductivity, is in series with the parallel combination of the sheath of high conductivity and the impedance of the interior region.

It might seem that the double-shell system is rather hypothetical. But a concrete example would be a dielectric spherical particle of permittivity ϵ_1 that exhibits surface conduction as characterized by a sheath with real conductance g. Then to account for a thin insulating film of capacitance C per square meter, we set $z = (i\omega C)^{-1}$. The surrounding medium could be a conducting electrolytic fluid of conductivity σ. The

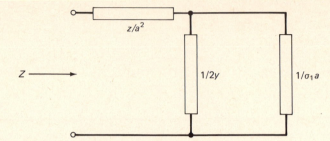

Figure 2.6   Equivalent circuit for double-shell model.

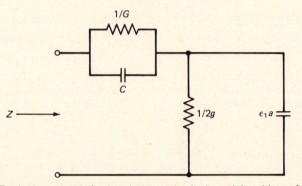

Figure 2.7   Equivalent circuit for insulating spherical particle with surface conductance g and leaky-condenser interface.

induced-dipole-moment factor χ is given by (2.48), but now

$$\delta = \frac{\sigma}{i\epsilon_1\omega + (2g/a)} + \frac{\sigma}{i\omega Ca} \tag{2.50}$$

A slightly more complicated case would be to represent the interface impedance by the form

$$z = (G + i\omega C)^{-1} \tag{2.51}$$

where G would account for the leakage of the capacitive layer. In this case

$$\delta = \frac{\sigma}{i\epsilon_1\omega + (2g/a)} + \frac{\sigma}{(G + i\omega C)a} \tag{2.52}$$

The corresponding equivalent circuit for this case is shown explicitly in Figure 2.7, where the input impedance is defined by (2.37); also the resistances and the condensers are identified (having dimensions of ohms and farads, respectively).

In dealing with electrochemical interface effects, it is sometimes desirable to employ a more complicated form for the frequency dependence of $z(i\omega)$. The so-called Warburg form [1] is to write

$$z(i\omega) = \text{constant} \times (i\omega)^{-1/2} \tag{2.53}$$

Other forms that tend to be rather empirical include the Cole-Cole type [1], whereby

$$z(i\omega) = \text{constant} \times (i\omega)^{-k} \tag{2.54}$$

where k, a number less than one, is chosen to fit observed data.

2.8 ENSEMBLE OF CONDUCTING SPHERICAL PARTICLES

At the risk of overemphasizing spherical-particle formulations, we would like to describe how an inhomogeneous medium can be modeled by an ensemble of such spherical particles. In particular, we wish to introduce the useful concept that a microscopically complicated region can be described by effective homogeneous medium parameters. This subject is one of current research using very complicated vector wave formulations [2, 3]. Here we restrict attention to quasi-static potential theory, which greatly simplifies the development. The present approach follows that introduced by Clerk Maxwell [5] and Maxwell-Garnett [6].

To deal with an ensemble of identical particles, we imagine they are located within a spherical region of radius r_0, as indicated in Figure 2.8. The total number within this volume is N. A uniform field $\mathbf{E_0}$ is now applied, which we assume acts on each particle. Neglecting the interaction between the particles, we then superimpose the fields produced by

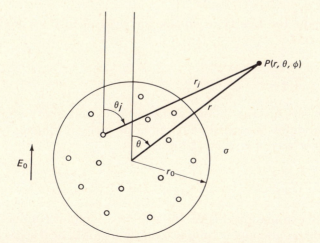

Figure 2.8   Ensemble of small spherical particles with reference sphere of radius r_0.

the N induced dipoles to yield the resultant potential Ψ at (r, θ, ϕ):

$$\Psi = -E_0 r \cos \theta + \sum_{j=1}^{N} \frac{E_0 a^3 \chi}{r^2} \cos \theta_j \tag{2.55}$$

As indicated in Figure 2.8, (r, θ, ϕ) are spherical coordinates with origin at the center of the reference volume, while (r_j, θ_j, ϕ_j) are centered at the individual particles, where $j = 1, 2, 3, \ldots, N$. Each particle has radius a, and the induced-dipole-moment factor is χ, as previously defined. We also assume that the particles are distributed randomly within the reference volume of radius r_0.

Also, for convenience we choose the polar axes of the coordinate systems (r, θ, ϕ) and (r_j, θ_j, ϕ_j) to be parallel with each other and with the direction of the applied primary field.

Now it is clear that at a sufficient distance from the ensemble (that is, when $r \gg r_0$), $r_j = r$ and $\theta_j = \theta$; and then Equation (2.55) can be approximated by

$$\Psi = -E_0 r \cos \theta + E_0 N a^3 \chi r^{-2} \cos \theta \tag{2.56}$$

The basic idea of the effective medium is to regard the volume within $r = r_0$ as a continuum. Also, because of our assumption about the randomness of the particle locations, the effective complex conductivity $\hat{\sigma}_e$ can be regarded as an average value for the continuum. Then using the known result for a single, homogeneous spherical body of radius r_0 and conductivity $\hat{\sigma}_e$, we write the equivalent expression for the potential at the exterior point P in the form

$$\Psi = -r E_0 \cos \theta + E_0 \frac{r_0^3}{r^2} \frac{\hat{\sigma}_e - \hat{\sigma}}{\hat{\sigma}_e + 2\hat{\sigma}} \cos \theta \tag{2.57}$$

We now take the liberty to equate the right-hand sides of (2.56) and (2.57) to yield

$$v\chi = \frac{\hat{\sigma}_e - \hat{\sigma}}{\hat{\sigma}_e + 2\hat{\sigma}} \tag{2.58}$$

where $v = Na^3/r_0^3$ is the volume of the particles per cubic meter. Usually, the dimensionless parameter v is called the volume loading.

Another way of writing (2.58) is

$$\frac{\hat{\sigma}_e}{\hat{\sigma}} = \frac{1 + 2v\chi}{1 - v\chi} \tag{2.59}$$

which is an explicit formula for the effective complex conductivity $\hat{\sigma}_e$ of the medium containing a volume loading v of identical spherical particles. When the particles are uncoated homogeneous spheres of complex conductivity $\hat{\sigma}_1$, we know that

$$\chi = \frac{\hat{\sigma}_1 - \hat{\sigma}}{\hat{\sigma}_1 + 2\hat{\sigma}} \tag{2.60}$$

This result is essentially the same result that was obtained by Maxwell [5] nearly a century ago, although he was considering only real conductivities. Another special case is when we are dealing with homogeneous, dielectric spherical particles of permittivity ϵ_1 in a host dielectric medium of permittivity ϵ. Now we replace $\hat{\sigma}_1$ by $i\epsilon_1\omega$, $\hat{\sigma}$ by $i\epsilon\omega$, and $\hat{\sigma}_e$ by $i\epsilon_e\omega$. The $i\omega$'s cancel in (2.59) and (2.60), so that

$$\frac{\epsilon_e}{\epsilon} = \frac{1 + 2v\chi}{1 - v\chi} \tag{2.61}$$

where

$$\chi = \frac{\epsilon_1 - \epsilon}{\epsilon_1 + 2\epsilon} \tag{2.62}$$

Equation (2.61) for the apparent permittivity ϵ_e is the celebrated Clausius-Mosotti formula, well known in dielectric theory [7].

As we see above, by introducing the concept of complex conductivity at the outset, we encompass both the classical direct current flow in conductors and the classic theory of dielectric polarization.

2.9 MORE GENERAL MIXTURE FORMULA

In using the Maxwell mixture type of formula, generalized to complex conductivities, such as given by (2.58), one should remember that the volume loading v has been assumed small. When v is greater than about 0.1 (that is, fractional volume of 10 percent), the interaction between the particles can be important. There is a vast literature on how the Maxwell or the equivalent Clausius-Mosotti mixture formulas can be generalized or extended to larger values of v. Here we will outline one particular approach that is very effective in spite of the lack of rigor.

We begin by writing (2.58) in a slightly simplified form:

$$\frac{\hat{\sigma}_e - \hat{\sigma}}{3\hat{\sigma}} \simeq v \frac{\hat{\sigma}_1 - \hat{\sigma}}{\hat{\sigma}_1 + 2\hat{\sigma}} \tag{2.63}$$

which is justified for very small v. Actually, (2.63) is in the form derived by Rayleigh [8] who made similar assumptions about the smallness of v.

Now following the ideas of Bruggeman [9], we argue that the Rayleigh type of formula can be applied sequentially to incremental changes in the volume loading of the composite medium. Thus we began by stating that for a given volume loading of spherical particles, the effective conductivity has a unique value $\hat{\sigma}_e$. We now augment every unit volume (such as cubic meter) of this particular loaded medium with a small volume du of additional spherical particles of conductivity $\hat{\sigma}_1$. The effective conductivity of this augmented medium is now $\hat{\sigma}_e + d\hat{\sigma}_e$. Now utilizing the Rayleigh formula (2.63), we may write in terms of differentials:

$$\frac{(\hat{\sigma}_e + d\hat{\sigma}_e) - \hat{\sigma}_e}{3\hat{\sigma}_e} = \frac{\hat{\sigma}_1 - \hat{\sigma}_e}{\hat{\sigma}_1 + 2\hat{\sigma}_e} \, du \tag{2.64}$$

There is now a tricky point here in recognizing that the augmented spherical particles of volume du have actually replaced particles already present. Thus the resultant or actual change dv of the volume loading is related to du as follows:

$$dv = (1 - v) \, du \tag{2.65}$$

Now we write (2.64) in the form

$$\frac{d\hat{\sigma}_e}{3\hat{\sigma}_e} + \frac{d\hat{\sigma}_e}{\hat{\sigma}_1 - \hat{\sigma}_e} - \frac{dv}{1 - v} = 0 \tag{2.66}$$

This expression is equivalent to

$$d\left[\hat{\sigma}_e^{1/3} \frac{1}{\hat{\sigma}_e - \hat{\sigma}_1}(1 - v) \right] = 0 \tag{2.67}$$

Thus

$$\hat{\sigma}_e^{1/3} \frac{1}{\hat{\sigma}_e - \hat{\sigma}_1}(1 - v) = \text{constant} \tag{2.68}$$

in so far as variations of v are concerned. This constant is determined by noting that $\hat{\sigma}_e$, for $v = 0$, must be $\hat{\sigma}$. That is,

$$\hat{\sigma}^{1/3} \frac{1}{\hat{\sigma} - \hat{\sigma}_1} = \text{constant} \tag{2.69}$$

On dividing (2.69) by (2.68), we obtain the remarkable Bruggeman [9] formula:

$$\left(\frac{\hat{\sigma}_e}{\hat{\sigma}}\right)^{1/3} \left(\frac{\hat{\sigma} - \hat{\sigma}_1}{\hat{\sigma}_e - \hat{\sigma}_1}\right)(1 - v) = 1 \tag{2.70}$$

Obviously, this equation is not an explicit equation for the desired $\hat{\sigma}_e$ in terms of the given $\hat{\sigma}$ and $\hat{\sigma}_1$. But a number of limiting cases emerge without additional work such as solving cubic equations. For example, if $|\hat{\sigma}_1| \ll |\hat{\sigma}|$ and $|\hat{\sigma}_e|$, we see that

$$\hat{\sigma}_e \approx \hat{\sigma}(1 - v)^{3/2} \tag{2.71}$$

This equation would correspond to the case where the particles or grains would be acting as insulators. Then (2.71) suggests that if $v = 1$, the effective conductivity is zero, while if $v = 0$, the conductivity is $\hat{\sigma}$, that of the host medium. There is no really good reason why a formula such as (2.71) should work when the particle-packing ratio v is near one. But there is convincing experimental evidence that the three-halves power law closely fits the experimental data for direct current flow in brine solution filled with glass beads [10] for v in the range from 0 to over 80 percent.

There is also a great deal of experimental data for brine-saturated sedimentary rocks to indicate that a formula of the type $\hat{\sigma}_e \simeq \hat{\sigma}(1 - v)^m$ is a good empirical fit for m adjusted to be in the range 1 to 4 or 5. At direct current this equation is known as Archie's law [11]. Recent theoretical developments in this subject have been reviewed by Wait [12], so the interested reader is referred to this source.

References

1. J. R. Wait: *Geo-Electromagnetism*, Academic Press, New York, 1982 (see chap. 2).
2. D. Stroud: "Generalized Effective-Medium Approach to the Conductivity of an Inhomogeneous Material," *Physical Reveiw B*, vol. 12, no. 8, 1975, pp. 3368–3373.
3. V. V. Varadan and V. K. Varadan: "Multiple Scattering of Electromagnetic Waves by Randomly Distributed and Oriented Dielectric Scatterers," *Physical Review D*, vol. 21, no. 2, 1980.
4. J. R. Wait: "A Phenomenological Theory of Induced Electrical Polarization," *Canadian Journal of Physics*, vol. 36, 1958, pp. 1634–1644.
5. J. C. Maxwell: *A Treatise on Electricity and Magnetism*, 3d. ed., part II, Clarendon Press, Oxford, England, 1891, pp. 314–315 (reprinted by Dover, New York, 1954).
6. J. C. Maxwell-Garnett: "Colours in Metal Glasses and Metal Films," *Philosophical Transactions A, Royal Society of London*, vol. 205, 1905, pp. 237–288.
7. V. Daniels: *Dielectric Relaxation*, Academic Press, London, 1967.
8. Lord Rayleigh (J. W. Strutt): "On the Influence of Obstacles Arranged in Rectangular Order upon the Properties of a Medium," *Philosophical Magazine*, vol. 34, 1892, pp. 481–502.
9. D. A. G. Bruggeman: "Berechnung Verschiedener Physikalischer Konstanten von Heterogenen Substanzen," *Annalen der Physik*, vol. 24, 1935, pp. 636–644.
10. P. N. Sen, C. Scala, and M. H. Cohen: "A Self Similar Model for Sedimentary Rocks," *Geophysics*, vol. 46, 1981, pp. 781–795.
11. G. E. Archie: "The Electrical Resistivity Log as an Aid in Determining Some Reservoir Characteristics," *Transactions AIME*, vol. 146, 1942, pp. 54–62.
12. J. R. Wait, Complex Conductivity of Disseminated Spheroidal Ore Grains, *Gerlands Beitrage fur Geophysik (Leipzig)*, vol. 92, 1983, pp. 49–69.
13. H. Looyenga: "Dielectric Constants of Mixtures," *Physica*, vol. 31, 1965, pp. 401–406.
14. C. J. F. Bottcher: *Theory of Electric Polarization*, Elsevier, Amsterdam, 1952, p. 419.

Appendix A. Alternative Derivation of Mixture Formulas

There is another way to derive the Bruggeman formula. Following Looyenga [13], we outline this method.

We consider a spherical region of radius b that is filled with homogeneous material of (complex) conductivity $\hat{\sigma}_b$ except for a small sphere of radius a at the center, which is homogeneous material of conductivity $\hat{\sigma}_a$. The surrounding region is homogeneous. We now place the system in a uniform electric field. Except for minor changes in notation, this is the same problem that was treated earlier.

It is not difficult to see that the field outside the large sphere is unchanged if we replace the configuration by a homogeneous sphere of radius b with conductivity $\hat{\sigma}$, where

$$\hat{\sigma} = \frac{1 - 2G}{1 + G} \, \hat{\sigma}_b \tag{A.1}$$

with

$$G = \frac{\hat{\sigma}_b - \hat{\sigma}_a}{2\hat{\sigma}_b + \hat{\sigma}_a} \left(\frac{a}{b}\right)^3 \tag{A.2}$$

Now, of course, if $\hat{\sigma}_a \neq \sigma_b$, $\hat{\sigma}$ differs from $\hat{\sigma}_b$. The change $\Delta\hat{\sigma}_e$ is

$$\Delta\hat{\sigma}_e = \hat{\sigma} - \hat{\sigma}_b = \frac{-3\hat{\sigma}_b G}{1 + G} \tag{A.3}$$

When $(a/b)^3 \ll 1$, this equation is reduced to

$$\Delta\hat{\sigma}_e \simeq -3\hat{\sigma}_b G = \frac{3\hat{\sigma}_b(\hat{\sigma}_a - \hat{\sigma}_b)}{(2\hat{\sigma}_b + \hat{\sigma}_a)}\left(\frac{a}{b}\right)^3 \tag{A.4}$$

We now apply the above result to a binary system containing a volume fraction v of material with conductivity $\hat{\sigma}_1$ and a volume fraction of $\phi = 1 - v$ of material with conductivity $\hat{\sigma}$. We wish to determine the effective conductivity $\hat{\sigma}_e(v)$ as a function of v.

We proceed by considering that the large sphere (of radius b) is filled with a mixture having a volume fraction v of material of conductivity $\hat{\sigma}_1$. We now place a small sphere of radius a of pure conductivity $\hat{\sigma}_1$ in the center of the large sphere. The resulting change in the apparent or effective conductivity is given by (A.4) with $\hat{\sigma}_a$ replaced by $\hat{\sigma}_1$ and $\hat{\sigma}_b$ by $\hat{\sigma}_e$. The resulting change in v in the large sphere is given by

$$\Delta v = (1 - v)\left(\frac{a}{b}\right)^3 \tag{A.5}$$

Thus

$$\frac{\Delta\hat{\sigma}_e}{\Delta v} = \frac{3\hat{\sigma}_e(\hat{\sigma}_1 - \hat{\sigma}_e)}{(2\hat{\sigma}_e + \hat{\sigma}_1)(1 - v)} \rightarrow \frac{d\hat{\sigma}_e}{dv} \tag{A.6}$$

This differential equation is precisely the same as (2.66), so the Bruggeman formula, as given by (2.69), is again obtained.

Using the same double-sphere model, Looyenga has also deduced the form of the celebrated Bottcher formula. The argument goes more or less as follows: Inside the large sphere of radius b with conductivity $\hat{\sigma}_e$, a small sphere of volume $(4\pi a^3/3)v$ is replaced by a pure material of conductivity $\hat{\sigma}_1$. Then another small sphere of volume $(4\pi a^3/3)(1 - v)$ is replaced by a pure substance of conductivity $\hat{\sigma}$. The resultant fractional loading of the mixture has not changed. Thus the resultant or effective conductivity $\hat{\sigma}_e$ has not changed. Then in view of the form of (A.4), we deduce that the sum of the two changes must be zero. Thus

$$\left[\frac{3\hat{\sigma}_e(\hat{\sigma}_1 - \hat{\sigma}_e)}{(2\hat{\sigma}_e + \hat{\sigma}_1)}v + \frac{3\hat{\sigma}_e(\hat{\sigma} - \hat{\sigma}_e)}{(2\hat{\sigma}_e + \hat{\sigma})}(1 - v)\right] = 0 \tag{A.7}$$

An equivalent statement is

$$v = \frac{(\hat{\sigma}_e - \hat{\sigma})(2\hat{\sigma}_e + \hat{\sigma}_1)}{3\hat{\sigma}_e(\hat{\sigma}_1 - \hat{\sigma})} \tag{A.8}$$

which was derived by Bottcher [14] from a consideration of the internal field for spherical particles.

Exercise: Derive an expression for the laplacian in cylindrical coordinates, given the form of the laplacian in rectangular coordinates.

SOLUTION: Now we have

$$\nabla^2 V = \frac{\partial^2 V}{\partial x^2} + \frac{\partial^2 V}{\partial y^2} + \frac{\partial^2 V}{\partial z^2}$$

in rectangular coordinates. In transforming this equation to cylindrical coordinates (ρ, ϕ, z), we note that $\rho = (x^2 + y^2)^{1/2}$ and $\tan \phi = y/x$. Also,

$$\frac{\partial \rho}{\partial x} = \frac{x}{\rho} \quad \text{and} \quad \frac{\partial \rho}{\partial y} = \frac{y}{\rho}$$

On the other hand,

$$\frac{\partial}{\partial x} \tan \phi = \frac{\partial}{\partial \phi} \tan \phi \frac{\partial \phi}{\partial x}$$

or

$$\frac{\partial}{\partial x} \frac{y}{x} = \frac{1}{\cos^2 \phi} \frac{\partial \phi}{\partial x}$$

or

$$\frac{-y}{x^2} = \frac{\rho^2}{x^2} \frac{\partial \phi}{\partial x}$$

Therefore

$$\frac{\partial \phi}{\partial x} = \frac{-y}{\rho^2}$$

Now

$$\frac{\partial V}{\partial x} = \frac{\partial V}{\partial \rho} \frac{\partial \rho}{\partial x} + \frac{\partial V}{\partial \phi} \frac{\partial \phi}{\partial x} \quad \text{and} \quad \frac{\partial V}{\partial y} = \frac{\partial V}{\partial \rho} \frac{\partial \rho}{\partial y} + \frac{\partial V}{\partial \phi} \frac{\partial \phi}{\partial y}$$

Thus

$$\frac{\partial V}{\partial x} = \frac{x}{\rho} \frac{\partial V}{\partial \rho} - \frac{y}{\rho^2} \frac{\partial V}{\partial \phi} \quad \text{and} \quad \frac{\partial V}{\partial y} = \frac{y}{\rho} \frac{\partial V}{\partial \rho} + \frac{x}{\rho^2} \frac{\partial V}{\partial \phi}$$

Now work out the next derivative to get

$$\frac{\partial^2 V}{\partial x^2} = \frac{1}{\rho} \frac{\partial V}{\partial \rho} - \frac{x^2}{\rho^3} \frac{\partial V}{\partial \rho} + \frac{x}{\rho} \frac{\partial^2 V}{\partial \rho} \frac{x}{\rho} + \frac{2xy}{\rho^3} \frac{\partial V}{\partial \phi} + \frac{y^2}{\rho^4} \frac{\partial^2 V}{\partial \phi^2}$$

and

$$\frac{\partial^2 V}{\partial y^2} = \frac{1}{\rho} \frac{\partial V}{\partial \rho} - \frac{y^2}{\rho^3} \frac{\partial V}{\partial \rho} + \frac{y}{\rho} \frac{\partial^2 V}{\partial \rho^2} \frac{y}{\rho} - \frac{2xy}{\rho^3} \frac{\partial V}{\partial \phi} + \frac{x^2}{\rho^4} \frac{\partial^2 V}{\partial \phi^2}$$

Add to get

$$\frac{\partial^2 V}{\partial x^2} + \frac{\partial^2 V}{\partial y^2} + \frac{\partial^2 V}{\partial z^2} = \frac{\partial^2 V}{\partial \rho^2} + \frac{1}{\rho} \frac{\partial V}{\partial \rho} + \frac{1}{\rho^2} \frac{\partial^2 V}{\partial \phi^2} + \frac{\partial^2 V}{\partial z^2}$$

$$= \frac{1}{\rho} \frac{\partial}{\partial \rho} \rho \frac{\partial V}{\partial \rho} + \frac{1}{\rho^2} \frac{\partial^2 V}{\partial \phi^2} + \frac{\partial^2 V}{\partial z^2} = \nabla^2 V$$

which is the desired result.

Chapter 3
Scalar Waves in One, Two, and Three Dimensions

3.1 ONE-DIMENSIONAL WAVE PROPAGATION

We will introduce the concept of wave propagation here in a somewhat unorthodox fashion. Essentially, the idea is to write down the wave equation for a physical parameter V that represents an observable quantity. It is remarkably similar to Laplace's equation, except that the term on the right-hand side is nonzero and proportional to V. Historically, such an equation arose in Helmholtz' studies of acoustic disturbances in solid, liquid, and gaseous media (see Hermann Helmholtz, *On the Sensations of Tone*, Dover, New York, Reprint 1954).

Frequently, the solutions of problems in electromagnetic theory involve the wave equation. In cartesian coordinates this equation would take the form

$$\frac{\partial^2 V}{\partial x^2} + \frac{\partial^2 V}{\partial y^2} + \frac{\partial^2 V}{\partial z^2} = \gamma^2 V \tag{3.1}$$

where V is some scalar function and γ is a propagation constant that we assume does not depend on x, y, or z. For example, if the problem were purely one-dimensional, $\partial/\partial y = \partial/\partial z = 0$, then we would have simply

$$\frac{\partial^2 V}{\partial x^2} - \gamma^2 V = 0 \tag{3.2}$$

or

$$\frac{d^2 V}{dx^2} - \gamma^2 V = 0$$

Clearly, solutions would be

$$V = \exp(\pm \gamma x) \tag{3.3}$$

and linear combinations thereof. For example,

$$V = Ae^{-\gamma x} + Be^{\gamma x} \tag{3.4}$$

where A and B are constants.

In wave propagation problems where the time factor is $\exp(i\omega t)$, with ω the angular frequency, the solution would be

$$v = \text{Re}(Ae^{-\gamma x} + Be^{+\gamma x})e^{i\omega t} \tag{3.5}$$

The quantity in parentheses is a phasor, where we note that A and B may also be complex. By convention, we take the real part of the product of the complex phasor and the factor $\exp(i\omega t)$ to represent the real-world (that is, physical) quantity.

To give further physical insight, we set $\gamma = \alpha + i\beta$, where α is the (real) attenuation factor and β is the (real) phase factor. The phasor V is then

$$V = Ae^{-\alpha x}e^{-i\beta x} + Be^{\alpha x}e^{i\beta x} \tag{3.6}$$

or

$$v = \text{Re}\left\{Ae^{-\alpha x}\exp\left[-i\beta\left(x - \frac{\omega}{\beta}t\right)\right] + Be^{\alpha x}\exp\left[i\beta\left(x + \frac{\omega}{\beta}t\right)\right]\right\} \tag{3.7}$$

It is clear that the two terms on the right-hand side of this equation represent traveling waves. The first term is a wave traveling in the positive x direction with an attenuation rate of α nepers per meter (Np/m) and a phase velocity of ω/β m/s. Conversely, the second term is a wave traveling in the negative x direction with the same attenuation rate and phase velocity.

This very simple and idealistic case is introduced because it illustrates beautifully the idea of special functions. Here $\exp(-\gamma x)$ and $\exp(+\gamma x)$ are the special functions for this rudimentary problem. Their properties are well understood, and a pocket calculator (such as the TI 58/59 type) can be used to obtain precise values even when γ is generally complex. In some cases, however, we might want to employ their series expansion, to wit

$$\exp(\mp \gamma x) = \sum_{n=0}^{\infty} (\mp)^n \frac{(\gamma x)^n}{n!} \tag{3.8}$$

where the summation is over all integers beginning with zero (for example, $n = 0, 1, 2, \ldots$).

Exercise: Specify that at $x = x_0$ we have $dV/dx = C = $ constant. What form does the solution take for the region $x_0 < x < \infty$, assuming that no sources exist in this region? *Hint:* Reject infinite physical quantities.

SOLUTION: Note that

$$\frac{dV}{dx} = -\gamma A e^{-\gamma x} + \gamma B e^{\gamma x} \tag{3.9}$$

Now as $x \to \infty$, the second term blows up on two accounts: (1) it becomes infinite since Re $\gamma = \alpha > 0$, and (2) it represents a plane wave traveling toward the source region at $x = x_0$. Thus we obtain the unknown A by insisting that

$$-\gamma A e^{-\gamma x_0} = C$$

or

$$A = -\gamma^{-1} e^{\gamma x_0} C$$

Then the desired solution is

$$V = -\gamma^{-1} C e^{-\gamma(x-x_0)} \tag{3.10}$$

Exercise: Consider the region $x_1 < x < x_2$. Assume that $V = V_1$ at $x = x_1$ and $dV/dx = V_2'$ at $x = x_2$. If there are no sources in this region, what form does the solution take?

SOLUTION: Now

$$A e^{-\gamma x_1} + B e^{\gamma x_1} = V_1 \tag{3.11}$$

and

$$-\gamma A e^{-\gamma x_2} + \gamma B e^{\gamma x_2} = V_2' \tag{3.12}$$

Solving for A and B is now purely algebra. Details are left to the reader.

Exercise: Another type of boundary condition is

$$\frac{dV}{dx} - V\delta = \hat{V} \tag{3.13}$$

at, say, $x = \hat{x}$. Here δ is a constant. If there are no sources in the region $\hat{x} < x < \infty$, what form would the solution take?

SOLUTION: Again, as in the first exercise, the solution of the form

$$V = Ae^{-\gamma x}$$

is finite (or zero) as $x \to \infty$, and it represents a wave traveling away from the source.

Now the boundary condition yields

$$-\gamma Ae^{-\gamma \hat{x}} - Ae^{-\gamma \hat{x}}\delta = \hat{V} \tag{3.14}$$

or

$$A = -\hat{V}e^{\gamma \hat{x}}(\delta + \gamma)^{-1} \tag{3.15}$$

Before leaving the one-dimensional case, we consider the static limit. Then γ tends to zero. Now the governing equation is

$$\frac{d^2V}{dx^2} = 0 \tag{3.16}$$

and the general solution is

$$V = ax + b \tag{3.17}$$

where a and b are constants. How does this solution square with Equation (3.4)? It would suggest that

$$V = A + B = \text{constant} \tag{3.18}$$

independent of x! But the apparent inconsistency is removed if we write

$$V = A(1 - \gamma x) + B(1 + \gamma x) \tag{3.19}$$

where terms of order $(\gamma x)^2$, $(\gamma x)^3$, . . . are neglected. Then clearly in the static limit we should have $a = \gamma(B - A)$ and $b = A + B$.

3.2 THE TRANSMISSION-LINE ANALOGY

In many of these wave problems it is fruitful to draw on the analogy with transmission-line theory. This is particularly the case when the problem involves various sections. We will illustrate our point by considering the simple one-dimensional model where transmission is now in the z direction.

The transmission line is idealized as two parallel, straight conductors. Such a line can be characterized by a series impedance $\hat{Z}$ per unit length and a series admittance $\hat{Y}$ per unit length.

A section of elementary length δz is illustrated in Figure 3.1. The current I changes by an amount $(dI/dz)\,\delta z$ across the section.

Now the voltage drop across the section must be balanced by the

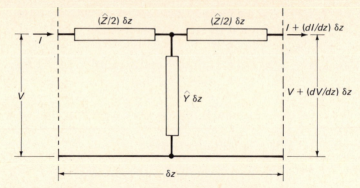

Figure 3.1   Small segment of uniform transmission line.

current times the incremental impedance. Similarly, the current change must be balanced by the voltage times the incremental admittance. Thus

$$\frac{dV}{dz} = -\hat{Z}I \tag{3.20}$$

and

$$\frac{dI}{dz} = -\hat{Y}V \tag{3.21}$$

On combining these equations, we immediately get

$$\frac{d^2V}{dz^2} - \gamma^2 V = 0 \tag{3.22}$$

and

$$\frac{d^2I}{dz^2} - \gamma^2 I = 0 \tag{3.23}$$

where $\gamma = (\hat{Z}\hat{Y})^{1/2}$. The general solution for the voltage is

$$V = Ae^{-\gamma z} + Be^{\gamma z} \tag{3.24}$$

which, by no coincidence, is the same as (3.4). Then in view of (3.20), we see that

$$I = -\frac{1}{\hat{Z}}\frac{dV}{dz} = \frac{\gamma}{\hat{Z}}Ae^{-\gamma z} - \frac{\gamma}{\hat{Z}}Be^{\gamma z} \tag{3.25}$$

or

$$I = K^{-1}Ae^{-\gamma z} - K^{-1}Be^{\gamma z} \tag{3.26}$$

where

$$K = \left(\frac{\gamma}{\hat{Z}}\right)^{-1} = \left(\frac{\hat{Z}}{\hat{Y}}\right)^{1/2} = \frac{\gamma}{\hat{Y}}$$

is termed the characteristic impedance of the line.

Now, clearly, if the line were of infinite length (that is, $0 < z < \infty$) and if there were no sources for $z > 0$, the wave solutions would have the form

$$V = Ae^{-\gamma z} \tag{3.27}$$

and

$$I = K^{-1}Ae^{-\gamma z} \tag{3.28}$$

Clearly, $V/I = K$ for all points $z > 0$. In this case the input impedance of the line is $K \ \Omega$.

To illustrate the usefulness of the concept, we consider the case where the line is terminated at $z = \ell$ in an impedance Z_ℓ as indicated in Figure 3.2. The problem is to determine the input impedance at $z = 0$. The solution is obtained by noting that

$$Z_\ell = \frac{V}{I}\bigg|_{z=\ell} = K\frac{Ae^{-\gamma\ell} + Be^{\gamma\ell}}{Ae^{-\gamma\ell} - Be^{\gamma\ell}} \tag{3.29}$$

This equation can be solved for the ratio B/A to give

$$\frac{B}{A} = \frac{Z_\ell - K}{Z_\ell + K}e^{-2\gamma\ell} \tag{3.30}$$

The input impedance at $x = 0$ is clearly

$$Z = \frac{V}{I}\bigg|_{x=0} = K\frac{1 + (B/A)}{1 - (B/A)} \tag{3.31}$$

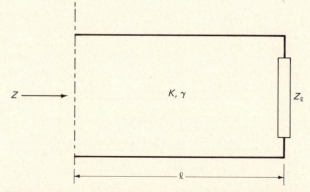

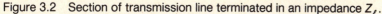

Figure 3.2   Section of transmission line terminated in an impedance Z_ℓ.

This equation can be written more explicitly as

$$Z = K \frac{(Z_\ell + K) + (Z_\ell - K)e^{-2\gamma\ell}}{(Z_\ell + K) - (Z_\ell - K)e^{-2\gamma\ell}} = K \frac{Z_\ell + K \tanh \gamma\ell}{K + Z_\ell \tanh \gamma\ell} \tag{3.32}$$

Consider two uniform sections of transmission lines connected in tandem and terminated in a semi-infinite line. What is the input impedance? The situation is illustrated in Figure 3.3.

The classical (physicist's) point of view will be described first. The coordinate z is the distance from the input of the line. To this end, we would write the following expressions for the voltage and current: For $0 < z < \ell_1$

$$V_1 = Ae^{-\gamma_1 z} + Be^{\gamma_1 z} \tag{3.33}$$

$$I_1 = K_1^{-1}Ae^{-\gamma_1 z} - K_1^{-1}Be^{\gamma_1 z} \tag{3.34}$$

For $\ell_1 < z < \ell_1 + \ell_2$

$$V_2 = Ce^{-\gamma_2 z} + De^{\gamma_2 z} \tag{3.35}$$

$$I_2 = K_2^{-1}Ce^{-\gamma_2 z} - K_2^{-1}De^{\gamma_2 z} \tag{3.36}$$

and for $\ell_1 + \ell_2 < z < \infty$

$$V_3 = Ee^{-\gamma_3 z} \tag{3.37}$$

$$I_3 = K_3^{-1}Ee^{-\gamma_3 z} \tag{3.38}$$

where $K_i = (\hat{Z}_i / \hat{Y}_i)^{1/2}$ is the characteristic impedance of the ith section ($i = 1, 2, 3$). The corresponding propagation constant of the ith section is γ_i.

To obtain the solution, we now enforce the continuity of current and voltage at the terminal planes. Thus

$$\begin{aligned} V_1 &= V_2 \\ I_1 &= I_2 \end{aligned} \quad \text{at } z = \ell_1 \tag{3.39}$$

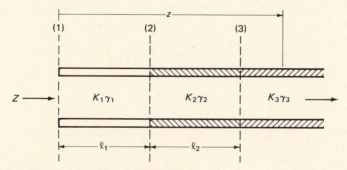

Figure 3.3 Three-section transmission line.

and

$$\begin{aligned} V_2 &= V_3 \\ I_2 &= I_3 \end{aligned} \quad \text{at } z = \ell_1 + \ell_2 \tag{3.40}$$

This result leads to four linear algebraic equations. Thus E, D, C, and B can be expressed in terms of A after successive elimination. The input impedance is then

$$Z_1 = \frac{V_1}{I_1}\bigg|_{z=0} = K_1 \frac{1 + (B/A)}{1 - (B/A)} \tag{3.41}$$

This equation can be written in the form

$$Z_1 = K_1 \frac{Z_2 + K_1 \tanh \gamma_1 \ell_1}{K_1 + Z_2 \tanh \gamma_1 \ell_1} \tag{3.42}$$

where

$$Z_2 = K_2 \frac{K_3 + K_2 \tanh \gamma_2 \ell_2}{K_2 + K_3 \tanh \gamma_2 \ell_2} \tag{3.43}$$

Our key observation is that this expression for Z_1 could be written directly only on the basis of the previous solution for the single section terminated in a known impedance. For example, on referring to Figure 3.3, we see that Z_2 is the input impedance at terminal plane (2) (that is, at $z = \ell_1$). In this case the terminating impedance is K_3 since the third section is a semi-infinite line. Then the form of Z_1 is appropriate for a uniform section of length ℓ_1 terminated in an impedance of Z_2.

At this point it is worthwhile to point out the symmetry of the basic equations for I and V. As a consequence, we could have just as well written the analog of (3.24) as

$$I = A'e^{-\gamma z} + B'e^{\gamma z} \tag{3.44}$$

when A' and B' are coefficients. Then clearly,

$$V = M^{-1}A'e^{-\gamma z} - M^{-1}A'e^{\gamma z} \tag{3.45}$$

which is a consequence of (3.21), where

$$M = K^{-1} = \left(\frac{\hat{Y}}{\hat{Z}}\right)^{1/2}$$

is the characteristic admittance of the line.

In the context of the previous example of two uniform sections of line terminated by a semi-infinite section, the desired expression for the *input admittance* Y_1 is given by

$$Y_1 = M_1 \frac{Y_2 + M_1 \tanh \gamma_1 \ell_1}{M_1 + Y_2 \tanh \gamma \ell} \tag{3.46}$$

where

$$Y_2 = M_2 \frac{M_3 + M_2 \tanh \gamma_2 \ell_2}{M_2 + M_3 \tanh \gamma_2 \ell_2} \tag{3.47}$$

Here M_i $(i = 1, 2, 3)$ is the characteristic admittance of the ith section.

3.3 N-SECTIONAL LINE

To obtain an expression for the input impedance of a transmission line composed of N sections is straightforward. The situation is depicted in Figure 3.4. The line parameters for the nth section are K_{n-1} and γ_{n-1}, where $n = 1, 2, 3, \ldots, N - 1, N$. The length of the nth section is ℓ_n.

The simplest way to handle this problem is to begin at the $(N - 1)$th terminal plane. Clearly, the input impedance is

$$Z_{N-1} = K_{N-1} \frac{K_N + K_{N-1} \tanh \gamma_{N-1} \ell_{N-1}}{K_{N-1} + K_N \tanh \gamma_{N-1} \ell_{N-1}} \tag{3.48}$$

Then the next step is to note that the input impedance at terminal plane $N - 2$ is

$$Z_{N-2} = K_{N-2} \frac{Z_{N-1} + K_{N-2} \tanh \gamma_{N-2} \ell_{N-2}}{K_{N-2} + Z_{N-1} \tanh \gamma_{N-2} \ell_{N-2}} \tag{3.49}$$

This process can be continued as needed. For example, the input impedance at the $(n - 1)$th terminal plane is

$$Z_{n-1} = K_{n-1} \frac{Z_n + K_{n-1} \tanh \gamma_{n-1} \ell_{n-1}}{K_{n-1} + Z_n \tanh \gamma_{n-1} \ell_{n-1}} \tag{3.50}$$

which is given in terms of the input impedance Z_n at the nth terminal plane. The final step in the iterative process is

$$Z_1 = K_1 \frac{Z_2 + K_1 \tanh \gamma_1 \ell_1}{K_1 + Z_2 \tanh \gamma_1 \ell_1} \tag{3.51}$$

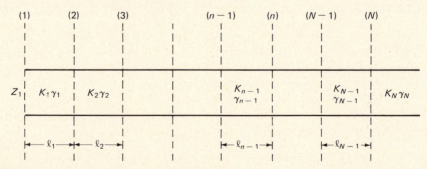

Figure 3.4   An N-sectional transmission line.

which, of course, is the desired result for the input impedance. The corresponding formulation in terms of admittances is directly analogous and hardly needs to be written out.

A word might be said here about the meaning of the line parameters $\hat{Z}$ and $\hat{Y}$. For time-harmonic fields [for the time factor $\exp(i\omega t)$] we may write

$$\hat{Z} = \hat{R} + i\omega\hat{L} \tag{3.52}$$

and

$$\hat{Y} = \hat{G} + i\omega\hat{C} \tag{3.53}$$

where

$\hat{R}$ = series resistance per unit length
$\hat{L}$ = series inductance per unit length
$\hat{G}$ = shunt conductance per unit length
$\hat{C}$ = shunt capacitance per unit length

In many wave propagation problems involving transmission lines or analogous systems, the losses are small and can be neglected. Then $\hat{Z} \simeq i\omega\hat{L}$ and $\hat{Y} \simeq i\omega\hat{C}$, whence

$$\gamma = i\beta$$

where $\beta = (\hat{L}\hat{C})^{1/2}\omega$ is the wave number. Then, for example,

$$\tanh \gamma\ell = i \tan \beta\ell \tag{3.54}$$

is purely imaginary.

The other extreme limiting case is when the frequency is sufficiently small that

$$\hat{Z} \simeq \hat{R} \qquad \text{and} \qquad \hat{Y} \simeq \hat{G}$$

so that everything becomes real. The transmission line then becomes a resistive network and the propagation constant γ is purely real [that is, $\gamma = (\hat{R}\hat{G})^{1/2}$]. This limiting case is a useful analogy to dc flow in layered conductors.

3.4 TWO-DIMENSIONAL WAVE PROPAGATION

We now return to (3.1) and allow variations in, say, two dimensions. Thus if $\partial/\partial z = 0$, we would have

$$\frac{\partial^2 V}{\partial x^2} + \frac{\partial^2 V}{\partial y^2} = \gamma^2 V \tag{3.55}$$

To illustrate the separation-of-variables technique, we seek a solution in the form

$$V = X(x)Y(y) \tag{3.56}$$

and thus

$$Y\frac{d^2X}{dx^2} + X\frac{d^2Y}{dy^2} = \gamma^2 XY \tag{3.57}$$

This equation is equivalent to

$$\frac{1}{X}\frac{d^2X}{dx^2} = \gamma_x^2 \quad \text{a constant} \tag{3.58}$$

and

$$\frac{1}{Y}\frac{d^2Y}{dy^2} = \gamma_y^2 \quad \text{a constant} \tag{3.59}$$

Furthermore, these two constants are constrained by

$$\gamma_x^2 + \gamma_y^2 = \gamma^2 \tag{3.60}$$

Thus a general solution of (3.54) must be made up as linear combinations of $\exp(\mp \gamma_x x)\exp(\mp \gamma_y y)$.

3.5 PLANAR-WAVEGUIDE MODEL

As an example to illustrate our point, let us find a solution for V that vanishes at $x = 0$ and $x = a$, and it should also vanish as $y \to \infty$. Clearly, the desired combination of $\exp(\mp \gamma_x x)$ is

$$X(x) = A\sinh(\gamma_x x) \tag{3.61}$$

where γ_x is constrained by

$$\sinh(\gamma_x a) = 0 \tag{3.62}$$

Here $\sinh(Z) = (e^Z - e^{-Z})/2$ denotes the hyberbolic sine function. If we write $\gamma_x = i\beta_{x,n}$, then

$$\sin\beta_{x,n}a = 0 \tag{3.63}$$

or

$$\beta_{xn} = \frac{n\pi}{a} \quad \text{where } n = 0, 1, 3, \ldots$$

The corresponding values of γ_y are obtained from

$$\gamma_y = (\gamma^2 - \gamma_x^2)^{1/2} \tag{3.64}$$

or

$$\gamma_{y,n} = \left[\gamma^2 + \left(\frac{n\pi}{a} \right)^2 \right]^{1/2} \tag{3.65}$$

By convention, we choose the radical such that $\mathrm{Re}\ \gamma_{y,n} > 0$. This equation now tells us that the desired form of the solution is

$$V = \sum_{n=0}^{\infty} A_n \sinh(i\beta_{x,n}) \exp(-\gamma_{y,n}y) \tag{3.66}$$

where n may be summed over all integers. More explicity, we would write

$$V(x, y) = i \sum_{n=1}^{\infty} A_n \left(\sin \frac{n\pi}{a} x \right) \exp\left\{ -\left[\gamma^2 + \left(\frac{n\pi}{a} \right)^2 \right]^{1/2} y \right\} \tag{3.67}$$

where the $n = 0$ term is identically zero. If $V(x, 0)$ is specified at the "input" surface $y = 0$, then A_n is to be determined from

$$V(x, 0) = i \sum_{n=1}^{\infty} A_n \sin \beta_{x,n}x \tag{3.68}$$

To this end, we multiply both sides by $\sin \beta_{x,m}x$ and integrate over x from 0 to a. That is,

$$\int_0^a V(x, 0) \sin \frac{m\pi}{a} x\ dx = i \sum_{n=1}^{\infty} A_n \int_0^a \sin \frac{n\pi}{a} x \sin \frac{m\pi}{a} x\ dx \tag{3.69}$$

The integration of the right-hand side is facilitated by noting that

$$\sin \frac{n\pi}{a} x \sin \frac{m\pi}{a} x = \frac{1}{2} \left[\cos \frac{\pi}{a}(n - m)x - \cos \frac{\pi}{a}(n + m)x \right]$$

Thus

$$\int_0^a V(x, 0) \sin \frac{m\pi}{a} x\ dx = i \sum_{n=1}^{\infty} A_n \frac{a}{2} \delta_{n,m} = i \frac{A_m}{2} a$$

where

$$\delta_{n,m} = \begin{cases} 1 & \text{if } m = n \\ 0 & \text{if } m \neq n \end{cases}$$

Thus

$$A_n = -\frac{2i}{a} \int_0^a V(x, 0)\sin \frac{n\pi}{a} x\ dx$$

is the desired value.

The fact that the integral on the right-hand side of (3.69) vanishes for $n \neq m$ is an example of mode orthogonality. This property is very useful, if not indispensable, in solving problems of this type.

Exercise: Take $V(x, 0) = V_0 \left[\sin\left(\dfrac{p\pi}{a}\right)x \right]$, where p is any positive integer. Then show that

$$V(x, y) = V_0 \sin\frac{p\pi}{a} x \, \exp\left\{-\left[\gamma^2 + \left(\frac{p\pi}{a}\right)^2\right]^{1/2} y\right\}$$

Exercise: Take

$$V(x, 0) = \begin{cases} 0 \text{ for} & 0 < x < x_0 \\ V_0 \text{ for} & x_0 > x > x_0 + \delta \\ 0 \text{ for} & x_0 + \delta < x < a \end{cases}$$

Then if δ is sufficiently small, show that the coefficient A_n in (3.67) is

$$A_n \simeq -\frac{2i}{a} V_0 \, \delta \sin\frac{\pi n x_0}{a}$$

3.6 CYLINDRICAL WAVE FUNCTIONS

Next, we consider wave propagation in cylindrical geometry, but for the moment we restrict attention to the case $\partial/\partial z = 0$. The governing equation for V is now

$$\frac{1}{\rho}\frac{\partial}{\partial\rho}\left(\rho\frac{\partial V}{\partial\rho}\right) + \frac{1}{\rho^2}\frac{\partial^2 V}{\partial\phi^2} = \gamma^2 V \tag{3.70}$$

as indicated in the exercise on page 74 of Chapter 2. Again we use the product solution

$$V = R(\rho)\Phi(\phi) \tag{3.71}$$

This solution leads to

$$\frac{\rho}{R}\frac{d}{d\rho}\left(\rho\frac{dR}{d\rho}\right) + \frac{1}{\Phi}\frac{d^2\Phi}{d\phi^2} = \gamma^2\rho^2 \tag{3.72}$$

Exercise: Show that (3.70) can be derived directly from the cartesian form

$$\frac{\partial^2 V}{\partial x^2} + \frac{\partial^2 V}{\partial y^2} = \gamma^2 V$$

where the variables are connected by $\rho = (x^2 + y^2)^{1/2}$ and $\tan\phi = y/x$. *Hint:* Note that

$$\frac{\partial V}{\partial x} = \frac{\partial V}{\partial\rho}\frac{\partial\rho}{\partial x} + \frac{\partial V}{\partial\phi}\frac{\partial\phi}{\partial x} \text{ (e.g. see Exercise, pg. 74)}$$

In equation (3.72), we set

$$\frac{1}{\Phi}\frac{d^2\Phi}{d\phi^2} = -m^2 \tag{3.73}$$

where $-m^2$ is a separation constant. Solutions of Φ are of the form $\exp(\mp im\phi)$. If these are single-valued functions of ϕ, we must have m an integer or zero. Then

$$\frac{\rho}{R}\frac{d}{d\rho}\left(\rho\frac{dR}{d\rho}\right) - \gamma^2\rho^2 - m^2 = 0 \tag{3.74}$$

or

$$\rho^2\frac{d^2R}{d\rho^2} + \rho\frac{dR}{d\rho} - (\gamma^2\rho^2 + m^2)R = 0 \tag{3.75}$$

If we change the variable to

$$x = i\gamma\rho$$

the equation for R becomes

$$x^2\frac{d^2R}{dx^2} + x\frac{dR}{dx} + (x^2 - m^2)R = 0 \tag{3.76}$$

This equation is Bessel's equation of order m, and solutions are designated $J_m(x)$ and $Y_m(x)$. The latter are termed the Bessel functions of the first and second kind of order m, respectively. The first of these functions is defined by the convergent series expansion

$$J_m(x) = \sum_{n=0}^{\infty}\frac{(-1)^n x^{m+2n}}{n!(m+n)!2^{m+2n}} \tag{3.77}$$

and the second by the limit process

$$Y_m(x) = \lim_{v\to m}\left(\frac{J_v(x)\cos v\pi - J_{-v}(x)}{\sin v\pi}\right) \tag{3.78}$$

Equations (3.77) and (3.80) are actually defining relations for noninte-gral order if m is replaced by v and $(n+m)!$ by $\Gamma(n+v+1)$.

Closely related functions would occur more naturally in this type of problem if we had changed the variable instead by the substitution

$$y = \gamma\rho$$

Then the governing equation for R would be

$$y^2\frac{d^2R}{dy^2} + y\frac{dR}{dy} - (y^2 + m^2)R = 0 \tag{3.79}$$

Independent solutions of this equation are the modified Bessel functions that are designated $I_m(y)$ and $K_m(y)$. Conventionally, they are defined by

$$I_m(y) = \sum_{n=0}^{\infty} \frac{y^{m+2n}}{n!(m+n)!2^{m+2n}} \tag{3.80}$$

and

$$K_m(y) = \lim_{v \to m} \frac{\pi}{2}\left[\frac{I_{-v}(y) - I_v(y)}{\sin v\pi}\right] \tag{3.81}$$

Of course, the modified forms are related to the J and Y forms in a clear-cut fashion. In fact, for any order v

$$I_v(iZ) = e^{iv\pi/2}J_v(Z) \tag{3.82}$$

and

$$K_v(iZ) = \frac{\pi}{2} e^{-i(v+1)\pi/2}[J_v(Z) - iY_v(Z)] \tag{3.83}$$

Thus, for example, solutions of (3.75) are $I_m(\gamma\rho)$ and $K_m(\gamma\rho)$, or, if we had defined $\gamma = i\beta$, then the choice, by convenience, would have been the pair $J_m(\beta\rho)$ and $Y_m(\beta\rho)$.

Two other functions that occur frequently are the Hankel functions of the first and second kind. They are defined by

$$H_v^{(1)}(y) = J_v(y) + iY_v(y) \tag{3.84}$$

and

$$H_v^{(2)}(y) = J_v(y) - iY_v(y) \tag{3.85}$$

In selecting a Bessel function for a particular physical problem, we usually need to have ready access to their properties for small and large values of the arguments. Some particularly useful results for small values of the argument are

$$J_v(x) \simeq \frac{x^v}{2^v v!} \qquad Y_v(x) \simeq -\frac{2^v(v-1)!}{\pi x^v} \qquad (v \neq 0)$$

$$J_0(x) \simeq 1 \qquad Y_0(x) \simeq \frac{2}{\pi}(\ln x + C - \ln 2) \tag{3.86}$$

where $C = 0.5773. \ldots$, and

$$I_v(y) \simeq \frac{y^v}{2^v v!} \qquad K_v(y) \simeq \frac{(v-1)!2^{v-1}}{y^v}$$

$$I_0(y) \simeq 1 \qquad K_0(y) \simeq -(\ln y + C - \ln 2) \tag{3.87}$$

The above expressions are the leading terms in the relevant convergent series expansions of the functions. They are valid for any value of v, but in the expression for K_v, Re $v > 0$. When the arguments of the functions are large, the leading terms of the relevant asymptotic expansions are also very valuable. They are

$$H_v^{(1)}(x) \simeq \left(\frac{2}{\pi x}\right)^{1/2} \exp\left[i\left(x - \frac{v\pi}{2} - \frac{\pi}{4}\right)\right] \tag{3.88}$$

$$H_v^{(2)}(x) \simeq \left(\frac{2}{\pi x}\right)^{1/2} \exp\left[-i\left(x - \frac{v\pi}{2} - \frac{\pi}{4}\right)\right] \tag{3.89}$$

$$I_v(y) \simeq \frac{e^y}{(2\pi y)^{1/2}} \tag{3.90}$$

$$K_v(y) \simeq \left(\frac{\pi}{2y}\right)^{1/2} e^{-y} \tag{3.91}$$

In general, we see that the ordinary Bessel functions J, Y, $H^{(1)}$, and $H^{(2)}$ have an oscillatory behavior of their real argument, while the modified Bessel functions I and K have an exponential growth or decay as a function of their real argument. When the arguments of the functions are generally complex, the choice again is one of convenience if custom has not dictated otherwise.*

---

* Many other useful properties of Bessel functions can be found from their defining series expansions. The following examples are noted:

$$J_n'(x) = \frac{dJ_n}{dx} = J_{n-1} - \frac{n}{x} J_n$$

which is also satisfied by Y_n, $H_n^{(1)}$, and $H_n^{(2)}$, all of argument x and order n. In particular, $J_0'(x) = -J_1(x)$. Also,

$$J_n Y_n' - J_n' Y_n = J_{n+1} Y_n - J_n Y_{n+1} = \frac{2}{\pi x}$$

the wronskian relation, has great utility. Corresponding results for modified Bessel functions are

$$I_n'(y) = \frac{dI_n}{dy} = I_{n-1} - \frac{n}{y} I_n$$

$$K_n'(y) = -K_{n-1} - \frac{n}{y} K_n$$

$$I_0'(y) = I_1(y) \qquad \text{and} \qquad K_0'(y) = -K_1(y)$$

Also,

$$K_n I_n' - K_n' I_n = I_{n+1} K_n + K_{n+1} I_n = \frac{1}{y}$$

where the arguments are y.

3.7 CYLINDRICAL RADIATOR

We consider the following exterior problem. Over the cylindrical surface $\rho = a$ we specify that the physical parameter $V(a, \phi)$ is specified for all values of ϕ and z. The external region is homogeneous, and V satisfies the equation

$$(\nabla^2 - \gamma^2)V = 0 \tag{3.92}$$

where

$$\nabla^2 = \frac{1}{\rho}\frac{\partial}{\partial \rho}\left(\rho \frac{\partial}{\partial \rho}\right) + \frac{1}{\rho^2}\frac{\partial^2}{\partial \phi^2} \tag{3.93}$$

Assuming that there are no sources outside the cylinder, what form should the solution take? We may set $\gamma = \alpha + i\beta$ and assume α and β are real for a time factor of $\exp(i\omega t)$.

The desired form of the solution must be

$$K_m(\gamma\rho)e^{\pm im\phi} \tag{3.94}$$

where m is an integer. Such solutions are periodic in ϕ and single-valued in the angular domain $0 < \phi < 2\pi$. Also, $K_m(\gamma\rho)$ has the proper asymptotic behavior as $\rho \to \infty$. The solution is now made up of a superposition of all such elementary forms; that is,

$$V(\rho, \phi) = \sum_{m=-\infty}^{+\infty} B_m K_m(\gamma\rho)e^{-im\phi}. \tag{3.95}$$

where the summation extends over both positive and negative integers. Now

$$V(a, \phi) = \sum_{m=-\infty}^{+\infty} B_m K_m(\gamma a)e^{-im\phi} \tag{3.96}$$

We now multiply both sides by $e^{in\phi}$ and integrate from 0 to 2π. Thus

$$\int_0^{2\pi} V(a, \phi)e^{in\phi}\,d\phi = \sum_{m=-\infty}^{+\infty} B_n K_n(\gamma a)\int_0^{2\pi} e^{i(n-m)\phi}\,d\phi$$

$$= \sum_{m=-\infty}^{+\infty} B_m K_m(\gamma a)2\pi\,\delta_{n,m} = B_n K_n(\gamma a)2\pi \tag{3.97}$$

Then we find that

$$B_m = \frac{1}{2\pi}\int_0^{2\pi} V(a, \phi)e^{im\phi}\,d\phi\,\frac{1}{K_m(\gamma a)} \tag{3.98}$$

For example, we might have specified that

$$V(a, \phi) = V_0\,\delta(\phi - \phi_0) \tag{3.99}$$

corresponding to a slit excitation at $\phi = \phi_0$ on the cylindrical surface.

Then

$$B_m = \frac{V_0}{2\pi} \frac{e^{im\phi_0}}{K_m(\gamma a)} \qquad (3.100)$$

The field expression for all space external to the cylinder is then

$$V(\rho, \phi) = \frac{V_0}{2\pi} \sum_{m=-\infty}^{+\infty} e^{im(\phi_0 - \phi)} \frac{K_m(\gamma\rho)}{K_m(\gamma a)} \qquad (3.101)$$

When $|\gamma\rho| \gg 1$, we can employ the far-field approximation given by (3.91); then

$$V(\rho, \phi) \simeq \frac{V_0 e^{-\gamma\rho}}{2(2\pi)^{1/2}(\gamma\rho)^{1/2}} P(\phi) \qquad (3.102)$$

where

$$P(\phi) = \sum_{m=-\infty}^{+\infty} \frac{e^{im(\phi_0 - \phi)}}{K_m(\gamma a)} \qquad (3.103)$$

is a pattern factor. Now, as indicated by (3.81),

$$K_m(\gamma a) = K_{-m}(\gamma a) \qquad (3.104)$$

when m is an integer. Thus the pattern factor can be written more conveniently as

$$P(\phi) = \sum_{m=0}^{\infty} \frac{\epsilon_m \cos m(\phi_0 - \phi)}{K_m(\gamma a)} \qquad \epsilon_0 = 1, \epsilon_m = 2 \ (m \neq 0) \qquad (3.105)$$

Now if the radius of the cylinder is electrically small, we would have $|\gamma a| \ll 1$. The series in (3.105) converges very quickly, and the first two terms would yield the approximation

$$P(\phi) \simeq P_0 + P_1 \cos(\phi - \phi_0) \qquad (3.106)$$

where

$$\frac{P_1}{4P_0} \simeq \frac{K_0(\gamma a)}{2K_1(\gamma a)} \simeq -\frac{\gamma a}{2}\left(\ln \frac{\gamma a}{2} + C\right)$$

which is also small.

Exercise: Repeat the foregoing analysis but instead specify $\partial V(\rho, \phi)'/\partial\rho$ at $\rho = a$.

3.8 ANOTHER CYLINDRICAL RADIATOR

Here we consider a slightly more complicated problem. The configuration, as shown in Fig. 3.5, is the same as in the above example but now we imagine there is a source located in a concentric cylindrical surface just

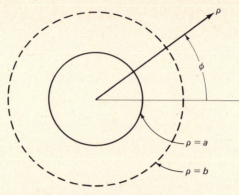

Figure 3.5   Cylindrical radiator with concentric source.

external to the cylinder that is ideally "soft." The "source" is such as to impart a jump in the normal derivative of the function $V(\rho, \phi)$ at $\rho = b$. Our objective is to deduce $V(\rho, \phi)$ everywhere for $\rho > a$ under the conditions

$$V'(b + 0, \phi) - V'(b - 0, \phi) = F(\phi) \tag{3.107}$$

and

$$V(a, \phi) = 0$$

for all ϕ. Also, we assume that $V(\rho, \phi)$ is continuous at $\rho = b$. Here $F(\phi)$ is a source function that is prescribed for $0 < \phi < 2\pi$. The function itself vanishes at the inner cylinder.

Actually, there are two distinct regions that must be considered. For $\rho > b$, clearly a suitable solution is of the form

$$V(\rho, \phi) = \sum_{m=-\infty}^{\infty} M_m e^{-im\phi} K_m(\gamma\rho) \tag{3.108}$$

where M_m is yet to be determined. Now in the region $a < \rho < b$ both functions I_m and K_m are permitted. This fact immediately tells us that a suitable form in this concentric region is

$$V(\rho, \phi) = \sum_{m=-\infty}^{+\infty} [N_m K_m(\gamma\rho) + T_m I_m(\gamma\rho)]e^{-im\phi} \tag{3.109}$$

Now when $V(a, \phi) = 0$ is applied to (3.109), term by term, we immediately deduce that

$$N_m K_m(\gamma a) + T_m I_m(\gamma a) = 0 \tag{3.110}$$

Thus within the concentric region $a < \rho < b$, we require that

$$V(\rho, \phi) = \sum_{m=-\infty}^{+\infty} \left\{ N_m \left[K_m(\gamma\rho) - \frac{K_m(\gamma a)}{I_m(\gamma a)} I_m(\gamma\rho) \right] \right\} e^{-im\phi} \tag{3.111}$$

As we see, there are two coefficients M_m and N_m yet to be determined. The continuity condition is met if

$$M_m K_m(\gamma b) = N_m \left[K_m(\gamma b) - \frac{K_m(\gamma a)}{I_m(\gamma a)} I_m(\gamma b) \right] \tag{3.112}$$

while the jump condition yields

$$\gamma M_m K'_m(\gamma b) - \gamma N_m \left[K'_m(\gamma b) - \frac{K_m(\gamma a)}{I_m(\gamma a)} I'_m(\gamma b) \right] = F_m \tag{3.113}$$

where

$$K'_m(\gamma b) = \frac{d}{dy} K_m(y) \Big|_{y=\gamma b} \tag{3.114}$$

$$I'_m(\gamma b) = \frac{d}{dy} I_m(y) \Big|_{y=\gamma b} \tag{3.115}$$

In the above equations we have expressed the source function in the form

$$F(\phi) = \sum_{m=-\infty}^{+\infty} F_m e^{-im\phi} \tag{3.116}$$

so that individual harmonic terms can be matched when implementing (3.113).

The two equations (3.112) and (3.113) can be solved for M_m and N_m. In particular, it is not difficult to deduce that

$$M_m = -bF_m \left[I_m(\gamma b) - \frac{I_m(\gamma a)}{K_m(\gamma a)} K_m(\gamma b) \right] \tag{3.117}$$

where we have used the wronskian identity, well known in Bessel function theory,

$$I_m(\gamma b) K'_m(\gamma b) - I'_m(\gamma b) K_m(\gamma b) = -\frac{1}{\gamma b}$$

The problem is thus formally solved. The explicit expression for the external field is

$$V(\rho, \phi) = -b \sum_{m=-\infty}^{+\infty} F_m e^{-im\phi} \frac{K_m(\gamma \rho)}{\Lambda_n(\gamma b)} \tag{3.118}$$

where

$$\frac{1}{\Lambda_m(\gamma b)} = I_m(\gamma b) - \frac{I_m(\gamma a)}{K_m(\gamma a)} K_m(\gamma b) \tag{3.119}$$

If $F(\phi)$ is specified, then we deduce F_m from

$$F_m = \frac{1}{2\pi} \int_0^{2\pi} F(\phi)e^{im\phi}\, d\phi \tag{3.120}$$

As in the previous example, the far-field behavior of the external field is embodied in (3.118) by replacing $K_m(\gamma\rho)$ by its asymptotic approximation as stated by (3.91). By definition, the corresponding "radiation pattern" is the ϕ dependence.

Exercise: Repeat the steps in the preceding exercise when (3.107) is replaced by

$$V(b + 0, \phi) - V(b - 0, \phi) = F(\phi) \quad \text{and} \quad V'(a, \phi) = 0$$

where now $V'(\rho, \phi)$ is continuous at $\rho = b$.

Exercise: Consider the following synthesis problem: If $V(\rho, \phi)$ is specified, indicate how the required excitation function $F(\phi)$ is to be determined.

3.9 CYLINDRICAL CAVITY RESONATOR

We can illustrate the important aspects of an internal type of problem by looking at the region inside a cylindrical tube. For example, assuming that $V(\rho, \phi)$ satisfies (3.92) everywhere inside the cylindrical surface $\rho = c$, what form would the solution take if, say, $V(a, \phi) = 0$? Now, clearly, the solution must be of the type

$$I_m(\gamma a) = 0 \tag{3.121}$$

Or if we set $\gamma = i\beta$, then the condition is equivalent to

$$J_m(\beta a) = 0 \tag{3.122}$$

For a given m (harmonic order) this equation yields a discrete set of solutions for β that are given by

$$\beta = \beta_{m,n} = \frac{x_{m,n}}{a}$$

where $x_{m,n}$ is the nth root of the equation

$$J_m(x) = 0$$

The roots are well known. For example,

$$
\begin{array}{lll}
x_{0,1} = 2.40 & x_{0,2} = 5.52 & x_{0,3} = 8.65 \\
x_{1,1} = 3.83 & x_{1,2} = 7.02 & x_{1,3} = 10.17 \\
x_{2,1} = 5.14 & x_{2,2} = 8.42 & x_{2,3} = 11.62
\end{array}
$$

and so on. Each one of these roots corresponds to a particular field configuration, and they are called modes.

Another class of modes would be obtained if we required that the normal derivative of V be zero at $\rho = a$. Then the model equation would be

$$J'_m(\beta a) = 0 \tag{3.123}$$

The roots $\beta'_{m,n}$ are not the same as in the previous example, but they are equally important in waveguide and cavity resonator theory.

3.10 THREE-DIMENSIONAL WAVE PROPAGATION

The case where the variation in all three coordinates is allowed presents some new complexities. In terms of the cartesian coordinates (x, y, z), we may proceed with the separation-of-variables technique by setting

$$V = X(x)Y(y)Z(z) \tag{3.124}$$

Then working with (3.1), we are readily led to

$$\left(\frac{d^2}{dx^2} - \gamma_x^2\right)X(x) = 0 \tag{3.125}$$

$$\left(\frac{d^2}{dy^2} - \gamma_y^2\right)Y(y) = 0 \tag{3.126}$$

$$\left(\frac{d^2}{dz^2} - \gamma_z^2\right)Z(z) = 0 \tag{3.127}$$

where γ_x, γ_y, and γ_z are the separation constants that are subject to the constraint

$$\gamma_x^2 + \gamma_y^2 + \gamma_z^2 = \gamma^2 \tag{3.128}$$

Clearly, solutions of (3.1) are of the form

$$\exp(\mp \gamma_x x)\exp(\mp \gamma_y y)\exp(\mp \gamma_z z) \tag{3.129}$$

3.11 RECTANGULAR CAVITY RESONATOR

As an example we will consider a rectangular enclosure bounded by $x = \pm a$, $y = \pm b$, and $z = \pm c$. We seek a general form for the field function $V(x, y, z)$ inside this enclosure subject to the condition that the normal derivative of V vanishes on all the walls. Also, we restrict attention to solutions where V is fully symmetrical about the origin (that is, the geometrical center).

A suitable form of the solution is seen to be

$$\cosh(\gamma_x x)\cosh(\gamma_y y)\cosh(\gamma_z z)$$

Now clearly,

$$\gamma_x = i\frac{\pi m}{a} \tag{3.130}$$

and

$$\gamma_y = i\frac{\pi n}{b} \tag{3.131}$$

where m and n are integers. Thus the form

$$\left(\cos\frac{\pi m}{a}x\right)\left(\cos\frac{\pi n}{a}y\right)(\cosh\gamma_z z)$$

where

$$\gamma_z^2 = \gamma^2 - (\gamma_x^2 + \gamma_y^2) = \gamma^2 + \left(\frac{\pi m}{a}\right)^2 + \left(\frac{\pi n}{b}\right)^2 \tag{3.132}$$

must now also be subject to

$$\gamma_z = i\frac{\pi p}{c}$$

where p is an integer. The so-called cavity resonator modes are then subject to the condition

$$\left(\frac{\pi m}{a}\right)^2 + \left(\frac{\pi n}{b}\right)^2 + \left(\frac{\pi p}{c}\right)^2 + \gamma^2 = 0 \tag{3.133}$$

If there are no losses in the enclosed medium, we can set $\gamma = i(2\pi/\lambda)$, where $\lambda, = \lambda_{m,n,p}$ designates the triply infinite, discrete wavelengths where the cavity is in resonance.

3.12 RECTANGULAR WAVEGUIDE

If instead of a closed cavity we were dealing with a rectangular channel of infinite extent in the z direction, the solution would be given by (3.133) subject only to the wall conditions at $x = \pm a$, $y = \pm b$. Thus for a specified value of γ, these would give the waveguide modes. So a general solution for the modal field, giving outgoing waves as $|z| \rightarrow \infty$, would be

$$V(x, y, z) = \sum_{m=0}^{\infty} \sum_{n=0}^{\infty} A_{m,n} \cos\frac{\pi m}{a}x \cos\frac{\pi n}{b}y \exp(-\gamma_{m,n}|z|) \tag{3.134}$$

where

$$\gamma_{m,n} = \left[\gamma^2 + \left(\frac{m\pi}{a}\right)^2 + \left(\frac{n\pi}{b}\right)^2\right]^{1/2} \tag{3.135}$$

The coefficient $A_{m,n}$ is to be obtained by the source conditions, such as the specified aperture field at $z = 0$.

This rather simple and idealistic example should point out to the reader the close relationship between cavity resonator and waveguide theories. As indicated above, when dealing with a *closed* cavity, we seek a source-free solution that yields a discrete set of wavelengths or frequencies. If there are losses in the system, we would find that such frequencies have a nonzero imaginary part. In the waveguide problem we first consider an infinitely extended uniform section and seek a discrete set of wave functions that individually satisfy the wall-boundary conditions and have a traveling-wave character.

3.13 THREE-DIMENSIONAL CYLINDRICAL PROBLEMS

A wide class of important wave propagation problems deal with cylindrical structures that can be regarded as infinite in axial extent. On the other hand, the associated fields may have variations in both the transverse and the axial direction. For example, the source of such fields is localized.

To obtain the desired form for the solutions, we again can use the separation-of-variables technique applied to the equation

$$(\nabla^2 - \gamma^2)V = 0 \tag{3.136}$$

In terms of cylindrical coordinates

$$\nabla^2 = \frac{1}{\rho}\frac{\partial}{\partial\rho}\rho\frac{\partial}{\partial\rho} + \frac{1}{\rho^2}\frac{\partial^2}{\partial\phi^2} + \frac{\partial^2}{\partial z^2} \tag{3.137}$$

is the laplacian operator. As a very straightforward extension of our previous work, we can easily deduce that the desired form of the solution is made up of linear combinations of

$$\begin{matrix} I_m(u\rho) \\ K_m(u\rho) \end{matrix} e^{\pm im\phi}\, e^{\pm i\lambda z}$$

where

$$u = (\gamma^2 + \lambda^2)^{1/2}$$

Here m is an integer if the solution is to be single-valued in ϕ (that is, periodic). In wedge-restricted regions this is not the case, and m is not necessarily an integer. In what follows, unless stated otherwise, we will assume m is an integer. The free parameter λ is unrestricted, and general solutions will involve an integration over this variable. The radical that defines u above calls for some comment. The branch of the square root is usually chosen such that $u \to \gamma$ as $|\lambda^2| \to 0$ and $u \to |\lambda|$ as $|\gamma^2| \to 0$. The use of the symbol λ here for the axial wave number is possibly unfortunate because it may become confused with λ_0, the symbol we use for the free-space wavelength. But the weight of custom is overbearing, so we will defer to Sommerfeld and retain this use.

3.14 GENERAL CYLINDRICAL RADIATOR

We consider the following exterior problem that is really an extension of Section 3.7. Over the cylindrical surface $\rho = a$, $-\infty < z < \infty$, we specify that

$$V(a, \phi, z) = F(\phi, z) \tag{3.138}$$

where $F(\phi, z)$ is a source function. What is the form for V for $\rho > a$? No sources exist outside this surface. We proceed as follows. We write

$$F(\phi, z) = \sum_{m=-\infty}^{\infty} \int_{-\infty}^{+\infty} F_m(\lambda) e^{-i\lambda z} \, d\lambda \, e^{-im\phi} \tag{3.139}$$

as an appropriate representation. To deduce the form for $F_m(\lambda)$, we utilize the following transform relations: If

$$F(z) = \int_{-\infty}^{+\infty} \hat{F}(\lambda) e^{-i\lambda z} \, d\lambda \tag{3.140}$$

then

$$\hat{F}(\lambda) = \frac{1}{2\pi} \int_{-\infty}^{+\infty} e^{i\lambda z} F(z) \, dz \tag{3.141}$$

according to Fourier integral theory. The integration over λ is along the real axis. Also, we know that if

$$F(\phi) = \sum_{m=-\infty}^{+\infty} F_m e^{-im\phi} \tag{3.142}$$

then

$$F_m = \frac{1}{2\pi} \int_0^{2\pi} e^{im\phi} F(\phi) \, d\phi \tag{3.143}$$

subject to the integrability of $F(\phi)$. Thus it follows that

$$F_m(\lambda) = \frac{1}{(2\pi)^2} \int_{\phi=0}^{2\pi} \int_{z=-\infty}^{+\infty} F(\phi, z) e^{i\lambda z} dz \, e^{im\phi} \, d\phi \tag{3.144}$$

This result immediately suggests that a suitable representation for the external field function is

$$V(\rho, \phi, z) = \sum_{-\infty}^{+\infty} \int_{-\infty}^{+\infty} V_m(\lambda) e^{-i\lambda z} e^{-im\phi} K_m(u\rho) \, d\lambda \tag{3.145}$$

The next step is very simple; we merely insist that (3.145) reduces to (3.135) as $\rho \rightarrow a$. This condition is met if

$$V_m(\lambda) = \frac{F_m(\lambda)}{K_m(u_m a)} \tag{3.146}$$

This step yields the desired formal integral solution for the problem, since $F_m(\lambda)$ is specified in terms of the given source function $F(\phi, z)$.

The evaluation of the infinite integral in the formal solution can be a formidable task, but there are some approximate methods that can be used to greatly simplify the task. This is particularly the case when the observer is in the far field.

3.15 CYLINDRICAL WAVEGUIDE

The interior cylindrical problem for a specified excitation is no less interesting. For example, we state the following objective: Determine the field function $V(\rho, \phi, z)$ for the region $\rho < a$ and $z_0 < z < \infty$ subject to the conditions that $V(a, \phi, z) = 0$ and $V(\rho, \phi, z_0) = S(\rho)$ is a specified excitation function, which is independent of ϕ.

The desired form of the solution must contain functions of the form

$$J_0(\nu \rho) e^{-i\lambda z}$$

where ν and λ are connected by

$$\nu^2 = \beta^2 - \lambda^2$$

Such functions satisfy the wave equation

$$\left(\frac{1}{\rho} \frac{\partial}{\partial \rho} \rho \frac{\partial}{\partial \rho} + \frac{\partial^2}{\partial z^2} + \beta^2 \right) V = 0 \tag{3.147}$$

where $\beta = -i\gamma = 2\pi/\text{wavelength}$.

The wall condition requires that

$$J_0(\nu a) = 0$$

This equation is satisfied if $\nu = x_m/a$, where x_m is the mth root of $J_0(x) = 0$. This result suggests that the general solution is of the form

$$V(\rho, z) = \sum_m G_m J_0(\nu_m \rho) e^{-i\lambda_m z} \tag{3.148}$$

where

$$\lambda_m = (\beta^2 - \nu_m^2)^{1/2}$$

We choose the root that has $\text{Re } \lambda_m > 0$ so that the waves are traveling in the positive z direction. However, we note that in the absence of losses

λ_m becomes purely imaginary if $\beta < v_m$. In that case we choose λ_m to be a negative imaginary quantity (that is, $i\lambda_m$ is positive and real). The modes in this case are cut off since the associated fields decay quickly.

To deal with the source condition, we are led to seek a representation in the form

$$S(\rho) = \sum_{m=1}^{\infty} s_m J_0\left(\frac{x_m \rho}{a}\right) \tag{3.149}$$

To obtain s_m, we utilize the orthogonality condition

$$\int_0^a \rho J_0\left(\frac{x_m \rho}{a}\right) J_0\left(\frac{x_n \rho}{a}\right) d\rho = \begin{cases} N_m^0 & \text{if } m = n \\ 0 & \text{if } m \neq n \end{cases} \tag{3.150}$$

Here

$$N_m^0 = \frac{a^2}{2} J_1^2(x_m) \tag{3.151}$$

is the normalization factor. We defer the proof of this result until later when we consider the ϕ-dependent case. The essential step is to multiply both sides of (3.149) by $\rho J_0(x_n \rho/a)$ and then integrate each side with respect to ρ from 0 to a. Clearly, this step gives the desired relation

$$s_m = \frac{1}{N_m^0} \int_0^a \rho S(\rho) J_0\left(\frac{x_m \rho}{a}\right) d\rho \tag{3.152}$$

By requiring that $V(\rho, z)$ given by (3.148) is the same as $S(\rho)$ as $z \rightarrow z_0$, we obtain

$$G_m = s_m e^{+i\lambda_m z_0}$$

Thus the final solution is

$$V(\rho, z) = \sum_{m=1}^{\infty} s_m J_0(v_m \rho) e^{-i\lambda_m(z - z_0)} \tag{3.153}$$

Consider the extension of the previous solution when the source function is specified to be

$$V(\rho, \phi, z_0) = S(\rho, \phi) \tag{3.154}$$

which can be a function of both ρ and ϕ. Again, we require that $V(a, \phi, z) = 0$.

We are now led to write

$$V(\rho, \phi, z) = \sum_{m} \sum_{n=-\infty}^{+\infty} G_{m,n} J_n(v_{n,m} \rho) e^{-in\phi} e^{-i\lambda_{n,m}z} \tag{3.155}$$

where

$$v_{n,m} = \frac{x_{n,m}}{a}$$

in terms of the roots $x_{n,m}$ of

$$J_n(x) = 0$$

Again,

$$\lambda_{n,m} = (\beta^2 - v_{n,m}^2)^{1/2} \tag{3.156}$$

The desired form of the source function is now

$$S(\rho, \phi) = \sum_{m=1}^{\infty} \sum_{n=-\infty}^{+\infty} s_{m,n} J_n\left(\frac{x_{n,m}\rho}{a}\right) e^{-in\phi} \tag{3.157}$$

To obtain $s_{m,n}$, we note that

$$\sum_{m=1}^{\infty} s_{m,n} J_n\left(\frac{x_{n,m}\rho}{a}\right) = \frac{1}{2\pi} \int_0^{2\pi} e^{in\phi} S(\rho, \phi)\, d\phi \tag{3.158}$$

The next step requires that we evaluate

$$\int_0^a \rho J_n\left(\frac{x_{n,m}\rho}{a}\right) J_n\left(\frac{x_{n,m'}\rho}{a}\right) d\rho = \Lambda_{m,m'}^n \tag{3.159}$$

This integral arises when we multiply both sides of (3.158) by $\rho J_n(x_{n,m'}\rho/a)$ and then integrate over ρ from 0 to a.

From Bessel function theory [1] we know that

$$\int_0^a J_n(k_1\rho)J_n(k_2\rho)\rho\, d\rho = \frac{a}{k_1^2 - k_2^2}\, [k_2 J_n(k_1 a)J_{n-1}(k_2 a)$$
$$- k_1 J_{n-1}(k_1 a)J_n(k_2 a)] \tag{3.160}$$

and

$$\int_0^a \rho[J_n(k_1 a)]^2\, d\rho = \frac{a^2}{2}\left\{\left[J_n'(k_1 a)\right]^2 + \left(1 - \frac{n^2}{k_1^2 a^2}\right)J_n^2(k_1 a)\right\} \tag{3.161}$$

The first integral above tells us immediately that $\Lambda_{m,m'}^n = 0$, if $m \neq m'$, when $k_1 a$ is identified with $x_{n,m}$. Thus we see that

$$s_{m,n} = \frac{1}{2\pi\Lambda_{m,m}^n} \int_0^{2\pi} e^{+in\phi} \int_0^a \rho S(\rho, \phi)J_n\left(\frac{x_{n,m}\rho}{a}\right) d\phi\, d\rho \tag{3.162}$$

where

$$\Lambda_{m,m}^n = \frac{a^2}{2}[J_n'(x_{n,m})]^2 \tag{3.163}$$

and where $x_{n,m}$ is the mth zero of the Bessel function $J_n(x)$ of order n.

The solution is obtained easily by requiring that

$$V(\rho, \phi, z_0) = S(\rho, \phi) \tag{3.164}$$

so that

$$G_{m,n} = s_{m,n} e^{i\lambda_{n,m}z_0} \tag{3.165}$$

Thus the final explicit result for the field function is

$$V(\rho, \phi, z) = \sum_{m=1}^{\infty} \sum_{n=-\infty}^{+\infty} s_{m,n} J_n(\nu_{n,m}\rho) e^{-in\phi} e^{-i\lambda_{n,m}(z-z_0)} \tag{3.166}$$

3.16 SPHERICAL WAVE FUNCTIONS

An important class of three-dimensional problems deals with spherical geometry.

The Helmholtz equation

$$\nabla^2 V = \gamma^2 V \tag{3.167}$$

is now written as follows, in terms of spherical coordinates (r, θ, ϕ):

$$\frac{1}{r^2}\frac{\partial}{\partial r} r^2 \frac{\partial V}{\partial r} + \frac{1}{r^2 \sin \theta}\frac{\partial}{\partial \theta}\left(\sin \theta \frac{\partial V}{\partial \theta}\right) + \frac{1}{r^2 \sin^2 \theta}\frac{\partial^2 V}{\partial \phi^2} = \gamma^2 V \tag{3.168}$$

Again we employ the separation-of-variables technique and set

$$V = R(r)T(\theta)\Phi(\phi)$$

This result is substituted into (3.168), and the partial differentiations are carried out. We then divide the result by $RT\Phi$ and multiply by $r^2 \sin^2 \theta$. The following equation then emerges:

$$\frac{\sin^2 \theta}{R}\frac{d}{dr}\left(r^2 \frac{dR}{dr}\right) + \frac{\sin \theta}{T}\frac{d}{d\theta}\left(\sin \theta \frac{dT}{d\theta}\right) + \frac{1}{\Phi}\frac{d^2\Phi}{d\phi^2}$$
$$- \gamma^2 r^2 \sin^2 \theta = 0 \tag{3.169}$$

We now set

$$\frac{1}{\Phi}\frac{d^2\Phi}{d\phi^2} = -m^2 \tag{3.170}$$

where m is a separation constant. Dividing the preceding equation by $\sin^2 \theta$ then yields

$$\frac{1}{R}\frac{d}{dr}\left(r^2 \frac{dR}{dr}\right) + \frac{1}{T \sin \theta}\frac{d}{d\theta}\left(\sin \theta \frac{dT}{d\theta}\right) - \frac{m^2}{\sin^2 \theta} = \gamma^2 r^2 \tag{3.171}$$

This equation clearly separates into r and θ dependence. Following convention, we now designate the second separation constant by $-(n+1)n$ and set

$$\frac{1}{T \sin \theta}\frac{d}{d\theta}\left(\sin \theta \frac{dT}{d\theta}\right) - \frac{m^2}{\sin^2 \theta} = -n(n+1) \tag{3.172}$$

and thus

$$\frac{1}{R}\frac{d}{dr}\left(r^2 \frac{dR}{dr}\right) - n(n+1) = \gamma^2 r^2 \tag{3.173}$$

Solutions of the three separated equations in the ϕ, θ, and r dimensions are of the form

$$\Phi = \exp(\pm im\phi) \tag{3.174}$$

$$T = \begin{cases} P_n^m(\cos\theta) \\ Q_n^m(\cos\theta) \end{cases} \tag{3.175}$$

and

$$R = \begin{cases} (\gamma r)^{-1/2} I_{n+1/2}(\gamma r) \\ (\gamma r)^{-1/2} K_{n+1/2}(\gamma r) \end{cases} \tag{3.176}$$

Linear combinations of any of these pairs of solutions are to be the basic building blocks to construct the general solution. The function designations used above call for some comment.

Clearly, if V is to be a single-valued function of ϕ, we may restrict m to be any integer including zero. In that case $P_n^m(\cos\theta)$ and $Q_n^m(\cos\theta)$ are associated Legendre polynomials of the order n and degree m. When $m = 0$, these terms are simply called the Legendre polynomials and designated $P_n(u)$ and $Q_n(u)$, where $u = \cos\theta$. The function of the first kind is of major interest because it is finite over the range $0 \le \theta \le \pi$. They can be defined as the finite sum

$$P_n(u) = \sum_{m=0}^{M} \frac{(-1)^m (2n - 2m)!}{2^n m! (n - m)! (n - 2m)!} u^{n-2m} \tag{3.177}$$

where $M = n/2$ or $(n-1)/2$, whichever is an integer. An alternative definition is Rodrigues' formula [3],

$$P_n(u) = \frac{1}{2^n n!} \frac{d^n}{du^n} (u^2 - 1)^n \tag{3.178}$$

The lower-order polynomials of the first kind are relatively simple [3]:

$$P_0(u) = 1 \qquad P_1(u) = u \qquad P_2 = \tfrac{1}{2}(3u^2 - 1)$$
$$P_3(u) = \tfrac{1}{5}(5u^3 - 3u)$$

and so on. In terms of θ:

$$P_0(\cos\theta) = 1 \qquad P_1(\cos\theta) = \cos\theta$$
$$P_2(\cos\theta) = \tfrac{1}{4}(3\cos 2\theta + 1)$$
$$P_3(\cos\theta) = \tfrac{1}{8}(5\cos 3\theta + 3\cos\theta)$$

and so on.

The Legendre polynomials $Q_n(u)$ of the second kind are infinite at $u = \pm 1$ (that is, at $\theta = 0$ and π). However, they may be formally defined by

$$Q_n(u) = \lim_{v \to n} \frac{\pi}{2} \frac{P_v(u)\cos v\pi - P_{-v}(-u)}{\sin v\pi} \tag{3.179}$$

where we would have to use the general form of $P_\nu(u)$ for any order ν. Without belaboring these classical derivations, we will simply note that for the lower-order Legendre polynomials of the second kind, we have

$$Q_0(u) = \frac{1}{2} \ln \frac{u+1}{u-1}$$

$$Q_1(u) = \frac{u}{2} \ln \frac{u+1}{u-1} - 1$$

$$Q_2(u) = \frac{3u^2 - 1}{4} \ln \frac{u+1}{u-1} - \frac{3u}{2}$$

and so on. As indicated, they have a singular behavior as $u \to \pm 1$.

The associated Legendre polynomials arise when we no longer have azimuthal symmetry. Then m is an integer. These functions by common convention are defined by

$$P_n^m(u) = (-1)^m (1 - u^2)^{m/2} \frac{d^m P_n(u)}{du^m} \tag{3.180}$$

and

$$Q_n^m(u) = (-1)^m (1 - u^2)^{m/2} \frac{d^m Q_n(u)}{du^m} \tag{3.181}$$

although sometimes the $(-1)^m$ factors are omitted. Our convention follows that used in the *Handbook of Mathematical Functions*.[*] An important general property is that $P_n^m(u) = 0$ for $m > n$. Some of the lower-order associated polynomials are as follows:

$$P_1^1(u) = -(1 - u^2)^{1/2}$$
$$P_2^1(u) = -3(1 - u^2)^{1/2}u$$
$$P_2^2(u) = 3(1 - u^2)$$
$$P_3^1(u) = \tfrac{3}{2}(1 - u^2)^{1/2}(1 - 5u^2)$$

and

$$Q_1^1 = -(1 - u^2)^{1/2}\left(\frac{1}{2} \ln \frac{1+u}{1-u} + \frac{u}{1-u^2} \right)$$

$$Q_2^1 = -(1 - u^2)^{1/2}\left(\frac{3}{2} \ln \frac{1+u}{1-u} + \frac{3u^2 - 2}{1-u^2} \right)$$

$$Q_2^2 = (1 - u^2)^{1/2}\left[\frac{3}{2} \ln \frac{1+u}{1-u} + \frac{5u - 3u^2}{(1-u^2)^3} \right]$$

and so on.

---

[*] M. Abramowitz and I. Stegen, *Handbook of Mathematical Functions*, Dover, New York, 1970.

Some specialized forms are particularly useful. For example, at $\theta = 0$ (that is, at $u = 1$)

$$P_n^m(1) = \begin{cases} 1 & \text{for } m = 0 \\ 0 & \text{for } m > 0 \end{cases} \tag{3.182}$$

and at $\theta = \pi/2$ (that is, at $u = 0$)

$$P_n^m(0) = \begin{cases} (-1)^{(n+m)/2} \dfrac{1 \cdot 3 \cdot 5 \cdots (n+m-1)}{2 \cdot 4 \cdot 6 \cdots (n-m)} & \\ & \text{for } n + m \text{ even} \\ 0 & \text{for } n + m \text{ odd} \end{cases} \tag{3.183}$$

$$Q_n^m(0) = \begin{cases} 0 & \text{for } n + m \text{ even} \\ (-1)^{(n+m+1)/2} \dfrac{2 \cdot 4 \cdot 6 \cdots (n+m-1)}{1 \cdot 3 \cdot 5 \cdots (n+m)} & \\ & \text{for } n + m \text{ odd} \end{cases} \tag{3.184}$$

3.17 SPHERICAL RADIATOR

To continue our discussions of spherical wave functions, we will deal with the external problem for a symmetrical situation. Specifically, we deal with a source-free region in which the wave function V is to be prescribed at a spherical surface $r = a$. That is,

$$V(a, \theta, \phi) = S(\theta) \tag{3.185}$$

which is independent of the azimuthal coordinate ϕ. For the region $r > a$

$$(\nabla^2 - \gamma^2)V = 0 \tag{3.186}$$

so that solutions will be of the form

$$\left(\frac{1}{\gamma r}\right)^{1/2} K_{n+1/2}(\gamma r)P_n(\cos \theta) \tag{3.187}$$

as indicated by (3.168) and (3.171). The solutions involving $I_{n+1/2}$ are not permitted because the field should remain finite as $r \to \infty$. Also, the function $Q_n(\cos \theta)$ is not allowed because it is singular (that is, blows up) at $\theta = 0$ and π. Now because n can be any integer, we are led to adopt the following representation for the field function:

$$V(r, \theta) = \sum_{n=0}^{\infty} d_n \frac{1}{\gamma r} \hat{K}_n(\gamma r)P_n(\cos \theta) \tag{3.188}$$

where we have defined the "new" spherical Bessel function according to

$$\hat{K}_n(\gamma r) = \left(\frac{2\gamma r}{\pi}\right)^{1/2} K_{n+1/2}(\gamma r)$$

in accordance with Schelkunoff [2]. For any argument x this function is given by the finite series

$$\hat{K}_n(x) = e^{-x} \sum_{p=0}^{n} \frac{(n+p)!}{p!(n-p)!(2x)^p} \tag{3.189}$$

where p is an integer. Thus, for example,

$$\hat{K}_0(x) = e^{-x} \qquad \hat{K}_1(x) = e^{-x}\left(1 + \frac{1}{x}\right)$$

$$\hat{K}_2(x) = e^{-x}\left(1 + \frac{3}{x} + \frac{3}{x^2}\right)$$

In general, we see that for $|\gamma r| \to \infty$

$$\hat{K}_n(\gamma r) \simeq e^{-\gamma r} \tag{3.190}$$

for any n.

We now proceed with the solution of our problem. Imposing the condition given by (3.185), we see from (3.188) that

$$\frac{1}{\gamma a} \sum_{n=0}^{\infty} d_n \hat{K}_n(\gamma a) P_n(\cos \theta) = S(\theta) \tag{3.191}$$

The unknown coefficient d_n is to be determined from this equation. This task would be straightforward if the excitation function $S(\theta)$ could be expressed in the form

$$S(\theta) = \sum_{n=0}^{\infty} a_n P_n(\cos \theta) \tag{3.192}$$

over the required range $0 \le \theta \le \pi$. To show that this representation can be done, we multiply both sides by $P_q(\cos \theta)\sin \theta$, where q is any integer, and integrate over θ from 0 to π. Thus

$$\int_0^{\pi} S(\theta)P_q(\cos \theta)\sin \theta \, d\theta = \sum_{n=0}^{\infty} a_n \int_0^{\pi} P_n(\cos \theta)P_q(\cos \theta)\sin \theta \, d\theta \tag{3.193}$$

Fortunately, we know from the theory of Legendre polynomials that the integral on the right-hand side vanishes if $n \ne q$. For the term $n = q$ we also know that

$$\int_0^{\pi} [P_n(\cos \theta)]^2 \sin \theta \, d\theta = \frac{2}{2n+1} \tag{3.194}$$

Equation (3.192) is a Legendre series and is fully analogous to the Fourier series involving exponential or trigonometric functions.

From (3.191) to (3.192) we see that

$$d_n = (\gamma a) \frac{a_n}{\hat{K}_n(\gamma a)} \tag{3.195}$$

Following (3.188) we then find that the expression for the external field function is given by

$$V(r, \theta) = \frac{a}{r} \sum_{n=0}^{\infty} a_n \frac{\hat{K}_n(\gamma r)}{\hat{K}_n(\gamma a)} P_n(\cos \theta) \tag{3.196}$$

where the coefficient a_n is specified by the excitation function in accordance with (3.192).

Using (3.190), we immediately deduce that the far-field form of (3.196) is

$$V(r, \theta) \simeq \frac{a}{r} e^{-\gamma r} G(\theta) \tag{3.197}$$

where the "radiation pattern" is

$$G(\theta) = \sum_{n=0}^{\infty} \frac{a_n}{\hat{K}_n(\gamma a)} P_n(\cos \theta) \tag{3.198}$$

Exercise: Consider an external spherical radiator of radius a as described above, where the excitation function on the surface is of the form

$$S(\theta) = S_0 \, \delta(\theta - \theta_0) \tag{3.199}$$

where θ_0 is the range from 0 to π. Hence $\delta(\theta - \theta_0)$ is the unit-impulse function, so we are dealing with a ring source. Show that the radiation pattern is

$$G(\theta) = S_0 \sum_{n=0}^{\infty} \frac{2n+1}{2} \frac{P_n(\cos \theta_0)\sin \theta_0}{\hat{K}_n(\gamma a)} P_n(\cos \theta) \tag{3.200}$$

If $|\gamma a| \ll 1$ (that is, small sphere), show that this expression reduces to

$$G(\theta) \propto \frac{S_0 \sin \theta_0}{2}(1 + 3x \cos \theta_0 \cos \theta + \text{terms in } x^2, x^3, \dots \tag{3.201}$$

where $x = \gamma a$.

We now extend the previous analysis of the spherical radiator to the case where the azimuthal symmetry no longer prevails. We again deal with a source-free region external to the spherical boundary at $r = a$. Our source or initial condition is now

$$V(a, \theta, \phi) = S(\theta, \phi) \tag{3.202}$$

The desired forms of the solution are now

$$\left(\frac{1}{\gamma r}\right)^{1/2} K_{n+1/2}(\gamma r) P_n^m(\cos \theta) e^{-im\phi} \tag{3.203}$$

where m and n are integers. By superimposing these elementary forms and changing the notation for the radial functions, we are led to write

$$V(r, \theta, \phi) = \sum_{n=0}^{\infty} \sum_{m=0}^{n} \frac{\hat{K}_n(\gamma r)}{\gamma r}$$
$$\times P_n^m(\cos \theta)(a_{mn} \cos m\phi + b_{mn} \sin m\phi) \tag{3.204}$$

where a_{mn} and b_{mn} are coefficients yet to be determined. The required representation for the aperture field is

$$S(\theta, \phi) = \sum_{n=0}^{\infty} \sum_{m=0}^{n} (s_{mn} \cos m\phi + t_{mn} \sin m\phi)P_n^m(\cos \theta) \tag{3.205}$$

Here it is not difficult to show that

$$s_{on} = \frac{2n+1}{4\pi} \int_0^{2\pi} \left[\int_0^{\pi} S(\theta, \phi)\sin \theta \, d\theta \right] d\phi \tag{3.206}$$

$$s_{mn} = \frac{2n+1}{2\pi} \frac{(n-m)!}{(n-m!}$$
$$\times \int_0^{2\pi} \left[\int_0^{\pi} S(\theta, \phi)P_n^m(\cos \theta)\sin \theta \, d\theta \right] \cos m\phi \, d\phi \tag{3.207}$$

and

$$t_{mn} = \frac{2n+1}{2\pi} \frac{(n-m)!}{(n+m)!}$$
$$\times \int_0^{2\pi} \left[\int_0^{\pi} S(\theta, \phi)P_n^m(\cos \theta)\sin \theta \, d\theta \right] \sin m\phi \, d\phi \tag{3.208}$$

Using (3.202), (3.204), and (3.205), we find that

$$a_{mn} = (\gamma a)\frac{s_{mn}}{\hat{K}_n(\gamma a)} \tag{3.209}$$

and

$$b_{mn} = (\gamma a)\frac{t_{mn}}{\hat{K}_n(\gamma a)} \tag{3.210}$$

This completes the general solution for the problem since the coefficients a_{mn} and b_{mn} in (3.204) are now specified in terms of the aperture function $S(\theta, \phi)$ via (3.206), (3.207), and (3.208).

Exercise: Consider an empty spherical enclosure of radius a with a boundary condition $V(a, \theta, \phi) = 0$. Determine the form of the mode expansion for the field at any interior point in the region $0 < r < a$. Show that free oscillations occur at frequencies $\omega_{m,n}$ determined by

$$J_{m+1/2}(x_{m,n}) = 0$$

where

$$x_{m,n} = \frac{\omega_{m,n} a}{c}$$

is the nth zero of the Bessel function of order m. Here c is the velocity of plane waves in the unbounded medium.

Exercise: Derive the expressions for $s_{m,n}$ and $t_{m,n}$ in (3.206) to (3.208). *Hint:* Multiply both sides of (3.205) by

$$(s_{pq} \cos q\phi + t_{pq} \sin q\phi) P_p^q(\cos \theta)(\sin \theta)$$

and integrate over θ from 0 to π and ϕ from 0 to 2π. Note that

$$\int_0^{2\pi} \sin m\phi \cos p\phi \, d\phi = 0$$

$$\int_0^{2\pi} \sin m\phi \sin p\phi \, d\phi = \int_0^{2\pi} \cos m\phi \cos p\phi \, d\phi$$

$$= \begin{cases} 0 & \text{if } m \neq p \\ \pi & \text{f } m = p \neq 0 \end{cases}$$

and

$$\int_0^\pi P_n^m(\cos \theta) P_q^m(\cos \theta) \sin \theta \, d\theta = \begin{cases} 0 & \text{if } n \neq q \\ \dfrac{2}{2n+1} \dfrac{(n+m)!}{(n-m)!} & \text{if } n = q \end{cases}$$

3.18 FURTHER WORKED EXAMPLES

Exercise: With reference to Equations (3.20) and (3.21), do the followng: Let $\hat{Z}$ be a function of z and $\hat{Y}$ independent of z. Then derive equations satisfied by V and I.

SOLUTION: Now, in general,

$$\frac{dV}{dz} = -\hat{Z}I \quad \text{and} \quad \frac{dI}{dz} = -\hat{Y}V$$

Differentiate the first equation with respect to z to get

$$\frac{d^2V}{dz^2} = -\frac{d\hat{Z}}{dz} I - \hat{Z} \frac{dI}{dz}$$

Use the second equation for dI/dz and the first equation for I to yield

$$\frac{d^2V}{dz^2} - \frac{1}{\hat{Z}} \frac{d\hat{Z}}{dz} \frac{dV}{dz} - \hat{Y}\hat{Z}V = 0$$

Then differentiate the second equation with respect to z to get

$$\frac{d^2I}{dz^2} = -\hat{Y}\frac{dV}{dz}$$

(noting that $d\hat{Y}/dz = 0$). Then use the first equation for dV/dz to yield

$$\frac{d^2I}{dz^2} - \hat{Y}\hat{Z}I = 0$$

Exercise: Consider a uniform transmission line of infinite length (extending from $-\infty < z < \infty$). What is the form for the current $I(z)$, for any z, when a voltage V_0 is applied at $z = z_0$?

SOLUTION: The voltage anywhere on the line is

$$V = \begin{cases} V_0 \exp[-\gamma(z - z_0)] & \text{for } z > z_0 \\ V_0 \exp[+\gamma(z - z_0)] & \text{for } z < z_0 \end{cases}$$

The corresponding current is

$$I = \begin{cases} \dfrac{V_0}{K}\exp[-\gamma(z - z_0)] & \text{for } z > z_0 \\[2mm] \dfrac{-V_0}{K}\exp[+\gamma(z - z_0)] & \text{for } z < z_0 \end{cases}$$

The driving-point impedance is clearly $V_0/2I_0 = K/2$.

Exercise: Design or specify how you would transform an impedance Z_1 to an impedance Z_2 by using a uniform section of transmission line.

SOLUTION: Connect the end of the line to the impedance Z_1; then the input impedance is given by

$$Z = K\frac{Z_1 + K\tanh\gamma\ell}{K + Z_1\tanh\gamma\ell}$$

where ℓ designates the length of the line, K is the characteristic impedance, and γ is the propagation constant. In principle, one can vary the length of the line to get Z to approximate Z_2 or, if necessary, vary K and γ. An important special case is when $\gamma\ell = i\pi/2$, whence $Z = K^2/Z_1 = Z_2$ is always possible if the line is lossless and K, Z_1, and Z_2 are real. The line is acting as an impedance transformer. Such operations can be implemented with the Smith chart even when loss is present (see Ramo, Whinnery and Van Duzer) [23].

Exercise: Consider a region $\infty > z > 0$ that is homogeneous. At $z = 0$ the function $V(x, z)$ vanishes [that is, $V(x, 0) = 0$]. For $\infty > z > 0$, V satisfies

$(\nabla^2 + k^2)V = 0$, where we may assume $\partial/\partial y = 0$. An incident wave has the form

$$V^{\text{inc}} = V_0 e^{ik(\cos\theta)z} e^{-ik(\sin\theta)x}$$

which is a plane wave making an angle of incidence θ with the z axis. What form does the reflected wave have? Repeat the exercise for the case where the boundary condition is $\partial V/\partial z = 0$ at $z = 0$.

SOLUTION: It should be obvious that the reflected wave has the form

$$V^{\text{refl}} = V_0 R e^{-ik(\cos\theta)z} e^{-ik(\sin\theta)x}$$

where $R = -1$ or $+1$ in the two respective cases.

Exercise: Consider a parallel-plate region that bounds a homogeneous space defined by $0 < x < a$ and $y > y_0$. Within this region

$$\left(\frac{\partial^2}{\partial x^2} + \frac{\partial^2}{\partial y^2} + \frac{\partial^2}{\partial z^2} - \gamma^2 \right) V = 0$$

At $y = y_0$ assume that $V = V_0 \exp(-ik_0 z)$ for $-\infty < z < \infty$, where V_0 is a constant. Also, we specify that $\partial V/\partial x = 0$ at $x = 0$ and $x = a$. What form does the solution take for $y > y_0$?

SOLUTION: Obviously, solutions would have the form

$$V = \frac{\cos}{\sin} (\beta_x x) \exp(\mp \Gamma_n y) \exp(-ik_0 z)$$

because the z dependence is fixed. These must satisfy

$$(\nabla^2 - \gamma^2)V = 0$$

Therefore

$$-\beta_x^2 + \Gamma_n^2 - k_0^2 - \gamma^2 = 0$$

or

$$\Gamma_n = (\gamma^2 + \beta_x^2 + k_0^2)^{1/2}$$

where $Re\ \Gamma_n > 0$. The field is to vanish as $y \to \infty$, so that the y dependence must be $\exp(-\Gamma_n y)$. Also, if $\partial V/\partial x = 0$ at $x = 0$ and $x = a$, clearly $\beta_x = n\pi/a$, where $n = 0, 1, 2, 3, \ldots$. Thus we are led to write for the region $0 < x < a$, $y > y_0$, $-\infty < z < \infty$,

$$V(x, y, z) = \sum_{n=0}^{\infty} A_n \cos \frac{\pi n x}{a} e^{-\Gamma_n y} e^{-ik_0 z}$$

At $y = y_0$ we have

$$V(x, y_0, z) = V_0(x)e^{-ik_0z} \sum_{n=0}^{\infty} A_n \cos \frac{\pi nx}{a} e^{-\Gamma_n y_0} e^{-ik_0z}$$

Then

$$\int_0^a V_0(x) \cos \frac{\pi nx}{a} dz = \sum_{n=0}^{\infty} A_n \int_0^a \cos \frac{\pi nx}{a} \cos \frac{\pi mx}{a} dx \, e^{-\Gamma_n y_0}$$

Now on the right-hand side

$$\int_0^a \cos \frac{\pi nx}{a} \cos \frac{\pi mx}{a} dx = \begin{cases} 0 & \text{if } n \neq m \\ a & \text{if } n = m = 0 \\ \dfrac{a}{2} & \text{if } n = m = 1, 2, 3, \ldots \end{cases}$$

Thus

$$\int_0^{\infty} V_0(x) \cos \frac{\pi mx}{a} dx = A_m \frac{a}{\epsilon_m} e^{-\Gamma_m y_0}$$

where $\epsilon_0 = 1$ and $\epsilon_m = 2 (m \neq 0)$, or

$$A_n = \frac{\epsilon_n}{a} \int_0^a V_0(x) \cos \frac{\pi nx}{a} dx \, e^{-\Gamma_n y_0}$$

where we have replaced m by n.

Now, of course, if $V_0(x) = V_0$ is independent of x, we see that $A_n = 0$ for $n = 1, 2, 3, \ldots$ and thus only the term for $n = 0$ survives. Then for $0 < x < a$, $y > y_0$, we have

$$V(x, y, z) = V_0 e^{-\Gamma_0(y - y_0)} e^{-ik_0z}$$

where

$$\Gamma_0 = (\gamma^2 + k_0^2)^{1/2}$$

Exercise: Repeat the above exercise for the case where the initial condition is $V = V_1 \cos(\pi x/a) \exp(-ik_1z)$ at $y = y_0$ for $0 < x < a$ and for $-\infty < z < \infty$.

SOLUTION: The only difference here is that we use $V_0(x) = V_1 \cos(\pi x/a)$ in the above problem and we replace k_0 by k_1. Then clearly,

$$A_n = \frac{\epsilon_n V_1}{a} \int_0^a \cos \frac{\pi x}{a} \cos \frac{\pi nx}{a} dx \, e^{\Gamma_n y_0}$$

which tells us that

$$A_n = 0 \quad \text{if } m \neq 1 \quad \text{and} \quad A_1 = V_1 \exp(\Gamma_1 y_0) \quad (\text{for } n = 1)$$

Exercise: If $\gamma = \alpha + i\beta$ in the preceding two exercises, where α and β are real, work out expressions for the attenuation rates and phase velocities of the waves in the region for both y and z directions. Assume that both k_0 and k_1 are real.

SOLUTION: We observe that V varies as $\exp(-\Gamma_n y)$. By definition, $\alpha_n = \text{Re } \gamma_n$ is the attenuation rate in the y direction and $\beta_n = \text{Re } \gamma_n$ is the corresponding phase constant or coefficient. Thus we write

$$e^{-\Gamma_n y} e^{i\omega t} = e^{-\alpha_n y} e^{-i(\beta_n y - \omega t)}$$
$$= e^{-\alpha_n y} e^{-i\omega[(y/v_n) - t]}$$

where $v_n = \omega/\beta_n$ is, by definition, the phase velocity in the y direction. Now in the case where $V_0(x) = $ constant, we are only concerned with the $n = 0$ term; thus

$$\Gamma_0 = \alpha_0 + i\beta_0 = [(\alpha + i\beta)^2 + k_0^2]^{1/2}$$

On squaring both sides and equating real and imaginary parts, we find that

$$\alpha_0^2 - \beta_0^2 = \alpha^2 - \beta^2 + k_0^2$$

and

$$\alpha_0 \beta_0 = \alpha\beta$$

Then

$$(\beta_0^2)^2 - (\beta^2 - \alpha^2 - k_0^2)\beta_0^2 - \alpha^2\beta^2 = 0$$

Thus we find that

$$2\beta_0^2 = (\beta^2 - \alpha^2 - k_0^2) \pm [(\beta^2 - \alpha^2 - k_0^2)^2 + 4\alpha^2\beta^2]^{1/2}$$

and

$$2\alpha_0^2 = -(\beta^2 - \alpha^2 - k_0^2) \pm [(\beta^2 - \alpha^2 - k_0^2) + 4\alpha^2\beta^2]^{1/2}$$

We choose the $+$ sign because $\alpha_0^2 > 0$ (α_0 is real). This gives α_0 and then the phase velocity

$$v_0 = \frac{\omega}{\beta_0} = \frac{\omega\alpha}{\alpha\beta}$$

(restricting attention to the case where $k_0 < \beta$). If $\alpha = 0$ (that is, lossless case) the situation is much simpler; then

$$\Gamma_0 = i\beta_0 = (k_0^2 - \beta^2)^{1/2}$$

whence $\alpha_0 = 0$ and

$$v_0 = \frac{\omega}{\beta_0} = \frac{\omega}{(\beta^2 - k_0^2)^{1/2}}$$

The phase velocity in the z direction is fixed by the $\exp(-ik_0z)$ dependence; it is ω/k_0.

In the case where $V_0(x)$ varies as $\cos(\pi x/a)$, we are concerned only with the $n = 1$ term. Then

$$\Gamma_1^2 = (\alpha_1 + i\beta_1)^2 = (\alpha + i\beta)^2 + k_1^2 + \left(\frac{\pi}{a}\right)^2$$

so that $v_1 = \omega/\beta_1$. Also, $2\beta_1^2$ and $2\alpha_1^2$ have the same form as $2\beta_0^2$ and $2\alpha_0^2$, respectively, if we merely replace k_1^2 by $[k_1^2 + (\pi/a)^2]$.

Exercise: Generalize the preceding exercises to the case where

$$V(x, y, z) = \int_{-\infty}^{+\infty} V_0(x, k_0)e^{-ik_0z}dk_0 = v_0(x, z)$$

at $y = y_0$. Note that

$$V_0(x, k_0) = \frac{1}{2\pi}\int_{-\infty}^{+\infty} v_0(x, z)e^{ik_0z}\, dz$$

is the inverse transform.

SOLUTION: The generalization can be obtained in the following fashion: As we indicated before, the form of the solution for an excitation of the spectral form $\exp(-ik_0z)$ can be written

$$V(x, y, z; k_0) = \sum_{n=0}^{\infty} A_n(k_0)\cos\frac{\pi n x}{a}\exp[-\Gamma_n(k_0)y]\exp(-ik_0z)$$

where the functional dependence on k_0 has been indicated explicitly. To allow for a general z variation, we integrate over all k_0; thus

$$V(x, y, z) = \int_{-\infty}^{+\infty}\left\{\sum_{n=0}^{\infty} A_n(k_0)\cos\frac{\pi n x}{a}\exp[-\Gamma_n(k_0)y]\right\}\exp(-ik_0z)\, dk_0$$

which must hold for $y_0 \leq y < \infty$. The quantity in braces is the Fourier transform of $V(x, y, z)$. Thus the inverse transform relation yields, for the case $y = y_0$,

$$\sum_{n=0}^{\infty} A_n(k_0)\cos\frac{\pi n x}{a}\exp[-\Gamma_n(k_0)y_0] = \frac{1}{2\pi}\int_{-\infty}^{+\infty} V(x, y_0, z)e^{ik_0z}\, dz$$

We finally note that the function of x on the right side of the above equation has the Fourier cosine series representation given on the left. Thus we can easily deduce that

$$A_n(k_0) = \frac{\epsilon_n}{2\pi a}\exp[\Gamma_n(k_0)y_0]\int_0^a\int_{-\infty}^{+\infty} V(x, y_0, z)\cos\frac{\pi n x}{a}\, dx\, e^{ik_0z}\, dz$$

which is the desired result.

References

1. G. N. Watson: *Theory of Bessel Functions,* Cambridge University Press, London, 1944. *The* authoritative book on Bessel functions.
2. S. A. Schelkunoff: *Electromagnetic Waves,* Van Nostrand, New York, 1943. Like a vintage wine, it improves with age; no other intermediate/advanced text on electromagnetic theory can match this classic for clarity and thoroughness.
3. P. M. Morse and H. Feshbach: *Methods of Theoretical Physics,* McGraw-Hill, New York, 1953. Without peer among general texts on analytical methods for dealing with boundary-value problems in wave transmission.
4. J. R. Wait: *Electromagnetic Radiation from Cylindrical Structures,* Pergamon Press, Elmsford, N.Y., 1959. Self-contained treatment of external radiation problems including slots in circular cylinders, wedges, and sheets.
5. J. R. Wait: *Wave Propagation Theory,* Pergamon Press, Elmsford, N.Y., 1981. An up-to-date collection of advanced topics including guided waves in nonuniform structures.
6. R. W. P. King: *Transmission Line Theory,* McGraw-Hill, New York, 1955. An exhaustive treatise of the subject.
7. H. Bremmer: *Propagation of Electromagnetic Waves, Handbuch der Physik,* vol. XVI, Springer-Verlag, Gottingen, 1958. Highly condensed but very clear.
8. T. M. MacRobert: *Spherical Harmonics,* 3d ed., Pergamon Press, Elmsford, N.Y., 1967. Excellent introduction to Legendre polynomials.
9. K. G. Budden: *The Waveguide Mode Theory of Wave Propagation,* Prentice-Hall (Logos Press), Englewood Cliffs, N.J., 1961. An advanced treatment of waveguide theory for electromagnetic and acoustic waveguides.
10. L. M. Brekhovskikh: *Waves in Layered Media,* 2d ed., Academic Press, New York, 1980. Covers a broad area of wave propagation with special emphasis on acoustics.
11. L. A. Chernov: *Wave Propagation in a Random Medium,* McGraw-Hill, New York, 1960. Translated from the Russian but still quite readable.
12. G. Tyras: *Radiation and Propagation of Electromagnetic Waves,* Academic Press, New York, 1969. Good coverage of boundary-value approaches at the intermediate/advanced level.
13. A. Ishimaru: *Wave Propagation and Scattering in Random Media,* parts I and II, Academic Press, New York, 1978. The latest and the best book on random-media effects, at the first-year graduate level.
14. H. R. L. Lamont: *Wave Guides,* 3d ed., Methuen, London, 1950. An unsurpassed masterpiece of brevity.
15. E. G. Jordan and K. G. Balmain: *Electromagnetic Waves and Radiating Systems,* 2d ed., Prentice-Hall, Englewood Cliffs, N.J., 1968. Best senior-level text.
16. R. E. Collin: *Field Theory of Guided Waves,* McGraw-Hill, New York, 1961. Contains good sections on wave propagation theory written clearly at the intermediate level.
17. J. R. Pierce: *Waves and Messages,* Doubleday, Garden City, N.Y., 1976. Enjoyable elementary introduction.
18. W. M. Ewing, W. S. Jardetzky, and F. Press: *Elastic Waves in Layered Media,* McGraw-Hill, New York, 1957. Exhaustive treatment of the analytical aspects with numerous examples in seismology.

19. I. Tolstoy: *Wave Propagation,* McGraw-Hill, New York, 1973. Very clear general introduction to ray and mode concepts.
20. C. S. Clay and H. Medwin: *Acoustical Oceanography,* Wiley, New York, 1977. Good survey with emphasis on sound waves in fluids.
21. V. L. Ginzburg: *The Propagation of Electromagnetic Waves in Plasmas,* 2d ed., Pergamon Press, Elmsford, N.Y., 1970. Very comprehensive coverage of a difficult subject.
22. J. M. Pearson: *A Theory of Waves,* Allyn & Bacon, Boston, 1966. Introductory level book on principles and theory.
23. S. Ramo, J. Whinnery, T. van Duzer: *Fields and Waves in Communication Electronics,* Wiley, New York, 1965.

Chapter 4
The Electromagnetic Field

4.1 INTRODUCTION

When dealing explicitly with electromagnetic fields, we need to employ Maxwell's equations [1]. These equations relate the time-varying electric and magnetic field quantities in a definitive sense. Historically, they are based on a set of postulates by J. C. Maxwell in order to encompass the earlier experimentally derived laws of Faraday and Ampere. The interested reader would be well advised to examine the elegant treatise first published by Maxwell in 1873 and the revised edition prepared by J. J. Thompson in 1891. Virtually hundreds of textbooks introduce Maxwell's equation in one form or another and present ever more elaborate discussions of their physical basis. At the risk of a certain bias, we would like to call attention to the writings of Sergei Schelkunoff [2] who spent many years as a mathematical consultant in the Bell Telephone Laboratories. His latest book is an extremely cogent and unadorned exposition of Maxwell's equations, with many later developments included. We warmly recommend this book for a modern-day account. An annotated list of other relevant texts are also given in the References [3–11].

4.2 MAXWELL'S EQUATIONS

Here we will not gild the lily but merely write down the differential form of Maxwell's equations for time-harmonic fields. We will also assume that the region under consideration has no sources. First of all, however, we should define our notation and conventions. These are fully consistent with those of earlier chapters where wave concepts were covered without the vector complexity of the electromagnetic field.

In the rationalized (MKS) system of units the electromagnetic properties of the medium are defined by ϵ, the (dielectric) permittivity, σ, the (electric) conductivity, and μ, the (magnetic) permeability. Since we will be dealing with linear phenomena, σ, ϵ, and μ will be assumed independent of the field strengths; for this reason they are "constant." But they will depend on the (angular) frequency ω except in certain idealized situations, such as in free space, where $\epsilon = \epsilon_0 = 8.854 \times 10^{-12}$ F/m, $\mu = \mu_0 = 4\pi \times 10^{-7}$ H/m, and $\sigma = 0$.

It is also possible that the "electromagnetic constants" of the medium will vary with position, that is, be a function of the coordinates. Except where stated explicitly, we will not deal with this case.

As before, we will adopt the common time factor $\exp(+i\omega t)$, where ω is the angular frequency and t is the time. Consequently, the actual electric field strength $e(t)$ and the actual magnetic field strength $h(t)$ are defined to be

$$e(t) = \mathrm{Re}(Ee^{i\omega t})$$

and

$$h(t) = \mathrm{Re}(He^{i\omega t})$$

where E and H are complex phasors. In general, $e(t)$ and $h(t)$ and their phasor counterparts are vectors in the coordinate space. Thus we designate the electric vector field strength as **E** with the understanding that the x component of the real-world physical counterpart $e_x(t)$ is the real part of $E_x \exp(i\omega t)$.

A preliminary step here is to state Ohm's law as it applies in this context. It reads

$$\mathbf{J} = (\sigma + i\omega\epsilon)\mathbf{E} \tag{4.1}$$

where **J** is the vector current density, in amperes per square meter, and **E** is vector electric field, in volts per meter. Here we should remind the reader that we are dealing with a source-free region, so **J** can be thought of as the driven current density that is proportional to the electric field. Here we are now free to define σ and ϵ to be real even though they, in general, are frequency dependent.

The analogous Ohm's law for magnetic quantities can be written

$$i\omega\mathbf{B} = i\omega\mu\mathbf{H} \tag{4.2}$$

or simply

$$\mathbf{B} = \mu\mathbf{H} \tag{4.2'}$$

where $\mathbf{B}$ is the vector magnetic induction, in webers per square meter (or teslas), and $\mathbf{H}$ is the vector magnetic field, in amperes per meter. It is useful to envisage $i\omega\mathbf{B}$ as the magnetic current associated with the time-varying flux. Not surprisingly, μ can be complex. It is useful to observe that the quantity $\sigma + i\epsilon\omega$ plays the same role in related electrical quantities as $i\mu\omega$ does in relating magnetic quantities.

The electric and magnetic field vectors are related by

$$\text{curl } \mathbf{E} = -i\mu\omega\mathbf{H} \tag{4.3}$$

and

$$\text{curl } \mathbf{H} = (\sigma + i\epsilon\omega)\mathbf{E} \tag{4.4}$$

These equations are Maxwell's equations for a source-free region.

An immediate consequence for homogeneous media is

$$\text{div } \mathbf{H} = 0 \tag{4.5}$$

and

$$\text{div } \mathbf{E} = 0 \tag{4.6}$$

because div curl of any vector is zero.

The corresponding integral forms of Maxwell's equations are obtained from (4.3) and (4.4) by integrating them over a volume and applying Stokes' theorem (see Chapter 1, Section 1.11) to get

$$\oint \mathbf{E} \cdot \mathbf{ds} = -i\omega \int B_n \, dS \tag{4.7}$$

and

$$\oint \mathbf{H} \cdot \mathbf{ds} = i\omega \int D_n dS \tag{4.8}$$

where

$$i\omega\mathbf{D} = (\sigma + i\epsilon\omega)\mathbf{E} = \mathbf{J}$$

is the current density written in terms of the electric induction vector $\mathbf{D}$. The integrations on the right are extended over *any* surface, while the integrations on the left are over the edge of these surfaces. The algebraic signs in these equations are determined by the right-hand rule, as indicated in Figure 4.1 for Equation (4.8). Here J_n denotes the component of the current density that points outward and is normal to the surface. We can write (4.8) in the form

$$\oint H_s \, ds = I_{\text{total}} \tag{4.9}$$

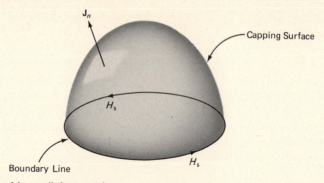

Figure 4.1   Maxwell-Ampere law.

where H_s is the component along the boundary line where ds is the element of length and I_{total} is the total current passing through it. Actually, (4.9) is a statement of Ampere's law. A precise description is embodied in (4.8), which is sometimes called the Ampere-Maxwell equation.

Corresponding remarks can be made about (4.7), where it can be rewritten

$$\oint E_s \, ds = -i\omega(\text{total magnetic flux}) \tag{4.10}$$

where E_s is the component of **E** along the boundary line. The right-hand side is the time rate of change (with a minus sign) of the total magnetic flux passing through. But this is a statement of Faraday's law of induction that dates back to the middle nineteenth century. A precise statement is embodied in (4.7), which can be aptly called the Faraday-Maxwell equation.

We should point out here that since we are dealing with time-harmonic fields, both div **D** and div **J** vanish even when the medium is inhomogeneous.

Maxwell's equations in the form given by (4.3) and (4.4) refer specifically to regions that are free of impressed currents or sources. But, of course, Ampere's law as stated by (4.9) applies even when the right-hand side includes the impressed current, such as that carried by a filamental wire that is enclosed by the integration path. The consequence of this fact is that Maxwell's equations can be generalized such that we may include the impressed current $\mathbf{J}_i$, expressed in amperes per square meter, on the right-hand side of both (4.4) and (4.8). Thus, for example,

$$\oint \mathbf{H} \cdot \mathbf{ds} = \int (\mathbf{J} + \mathbf{J}_i)_n dS \tag{4.11}$$

is the integral form of the Ampere-Maxwell equation to allow for both driven and impressed currents.

We will deal specifically with Maxwell's equations written in differential form:

$$\text{curl } \mathbf{E} = -i\mu\omega\mathbf{H} \tag{4.12}$$

and

$$\text{curl } \mathbf{H} = (\sigma + i\epsilon\omega)\mathbf{E} + \mathbf{J}_i \tag{4.13}$$

To be explicit, in cartesian coordinates (x, y, z) these equations would read

$$\frac{\partial E_z}{\partial y} - \frac{\partial E_y}{\partial z} = -i\mu\omega H_x \tag{4.14}$$

$$\frac{\partial E_x}{\partial z} - \frac{\partial E_z}{\partial x} = -i\mu\omega H_y \tag{4.15}$$

$$\frac{\partial E_y}{\partial x} - \frac{\partial E_x}{\partial y} = -i\mu\omega H_z \tag{4.16}$$

and

$$\frac{\partial H_z}{\partial y} - \frac{\partial H_y}{\partial z} = (\sigma + i\epsilon\omega)E_x + (J_i)_x \tag{4.17}$$

$$\frac{\partial H_x}{\partial z} - \frac{\partial H_z}{\partial x} = (\sigma + i\epsilon\omega)E_y + (J_i)_y \tag{4.18}$$

$$\frac{\partial H_y}{\partial x} - \frac{\partial H_x}{\partial y} = (\sigma + i\epsilon\omega)E_z + (J_i)_z \tag{4.19}$$

We stress, once again, that these forms are valid for time-harmonic time variation.*

---

* A word might be said here about the time-harmonic assumption where we have adopted the factor $\exp(i\omega t)$. This assumption in no way restricts the generality of the results. For example, if we wished to write the time-dependent form of Maxwell's equations, we could employ the forms given by (4.14) and (4.19), where $i\omega$ is merely replaced by the operator $\partial/\partial t$. Then we regard the quantities E_x, E_y, E_z, H_x, H_y, H_z, and J_i as real functions of time t. To be more explicit, we would identify $E_x(i\omega)$, for example, as the transform of $E_x(t)$, in the sense that

$$E_x(i\omega) = \int_0^\infty E_x(t)e^{-i\omega t}\, dt$$

where $E_x(t) = 0$ for $t < 0$. The corresponding inverse transform is

$$E_x(t) = \frac{1}{2\pi}\int_{-\infty}^{+\infty} E_x(i\omega)e^{i\omega t}\, d\omega$$

In this book we do not deal explicitly with transient excitation.

4.3 PLANE WAVE TRANSMISSION

In order to touch base with our earlier discussion of plane wave transmission, we will first consider the case where variations only in the z direction are permitted. Thus with reference to a cartesian coordinate system (x, y, z) we have $\partial/\partial x = \partial/\partial y = 0$. Furthermore, we deal with source-free regions. Then (4.15) and (4.17) reduce to

$$\frac{dE_x}{dz} = -i\mu\omega H_y \tag{4.20}$$

and

$$\frac{dH_y}{dz} = -(\sigma + i\epsilon\omega)E_x \tag{4.21}$$

Clearly, these equations have the same form as the transmission-line equations

$$\frac{dV}{dz} = -ZI \tag{4.22}$$

and

$$\frac{dI}{dz} = -YV \tag{4.23}$$

where V and I are identified with E_x and H_y, respectively, Z is identified with $i\mu\omega$, and Y is identified with $\sigma + i\epsilon\omega$. Here we note that E_x, H_y, and z form a right-handed triad, as indicated in Figure 4.2.

In addition to (4.20) and (4.21), we also have the pair that are reduced from (4.14) and (4.18):

$$\frac{dE_y}{dz} = -i\mu\omega(-H_x) \tag{4.24}$$

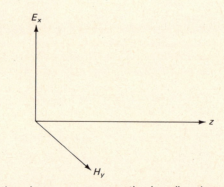

Figure 4.2   Depicting plane wave propagating in z direction when **E** is in x direction.

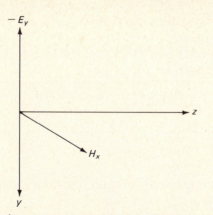

Figure 4.3 Depicting plane wave propagating in *z* direction when **H** is in *x* direction.

and

$$\frac{d(-H_x)}{dz} = -(\sigma + i\epsilon\omega)E_y \tag{4.25}$$

These equations also have the same form as the transmission-line equations (4.22) and (4.23), but now V and I are identified with E_y and $-H_x$, respectively. The prefixed negative sign on H_x is consistent with the right-handed triad formed by $-E_y$, H_x, and z, as indicated in Figure 4.3.

In the above discussion we have not restricted the medium properties to be independent of z. Now we do assume $d\sigma/dz$, $d\epsilon/dz$, and $d\mu/dz = 0$. Then it is a simple matter to show, from (4.20) and (4.21), that

$$\left(\frac{d^2}{dz^2} - \gamma^2\right)E_x = 0 \tag{4.26}$$

where $\gamma = [i\mu\omega(\sigma + i\epsilon\omega)]^{1/2}$, by definition, is the propagation constant, and the radical is always chosen such that Re $\gamma > 0$. It is also easy to see that H_z, E_z, and H_x also satisfy (4.26), bearing in mind that we are restricting attention to source-free regions.

Now clearly, elementary solutions of (4.26) are of the form $\exp(\mp\gamma z)$. Thus a general solution for E_x takes the form

$$E_x = Ae^{-\gamma z} + Be^{\gamma z} \tag{4.27}$$

where A and B are (z-independent) arbitrary constants. Then according to (4.20), we have

$$H_y = \frac{1}{\eta}\left(Ae^{-\gamma z} - Be^{\gamma z}\right) \tag{4.28}$$

where $\eta = i\mu\omega/\gamma = [i\mu\omega/(\sigma + i\epsilon\omega)]^{1/2}$. Because of the close analogy with the transmission-line forms, we define η to be the characteristic impedance of the medium.

Working with the pair (4.24) and (4.25), we would also be led to write general solutions as

$$-E_y = Ce^{-\gamma z} + De^{\gamma z} \tag{4.29}$$

and

$$H_x = \frac{1}{\eta}\left(Ce^{-\gamma z} - De^{\gamma z} \right) \tag{4.30}$$

where C and D are generally different arbitrary constants.

4.4 CURRENT SHEET EXCITATION

We will consider a simple example of plane wave excitation in a homogeneous medium to illustrate basic principles. At the plane $z = z_0$ a uniform current sheet of j_y A/m flows in the positive y direction for all values of x. Now according to the Ampere-Maxwell equation, as given by (4.11), we can assert that

$$\lim_{\delta \to 0}\left[H_x(z = z_0 + \delta) - H_x(z = z_0 - \delta) \right] = j_y \tag{4.31}$$

This equation is a statement that the magnetic field tangential to (an impressed) current sheet is discontinuous by an amount equal to the surface current density. Then we also state

$$\lim_{\delta \to 0}\left[E_y(z = z_0 + \delta) - E_y(z = z_0 - \delta) \right] = 0 \tag{4.32}$$

which follows from (4.7) or (4.10). In other words, E_y is continuous through a postulated electric current sheet. Also, for this idealized, two-dimensional problem it easily follows that E_x and H_y would both be continuous through the y-directed current sheet.

For the above example we then are led to write

$$H_x(z) = f \exp[-\gamma(z - z_0)] \qquad \text{for } z > z_0 \tag{4.33}$$

and

$$H_x(z) = g \exp[+\gamma(z - z_0)] \qquad \text{for } z < z_0 \tag{4.34}$$

bearing in mind that no other sources are present. The constants f and g have yet to be determined. Now according to (4.25), we must have that

$$E_y(z) = -\eta f \exp[-\gamma(z - z_0)] \qquad \text{for } z > z_0 \tag{4.35}$$

and

$$E_y(z) = +\eta g \exp[+\gamma(z - z_0)] \qquad \text{for } z < z_0 \tag{4.36}$$

From (4.31) and (4.32) it is now a simple matter to show that

$$f = -g = \frac{j_y}{2} \tag{4.37}$$

which is the desired solution of the excitation problem. Obviously, for this case the wave associated with E_x and H_y would be zero.

Exercise: Obtain the corresponding field expressions for the case where the impressed current sheet at $z = z_0$ is j_x A/m in the x direction for all values of y.

4.5 REFLECTION OF PLANE WAVES AT NORMAL INCIDENCE FOR SINGLE INTERFACE

A very basic problem in electromagnetics is to consider reflection of a plane wave from a plane interface separating two homogeneous regions. The situation is illustrated in Figure 4.4, where the region $z < 0$ is homogeneous with electromagnetic constants σ_1, ϵ_1, and μ_1 and where the region $z > 0$ has electromagnetic constants σ_2, ϵ_2, and μ_2. The incident wave from the left is characterized by

$$E_x^{\text{inc}} = Ae^{-\gamma_1 z} \tag{4.38}$$

and

$$H_y^{\text{inc}} = \eta_1^{-1} Ae^{-\gamma_1 z} \tag{4.39}$$

where $\gamma_1 = [i\mu_1\omega(\sigma_1 + i\epsilon_1\omega)]^{1/2}$, $\eta_1 = i\mu_1\omega/\gamma_1$, and A is a specified constant. Since there are no variations in the x and y directions, we are

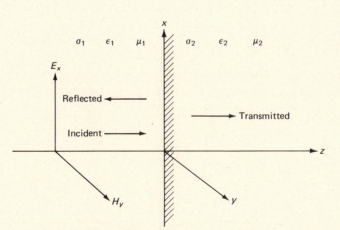

Figure 4.4 Plane wave incident onto the planar interface between two homogeneous regions of semi-infinite extent.

immediately led to write the corresponding forms for the reflected wave in the region $z < 0$:

$$E_x^{\text{refl}} = Be^{\gamma_1 z} \tag{4.40}$$
$$H_y^{\text{refl}} = -\eta_1^{-1}Be^{\gamma_1 z} \tag{4.41}$$

These fields clearly satisfy Maxwell's equations, and they vanish as $z \to -\infty$. Futhermore, they represent a plane wave traveling in the negative z direction away from the interface from whence they came.

The transmitted wave in the region $z > 0$ clearly must have the form

$$E_x^{\text{trans}} = Ce^{-\gamma_2 z} \tag{4.42}$$

and

$$H_y^{\text{trans}} = \eta_2^{-1}Ce^{-\gamma_2 z} \tag{4.43}$$

These vanish as $z \to +\infty$, and they represent a plane wave traveling away from the interface.

The constants B and C are not yet known, and they must be determined from boundary conditions at the interface $z = 0$. These conditions require that the tangential field components be continuous. Such a statement is consistent with Maxwell's equations as written in integral form in (4.7) and (4.8). For example, as indicated in Figure 4.5, we may choose a slender, rectangular-shaped contour contained in the (x, z) plane. The line integral of the electric field around this contour must vanish in the case where the breadth 2δ tends to zero. Thus the electric field E_x at $z = +\delta$ must tend to the electric field at $z = -\delta$ in the limit $\delta \to 0$. The same argument applies to the tangential magnetic fields, which are also continuous across the interface.

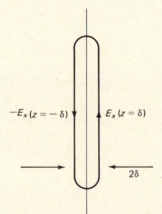

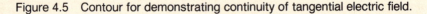

Figure 4.5 Contour for demonstrating continuity of tangential electric field.

In the context of the present one-dimensional problem our boundary conditions are thus

$$\left. \begin{array}{l} E_x^{\text{inc}} + E_x^{\text{refl}} = E_x^{\text{trans}} \\[6pt] H_y^{\text{inc}} + H_y^{\text{refl}} = H_y^{\text{trans}} \end{array} \right\} \quad \text{at } z = 0 \qquad \begin{array}{l} (4.44) \\[12pt] (4.45) \end{array}$$

As a consequence,

$$A + B = C \qquad (4.46)$$

$$\eta_1^{-1}(A - B) = \eta_2^{-1}C \qquad (4.47)$$

Solving these algebraic equations leads to

$$\frac{B}{A} = \frac{\eta_2 - \eta_1}{\eta_2 + \eta_1} = R \qquad (4.48)$$

and

$$\frac{C}{A} = \frac{2\eta_2}{\eta_2 + \eta_1} = T \qquad (4.49)$$

where we define R and T as reflection and transmission coefficients, respectively.

The complete expressions for the fields are thus

$$E_x = A(e^{-\gamma_1 z} + Re^{\gamma_1 z}) \qquad \text{for } z < 0 \qquad (4.50)$$

$$H_y = A\eta_1^{-1}(e^{-\gamma_1 z} - Re^{\gamma_1 z}) \qquad \text{for } z < 0 \qquad (4.51)$$

$$E_x = ATe^{-\gamma_2 z} \qquad \text{for } z > 0 \qquad (4.52)$$

$$H_y = A\eta_2^{-1}Te^{-\gamma_2 z} \qquad \text{for } z > 0 \qquad (4.53)$$

where now everything is specified in terms of the factor A, which characterizes the incident field.

For the present problem the "surface impedance" Z_s is given exactly by

$$Z_s = \frac{E_x}{H_y}\Bigg]_{z=0} = \eta_2 \qquad (4.54)$$

whether we use the pair (4.50) and (4.51) or the pair (4.52) and (4.53).

The transmission-line analog for the above problem is shown in Figure 4.6. The two half spaces of the posed problem are now represented by two semi-infinite transmission lines that have a common junction or connection at the terminal plane $z = 0$. The propagation constants of these two lines are γ_1 and γ_2, and the characteristic impedances are η_1 and η_2. The voltage V on the lines and the current I on the lines are identified with E_x and H_y, respectively.

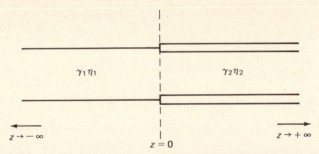

Figure 4.6 Equivalent transmission-line circuit for normally incident plane wave on single planar interface between homogeneous regions.

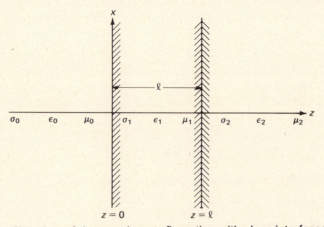

Figure 4.7 Side view of three-region configuration with plane interfaces at $z = 0$ and ℓ.

Exercise: The same problem as shown in Figure 4.4 is posed, but now we specify that $E_y^{\text{inc}} = \hat{A}e^{-\gamma_1 z}$ when $\hat{A}$ characterizes the incident field coming from $z = -\infty$. Show that the solution has the same form as $(4.50) - (4.53)$ if $\hat{A}$ replaces A, E_y replaces E_x, and $-H_x$ replaces H_y. Note that the exact surface impedance is now defined by

$$Z_s = \frac{-E_y}{H_x}\bigg]_{z=0} = \eta_2$$

4.6 MULTIPLE INTERFACES

We are now in the position to consider more complicated geometries. But for the moment we will stick with normal incidence. To this end, we consider a three-region problem, as illustrated in Figure 4.7. The incident plane wave coming from the left is specified to be

$$E_x^{inc} = E_0 e^{-\gamma_0 z} \tag{4.55}$$

where $\gamma_0 = [i\mu_0\omega(\sigma_0 + i\epsilon_0\omega)]^{1/2}$ in terms of the electromagnetic constants σ_0, ϵ_0, and μ_0 in the region $z < 0$. Thus the associated magnetic field must be given by

$$H_y^{inc} = \eta_0^{-1} E_0 e^{-\gamma_0 z} \tag{4.56}$$

where $\eta_0 = i\mu_0\omega/\gamma_0$. Actually, the region $z < 0$ could be free space, in which case $\sigma_0 = 0$ and $\gamma_0 = i(\epsilon_0\mu_0)^{1/2}\omega$ and $\eta_0 = (\mu_0/\epsilon_0)^{1/2}$, but we do not need to make this specialization at this stage.

On the basis of the previous example we can anticipate the form of the final solution in the region $z < 0$. Thus we would write for this region

$$E_x = E_0(e^{-\gamma_0 z} + R e^{\gamma_0 z}) \qquad \text{for } z < 0 \tag{4.57}$$

and

$$H_y = E_0 \eta_0^{-1}(e^{-\gamma_0 z} - R e^{\gamma_0 z}) \qquad \text{for } z < 0 \tag{4.58}$$

where R is yet to be determined. In the intermediate region we have

$$E_x = E_0(C e^{-\gamma_1 z} + D e^{+\gamma_1 z}) \qquad \text{for } 0 < z < \ell \tag{4.59}$$

$$H_y = E_0 \eta_1^{-1}(C e^{-\gamma_1 z} - D e^{+\gamma_1 z}) \qquad \text{for } 0 < z < \ell \tag{4.60}$$

where C and D are also unknown. Finally, in the transmitted region we have

$$E_x = E_0 F e^{-\gamma_2 z} \qquad \text{for } \ell < z < \infty \tag{4.61}$$

$$H_y = E_0 \eta_2^{-1} F e^{-\gamma_2 z} \qquad \text{for } \ell < z < \infty \tag{4.62}$$

where F is unknown.

We now impose the continuity conditions at $z = \ell$. Thus we find that

$$F = C e^{\gamma_2 \ell} e^{-\gamma_1 \ell} \frac{2\eta_2}{\eta_2 + \eta_1} \tag{4.63}$$

$$\frac{D}{C} = R_1 e^{-2\gamma_1 \ell} \qquad \text{where } R_1 = \frac{\eta_2 - \eta_1}{\eta_2 + \eta_1} \tag{4.64}$$

Imposing the continuity conditions at $z = 0$ leads to

$$R = \frac{Z_1 - \eta_0}{Z_1 + \eta_0} \tag{4.65}$$

where

$$Z_1 = \eta_1 \frac{1 + R_1 e^{-2\gamma_1 \ell}}{1 - R_1 e^{-2\gamma_1 \ell}}$$

$$= \eta_1 \frac{\eta_2 + \eta_1 \tanh \gamma_1 \ell}{\eta_1 + \eta_2 \tanh \gamma_1 \ell} \tag{4.66}$$

Then, finally, we find that

$$C = \frac{2\eta_1}{\eta_0 + \eta_1} \frac{1}{1 - R_0 R_1 \exp(-2\gamma_1 \ell)} \tag{4.67}$$

where

$$R_0 = \frac{\eta_0 - \eta_1}{\eta_0 + \eta_1} \tag{4.68}$$

Using (4.63) and (4.67), we see that (4.61) can be written

$$E_x = E_0 \frac{[2\eta_1/(\eta_1 + \eta_0)][2\eta_2/(\eta_2 + \eta_1)]e^{-\gamma_1 \ell}e^{-\gamma_2(z - \ell)}}{1 - [(\eta_0 - \eta_1)/(\eta_0 + \eta_1)][(\eta_2 - \eta_1)/(\eta_2 + \eta_1)]e^{-2\gamma_1 \ell}} \tag{4.69}$$

The factors in this expression have a clear physical interpretation. The term $2\eta_1/(\eta_1 + \eta_0)$ is the transmission coefficient for the electric field as it crosses the interface at $z = 0$ between media 0 and 1. The term $2\eta_2/(\eta_2 + \eta_1)$ is the transmission coefficient for the electric field as it crosses the interface $z = \ell$ between media 1 and 2. Then $\exp(-\gamma_1 \ell)$ is the transmission factor for propagation across the (sectionally) homogeneous medium from $z = 0$ to ℓ. The factor $\exp[-\gamma_2(z - \ell)]$ is the corresponding transmission factor for propagation from $z = \ell$ to z. Now if Re $\gamma_1 \ell \gg 1$, the denominator in (4.69) would be essentially one, and the physical situation would correspond to noninteraction of the two interfaces. However, if the denominator is different from one, we should interpret this result as the effect of multiple interactions of forward- and backward-traveling waves within the intermediate section. This point can be readily confirmed if the denominator is expanded in the manner

$$(1 - R_0 R_1 e^{-2\gamma_1 \ell})^{-1} = 1 + R_0 R_1 e^{-2\gamma_1 \ell} + (R_0 R_1)^2 e^{-4\gamma_1 \ell} + \cdots \tag{4.70}$$

Each one of the terms on the right can be identified with progressively increasing orders of internal reflections.

The equivalent transmission line for the three-region problem is shown in Figure 4.8. Basically, we now have two semi-infinite lines connected by a finite section of line of length ℓ. The characteristic impedances η_0, η_1, η_2 are as shown in Figure 4.8. The voltage V and current I on the lines are to be identified with E_x and H_y as before.

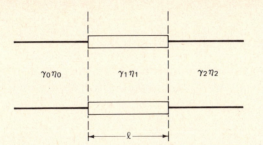

Figure 4.8   Equivalent transmission-line circuit for three-region problem for normal incidence.

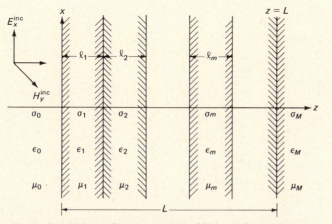

Figure 4.9   Normally incident plane wave onto M-layered structure.

Exercise: For the model shown in Figure 4.7, show that no reflection takes place if $\gamma_1 \ell = i(2n + 1)\pi/2$ for $n = 0, 1, 2, \ldots$ and $\eta_1^2 = \eta_0 \eta_2$. Note that for $n = 0$ the thickness of the layer is one-quarter wavelength if the conductivity σ_1 is zero.

We now consider the normal incidence of a plane wave onto a layered medium of M homogeneous layers. The mth layer of thickness ℓ_m has properties σ_m, ϵ_m, and μ_m. The situation is illustrated in Figure 4.9. The incident plane wave is again taken to have the form

$$E_x^{\text{inc}} = E_0 e^{-\gamma_0 z} \tag{4.71}$$

and

$$H_y^{\text{inc}} = \eta_0^{-1} E_0 e^{-\gamma_0 z} \tag{4.72}$$

The resultant fields in the region $z < 0$ again have the form

$$E_x = E_0(e^{-\gamma_0 z} + R e^{\gamma_0 z}) \tag{4.73}$$

and

$$H_y = E_0 \eta_0^{-1}(e^{-\gamma_0 z} - R e^{\gamma_0 z}) \tag{4.74}$$

where R is the reflection coefficient. In the region $z > L$, where the medium has properties σ_M, ϵ_M, and μ_M, the transmitted fields are written

$$E_x = E_0 T e^{-\gamma_M(z-L)} \tag{4.75}$$

and

$$H_y = E_0 \eta_M^{-1} T e^{-\gamma_M(z-L)} \tag{4.76}$$

Here T is the overall transmission coefficient and $L = \ell_1 + \ell_2 + \cdots + \ell_M + \cdots + \ell_{M-1}$.

Within the mth region, where m varies from 1 to $M-1$, the fields would have the form

$$E_x = A_m e^{-\gamma_m z} + B_m e^{\gamma_m z}$$

and

$$H_y = \eta_m^{-1}(A_m e^{-\gamma_m z} - B_m e^{\gamma_m z}) \tag{4.77}$$

where

$$\gamma_m = [i\mu_m\omega(\sigma_m + i\epsilon_m\omega)]^{1/2}$$

and

$$\eta_m = \frac{i\mu_m\omega}{\gamma_m}$$

Now clearly, for any specified value of M we could solve directly for the coefficients R, T, A_m, and B_m ($m = 1, 2, \ldots, M-1$) by matching E_x and H_y across each of the M interfaces. But as we have indicated, the problem can be recast in terms of transmission lines. Clearly, for this problem we are dealing with the equivalent circuit shown in Figure 4.10. Again, the voltage can be identified with E_x and the current with H_y.

The above analogy or equivalence with transmission-line circuitry allows us to write the following expression for the reflection coefficient:

$$R = \frac{Z_1 - \eta_0}{Z_1 + \eta_0} \tag{4.78}$$

where

$$Z_1 = \eta_1 \frac{Z_2 + \eta_1 \tanh \gamma_1 \ell_1}{\eta_1 + Z_2 \tanh \gamma_1 \ell_1} \tag{4.79}$$

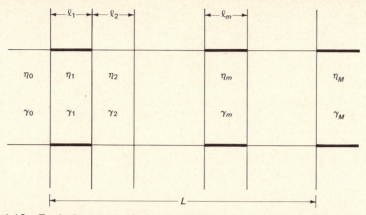

Figure 4.10   Equivalent transmission-line circuit for normally incident plane wave on *M*-layered structure.

$$Z_2 = \eta_2 \frac{Z_3 + \eta_2 \tanh \gamma_2 \ell_2}{\eta_2 + Z_3 \tanh \gamma_2 \ell_2} \tag{4.80}$$

.
.
.

$$Z_m = \eta_m \frac{Z_{m+1} + \eta_m \tanh \gamma_m \ell_m}{\eta_m + Z_{m+1} \tanh \gamma_m \ell_m} \tag{4.81}$$

.
.
.

$$Z_{M-1} = \eta_{M-1} \frac{\eta_M + \eta_{M-1} \tanh \gamma_{M-1} \ell_{M-1}}{\eta_{M-1} + \eta_M \tanh \gamma_{M-1} \ell_{M-1}} \tag{4.82}$$

On the other hand, the transmission coefficient is a product in the form

$$T = \left(\frac{2\eta_1}{\eta_1 + \eta_0} \frac{2\eta_2}{\eta_2 + \eta_1} \cdots \frac{2\eta_m}{\eta_m + \eta_{m-1}} \cdots \frac{2\eta_M}{\eta_M + \eta_{M-1}} \right) D^{-1}$$
$$\times \exp(-\gamma_1 \ell_1 - \gamma_2 \ell_2 - \cdots - \gamma_m \ell_m - \cdots - \gamma_{M-1} \ell_{M-1}) \tag{4.83}$$

where

$$D = \left[1 - \left(\frac{\eta_0 - \eta_1}{\eta_0 + \eta_1} \right) \left(\frac{Z_2 - \eta_1}{Z_2 + \eta_1} \right) e^{-2\gamma_1 \ell_1} \right]$$
$$\times \left[1 - \left(\frac{\eta_1 - \eta_2}{\eta_1 + \eta_2} \right) \left(\frac{Z_3 - \eta_2}{Z_3 + \eta_2} \right) e^{-2\gamma_2 \ell_2} \right] \cdots$$

$$\times \left[1 - \left(\frac{\eta_{m-1} - \eta_m}{\eta_{m-1} + \eta_m} \right) \left(\frac{Z_{m+1} - \eta_m}{Z_{m+1} + \eta_m} \right) e^{-2\gamma_m \ell_m} \right] \cdots$$

$$\times \left[1 - \left(\frac{\eta_{M-2} - \eta_{M-1}}{\eta_{M-2} + \eta_{M-1}} \right) \left(\frac{\eta_M - \eta_{M-1}}{\eta_M + \eta_{M-1}} \right) e^{-2\gamma_{M-1} \ell_{M-1}} \right] \tag{4.84}$$

Exercise: Consider the four-region problem (that is, $M = 3$); show that the transmission factor is

$$T = \frac{2\eta_1}{\eta_1 + \eta_0} \frac{2\eta_2}{\eta_2 + \eta_1} \frac{2\eta_3}{\eta_3 + \eta_2} D^{-1} \exp(-\gamma_1 \ell_1 - \gamma_2 \ell_2) \tag{4.85}$$

so that the transmitted field is

$$E_x = E_0 T \exp[-\gamma_3 (z - \ell_1 - \ell_2)] \tag{4.86}$$

Also, show for this case that

$$D = \left[1 - \left(\frac{\eta_0 - \eta_1}{\eta_0 + \eta_1} \right) \left(\frac{\eta_2 - \eta_1}{\eta_2 + \eta_1} \right) e^{-2\gamma_1 \ell_1} \right.$$
$$- \left(\frac{\eta_0 - \eta_1}{\eta_0 + \eta_1} \right) \left(\frac{\eta_3 - \eta_2}{\eta_3 + \eta_2} \right) e^{-2\gamma_2 \ell_2} e^{-2\gamma_1 \ell_1}$$
$$\left. - \left(\frac{\eta_1 - \eta_2}{\eta_1 + \eta_2} \right) \left(\frac{\eta_3 - \eta_2}{\eta_3 + \eta_2} \right) e^{-2\gamma_2 \ell_2} \right] \tag{4.87}$$

It is useful to note that for the present problem the following impedance relations are exact:

$$\left[\frac{E_x}{H_y} \right]_{z=0} = Z_1 \tag{4.88}$$

where Z_1 is given by (4.79), and

$$\left[\frac{E_x}{H_y} \right]_{z \geq L} = \eta_M \tag{4.89}$$

Of course, Z_1 is just the surface impedance, by definition. Thus if Z_1 were known at the outset, we could proceed by postulating the forms (4.73) and (4.74) and then deducing directly that R was given by (4.78) by imposing (4.88).

Actually, in general, we have that

$$\left[\frac{E_x}{H_y} \right]_{z=\ell_1 + \ell_2 + \cdots + \ell_{m-1}} = Z_m \tag{4.90}$$

where $m = 1, 2, \ldots, M$. This includes the case where $m = M$ and $z = L$; whence $Z_M = \eta_M$.

We might have chosen the incident plane wave to be polarized such that the electric vector had only an E_y component. Maxwell's equations for this case would be given explicitly by (4.24) and (4.25). Bearing in mind the preceding discussion, the fields in the region $z < 0$ take the form

$$E_y = \hat{E}_0(e^{-\gamma_0 z} + Re^{\gamma_0 z}) \tag{4.91}$$

$$H_x = -\hat{E}_0(e^{-\gamma_0 z} - Re^{\gamma_0 z})\eta_0^{-1} \tag{4.92}$$

where $\hat{E}_0$ is the strength of the incident plane wave. The reflection coefficient as defined is still given by (4.78). The corresponding impedances Z_m have precisely the same form [that is, as given by (4.79)–(4.82)]. Of course, now the surface impedance Z_1 is given by

$$Z_1 = \frac{-E_y}{H_x}\bigg]_{z=0} \tag{4.93}$$

4.7 REFLECTION OF ELLIPTICALLY POLARIZED PLANE WAVES: NORMAL INCIDENCE

The most general type of incident plane is one that is elliptically polarized. In other words, we are saying that

$$\mathbf{E}^{\text{inc}} = (E_x^{\text{inc}}, E_y^{\text{inc}}, 0) \tag{4.94}$$

where

$$E_x^{\text{inc}} = E_0 e^{-\gamma_0 z} \tag{4.95}$$

and

$$E_y^{\text{inc}} = \hat{E}_0 e^{-\gamma_0 z} \tag{4.96}$$

Now we may choose our time origin so that E_0 is real; then the actual time-varying quantities are

$$e_x^{\text{inc}} = \text{Re } E_0 e^{-\gamma_0 z} e^{i\omega t} \tag{4.97}$$

and

$$e_y^{\text{inc}} = \text{Re}|\hat{E}_0|e^{i\phi_0}e^{i\omega t}e^{-\gamma_0 z} \tag{4.98}$$

where ϕ_0 is the phase of $\hat{E}_0$. At $z = 0$ we note that

$$e_x^{\text{inc}} = E_0 \cos \omega t \tag{4.99}$$

and

$$e_y^{\text{inc}} = |\hat{E}_0|\cos(\omega t + \phi_0) \tag{4.100}$$

This is just the parametric description of an ellipse that describes the tip of the electric field vector. For the present problem the reflected electric fields at $z = 0$ are given by

$$e_x^{\text{refl}} = E_0|R|\cos(\omega t + \phi_r) \tag{4.101}$$

and

$$e_y^{\text{refl}} = |\hat{E}_0||R| \cos(\omega t + \phi_0 + \phi_r) \tag{4.102}$$

This is also an ellipse with the same orientation and shape as for the incident wave, but there is a shift in phase of the tip of the electric field vector.

An important special case is when $\phi_0 = \pi/2$ and $|\hat{E}_0| = |E_0|$, which corresponds to a circular polarization. Our analysis shows that even after normal reflection from the layered structure, the polarization would remain circular. Of course, it is important to note that the sense of rotation of the field vector is reversed with respect to the direction of propagation.

Finally, we might note that the polarization characteristics of the transmitted wave remains unchanged from the normally incident wave.

4.8 TRANSMISSION OF POWER

As indicated in Appendix B of this chapter, the power flow or energy transport can be calculated from Poynting's theorem for electromagnetic waves. In the context of reflection and transmission of plane waves for layered media, we are normally interested in knowing the division of power among the incident, reflected, transmitted, and absorbed fields.

The average power flow P_x in the incident field at $z = 0$, for the electric field polarized in the x direction, is

$$P_x^{\text{inc}} = \tfrac{1}{2} \operatorname{Re} E_x^{\text{inc}}(H_y^{\text{inc}})^*]_{z=0} \tag{4.103}$$

where $(H_y^{\text{inc}})^*$ is the conjugate of H_y^{inc}. The units of P_x are watts per square meter. Now clearly, in view of (4.71) and (4.72),

$$P_x^{\text{inc}} = \frac{|E_0^2|}{2} (\operatorname{Re} \eta_0^{-1}) \tag{4.104}$$

The power flow P_x^{refl} in the reflected field is easily shown to be given by

$$P_x^{\text{refl}} = \frac{|E_0|^2}{2} |R|^2 (\operatorname{Re} \eta_0^{-1}) \tag{4.105}$$

When dealing with the transmitted power P_x^{trans}, we work with

$$P_x^{\text{trans}} = \tfrac{1}{2} \operatorname{Re} E_x H_y^*]_{z=L} \tag{4.106}$$

where E_x and H_y, of course, refer to the total fields at the interface. In view of (4.89) it is clear that

$$P_x^{\text{trans}} = \frac{|E_0|^2}{2} |T|^2 (\operatorname{Re} \eta_M^{-1}) \tag{4.107}$$

The simplest method of calculating the total power dissipated P_x^{abs} within the multilayer structure (bounded by $0 < z < L$) is to exploit energy balance (that is, conservation of energy). Thus

$$P_x^{abs} = P_x^{inc} - P_x^{refl} - P_x^{trans} \tag{4.108}$$

$$= \frac{|E_0|^2}{2} [(\text{Re } \eta_0^{-1})(1 - |R|^2) - (\text{Re } \eta_M^{-1})|T|^2]$$

If we were dealing with free space on either side of the layered structure, we could set

$$\text{Re } \eta_0^{-1} = \text{Re } \eta_M^{-1} = \frac{1}{120\pi}$$

Then

$$P_x^{abs} = \frac{|E_0|^2}{240\pi} (1 - |R|^2 - |T|^2) \tag{4.109}$$

Exercise: Consider normal incidence from free space onto a homogeneous slab with electrical properties σ, ϵ, and μ_0 and with thickness ℓ. Show that the transmission coefficient (relating tangential E fields on the two sides of the slab) is given by

$$T = \frac{[2\eta/(\eta + \eta_0)][2\eta_0/(\eta_0 + \eta)]e^{-\gamma\ell}}{1 - [(\eta_0 - \eta)/(\eta_0 + \eta)]^2 e^{-2\gamma\ell}} \tag{4.110}$$

Then show that if the slab is electrically thin (that is, $|\gamma\ell| \ll 1$) and highly conducting (that is, $|\eta/\eta_0| \ll 1$), we can obtain the following approximate formula:

$$T \simeq \frac{1}{1 + \Delta} \tag{4.111}$$

where $2\Delta = \sigma\ell(\mu_0/\epsilon_0)^{1/2} \simeq 120\pi\sigma\ell$. Then show that the input or surface impedance Z of the slab (on the incident side) is approximately given by

$$\frac{1}{Z} = \frac{1}{\eta_0} + \frac{\gamma\ell}{\eta} \tag{4.112}$$

Then, finally, show that the reflection coefficient R of the thin slab under the same limitations is given by

$$R \simeq -\frac{\Delta}{1 + \Delta} \tag{4.113}$$

Exercise: Solve the thin-conductive-slab problem as in the preceding exercise by using a direct boundary-value approach. *Hint:* Apply a "jump" boundary condition that could be written

$$\lim_{\delta \to 0} [H_y(z = \delta) - H_y(z = -\delta)] = -\sigma E_x(0)\ell$$

and further note that E_x is approximately continuous through the thin slab. Hence obtain $R = -\Delta/(1 + \Delta)$ and $T = 1/(1 + \Delta)$. Furthermore, show that the power absorbed P^{abs} in the thin slab can be deduced directly from

$$P_x^{\text{abs}} = \frac{|E_x(0)|^2 \sigma \ell}{2} \tag{4.114}$$

to yield

$$P^{\text{abs}} = \frac{1}{240\pi} |E_0|^2 \frac{\Delta}{(1 + \Delta)^2} \tag{4.115}$$

and, finally, confirm that the total power is conserved, that is,

$$P_x^{\text{abs}} = \frac{1}{240\pi} |E_0|^2 (1 - |R|^2 - |T|^2) \tag{4.116}$$

4.9 REFLECTION OF OBLIQUELY INCIDENT PLANE WAVES

When plane waves impinge obliquely on the plane interface between two contrasting media, the situation becomes interesting. To deal with this case, we consider initially two homogeneous regions with electrical characteristics σ_0, ϵ_0, μ_0 and σ_1, ϵ_1, μ_1, separated by a plane interface at $z = 0$. The situation is illustrated in Figure 4.11.

The incident wave is polarized such that the magnetic field vector is parallel to the interface. Thus we choose our cartesian coordinate system such that the magnetic field vector has only a y component, H_y^{inc}. This component is written in the form

$$H_y^{\text{inc}} = H_0 e^{-\gamma_0 \cos\theta z} e^{-\gamma_0 \sin\theta x} \tag{4.117}$$

or if the medium is lossless (that is, $\sigma_0 = 0$), we can write

$$H_y^{\text{inc}} = H_0 e^{-i\beta z \cos\theta} e^{-i\beta x \sin\theta} \tag{4.118}$$

where $\beta = -i\gamma_0 = (\epsilon_0 \mu_0)^{1/2} \omega = \omega/c$. Here θ can be identified as the angle of incidence. As a consequence, we see that for a fixed level (that is, at $z = z_0$), the value of H_y^{inc} remains the same except for the factor $e^{-i\beta x \sin\theta}$. In fact, the phase velocity in the x direction is $c/(\sin\theta)$, where $c = (\epsilon_0 \mu_0)^{-1/2} = \omega/\beta$. It is also useful to note that planes of constant phase subtend an angle θ with the interface. Furthermore, the phase velocity in the positive z direction is $c/\cos\theta$, which, of course, becomes merely c at normal incidence (that is, at $\theta = 0$).

On physical grounds we can easily predict the form of the reflected field in the region $z < 0$. Clearly, the phase velocity in the x direction

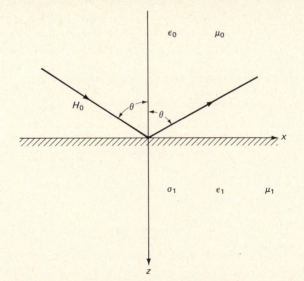

Figure 4.11 Plane were incident obliquely on to homogeneous half space.

should remain the same, and the phase velocity in the z direction should be directed away from the interface. Thus the magnetic field H_y^{refl} would have the form

$$H_y^{\text{refl}} = H_0 R e^{i\beta z\cos\theta} e^{-i\beta x\sin\theta} \qquad (4.119)$$

where, by definition, R is a reflection coefficient.

It is useful to note that

$$(\nabla^2 + \beta^2)\, \genfrac{}{}{0pt}{}{H_y^{\text{inc}}}{H_y^{\text{refl}}} = 0 \qquad (4.120)$$

for the region $z < 0$, where

$$\nabla^2 = \frac{\partial^2}{\partial x^2} + \frac{\partial^2}{\partial z^2} \qquad (4.121)$$

is the laplacian operator.

The total magnetic field H_y in the region $z < 0$ is now given by

$$H_y = H_0(e^{-i\beta z\cos\theta} + R e^{i\beta z\cos\theta})e^{-i\beta x\sin\theta} \qquad (4.122)$$

The corresponding electric field components in the region $z < 0$ are obtained from the Maxwell-Ampere equation:

$$i\epsilon_0\omega E = \text{curl } H \qquad (4.123)$$

or, in the present case,

$$i\epsilon_0\omega E_x = -\frac{\partial H_y}{\partial z} \qquad (4.124)$$

$$i\epsilon_0 \omega E_z = \frac{\partial H_y}{\partial x} \tag{4.125}$$

$$i\epsilon_0 \omega E_y = 0$$

Thus on using (4.124) and (4.125), we see that

$$E_x = H_0 \eta_0 \cos \theta \, (e^{-i\beta z \cos\theta} - R e^{i\beta z \cos\theta}) e^{-i\beta x \sin\theta} \tag{4.126}$$

and

$$E_z = -H_0 \eta_0 \sin \theta \, (e^{-i\beta z \cos\theta} + R e^{i\beta z \cos\theta}) e^{-i\beta x \sin\theta} \tag{4.127}$$

where

$$\eta_0 = \frac{\beta}{\epsilon_0 \omega} = \left(\frac{\mu_0}{\epsilon_0}\right)^{1/2}$$

For further consideration of the problem we change our notation to a more compact form. Thus we write the tangential field components in the regions $z < 0$ as follows:

$$H_y = H_0(e^{-u_0 z} + R e^{u_0 z}) e^{-i\lambda x} \tag{4.128}$$

and

$$E_x = H_0 K_0 (e^{-u_0 z} - R e^{u_0 z}) e^{-i\lambda x} \tag{4.129}$$

where

$$\lambda = \beta \sin \theta$$

$$u_0 = (\lambda^2 - \beta^2)^{1/2} = i\beta \cos \theta$$

and

$$K_0 = \frac{u_0}{i\epsilon_0 \omega} = \eta_0 \cos \theta$$

Clearly, $i\lambda$ is the effective propagation constant in the horizontal direction, while u_0 is the effective propagation in the vertical direction. Also, we see that K_0 plays the role of an effective wave impedance for the transverse field components. For example, it may be noted that

$$\frac{E_x^{\text{inc}}}{H_y^{\text{inc}}} = K_0 \tag{4.130}$$

and

$$\frac{E_x^{\text{refl}}}{H_y^{\text{refl}}} = -K_0 \tag{4.131}$$

We now deal with the field expressions in the region $z > 0$. It should be evident that the x dependence of the fields must be according to the factor $\exp(-i\lambda x)$. Thus we are led to write

$$H_y = f(z) \exp(-i\lambda x) \tag{4.132}$$

for the region $z > 0$. Now we note that

$$(\nabla^2 - \gamma_1^2)H_y = 0 \tag{4.133}$$

where

$$\gamma_1^2 = i\mu_1\omega(\sigma_1 + i\epsilon_1\omega)$$

Then it is seen that $f(z)$ must satisfy

$$\left(\frac{d^2}{dz^2} - \lambda^2 - \gamma_1^2\right)f(z) = 0 \tag{4.134}$$

which has solutions $\exp(\pm u_1 z)$, where $u_1 = (\lambda^2 + \gamma_1^2)^{1/2}$ and where we define Re $u_1 > 0$. Thus we find that the desired form of the solution is

$$H_y = Ae^{-u_1 z}e^{-i\lambda x} \tag{4.135}$$

which decays appropriately as $z \to \infty$. The appropriate electric field components are obtained from

$$(\sigma_1 + i\epsilon_1\omega)E_x = \frac{-\partial H_y}{\partial z} \tag{4.136}$$

$$(\sigma_1 + i\epsilon_1\omega)E_z = \frac{\partial H_y}{\partial x} \tag{4.137}$$

or

$$E_x = \frac{u_1}{\sigma_1 + i\epsilon_1\omega} Ae^{-u_1 z}e^{-i\lambda x} \tag{4.138}$$

and

$$E_z = -\frac{i\lambda}{\sigma_1 + i\epsilon_1\omega} Ae^{-u_1 z}e^{-i\lambda x} \tag{4.139}$$

for all $z > 0$.

The boundary conditions require that E_x and H_y be continuous on the two sides of the interface at $z = 0$. Thus we obtain the following two equations to determine the unknown coefficients A and R:

$$H_0(1 + R) = A \tag{4.140}$$

$$H_0 K_0(1 - R) = K_1 A \tag{4.141}$$

where

$$K_1 = \frac{u_1}{\sigma_1 + i\epsilon_1\omega}$$

Solving for R and A yields

$$R = \frac{K_0 - K_1}{K_0 + K_1} \tag{4.142}$$

and

$$A = H_0 \frac{2K_0}{K_0 + K_1} \tag{4.143}$$

The factor K_1 as defined above is an effective impedance (looking downward) in the lower half space, that is,

$$\left. \frac{E_x}{H_z} \right]_{z \geq 0} = K_1 \tag{4.144}$$

It is useful to note that this relation can be regarded as a boundary condition at $z = 0$; whence (4.128) and (4.129) can be solved directly to get R in accordance with (4.142).

In the important special case where we are dealing with two dielectric half spaces with respective permittivities ϵ_0 and ϵ_1, it follows that

$$R_\parallel = \frac{N_r^2 \cos \theta - (N_r^2 - \sin^2 \theta)^{1/2}}{N_r^2 \cos \theta + (N_r^2 - \sin^2 \theta)^{1/2}} \tag{4.145}$$

where

$$N_r = \left(\frac{\epsilon_1}{\epsilon_0} \right)^{1/2}$$

is, by definition, the relative refractive index. This particular form is the Fresnel reflection coefficient for plane waves that are polarized in the plane of incidence (that is, the $x - z$ plane). In fact, this is the main justification for adding a subscript $\parallel$ to R in any further considerations of this field configuration.

Two special cases of the Fresnel reflection are of special interest. At normal incidence $R_\parallel = (N_r - 1)/(N_r + 1)$, and at a special angle $\theta = \theta_c$, $R_\parallel$ actually vanishes. This so-called Brewster angle occurs when $\tan \theta_c = N_r$. It is also useful to note that

$$\lim_{\theta \to \pi/2} R_\parallel = -1$$

which is the grazing limit. But also note that

$$\lim_{N_r \to \infty} R_\parallel = +1$$

for all angles of incidence! This apparent paradox at grazing incidence is merely a product of the nonpreciseness in defining limiting conditions.

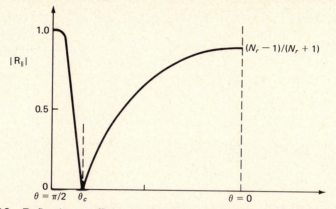

Figure 4.12 Reflection coefficient as function of angle of incidence, illustrating Brewster angle effect.

What we really should be saying is that *provided N_r is finite, $R_\parallel$ approaches -1 as $\theta \to \pi/2$.*

The behavior of the reflection coefficient $R_\parallel$ for the dielectric interface is shown in Figure 4.12. Clearly, as can be seen, the reflection coefficient has a rapid variation in the vicinity of $\theta = \pi/2$ when the refractive index is large.

Exercise: Consider the same geometry as in Figure 4.11 but now choose the electric vector of the incident wave to be polarized in the y direction, that is,

$$E_y^{inc} = E_0 e^{-i\beta z \cos\theta} e^{-i\beta x \sin\theta} \tag{4.146}$$

Show that the fields in the region $z < 0$ are now given by

$$E_y = E_0(e^{-u_0 z} + R_\perp e^{u_0 z})e^{-i\lambda x} \tag{4.147}$$

and

$$-H_x = E_0 N_0 (e^{-u_0 z} - R_\perp e^{u_0 z})e^{-i\lambda x} \tag{4.148}$$

where

$$R_\perp = \frac{N_0 - N_1}{N_0 + N_1} \tag{4.149}$$

and where

$$N_0 = \frac{u_0}{i\mu_0 \omega} \quad \text{and} \quad N_1 = \frac{u_1}{i\mu_1 \omega}$$

Show that $R_\perp = -R_\parallel$ at normal incidence.

For the pure dielectric model of the lower half-space, show that the corresponding Fresnel reflection coefficient is given by

$$R_\perp = \frac{\cos\theta - (N_r^2 - \sin^2\theta)^{1/2}}{\cos\theta + (N_r^2 - \sin^2\theta)^{1/2}} \qquad (4.150)$$

where

$$N_r = (\epsilon_1/\epsilon_0)^{1/2}$$

In our discussion above we have specifically referred to the case where the upper region is air. In this case the reflected wave has the same character as the incident wave, except that the direction of propagation in the vertical z direction is reversed. In both cases the planes of constant phase and planes of constant amplitude are parallel to each other. However, in the lower region the transmitted plane waves are often called inhomogeneous plane waves, because, in general, the planes of constant amplitude are not parallel to the planes of constant phases. This fact is certainly obvious when we note that the amplitude-equivalue planes for $z > 0$ are parallel to the $z = 0$ interface when $\lambda = \beta\sin\theta$ is real. However, the phase varies in both the x and the z directions unless, of course, $\theta = 0$ (that is, normal incidence).

4.10 TOTAL REFLECTION

Another closely related situation that warrants discussion is when a plane wave is incident from the direction of the denser medium. We will discuss this situation in the context posed in Figure 4.13, but now we will

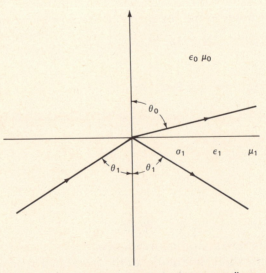

Figure 4.13 Incident plane wave from relatively denser medium.

let the incident wave come from below, as indicated in Figure 4.13. The field in the lower half space for the case where the magnetic field vector is along the y axis can be written

$$H_y = H_0(e^{-u_1 z} + R_{\parallel} e^{u_1 z})e^{-i\lambda x} \qquad (4.151)$$

where

$$R_{\parallel} = \frac{K_1 - K_0}{K_1 + K_0} \qquad (4.152)$$

Now if we were dealing with pure dielectric regions only ($\sigma_1 = 0$, $\mu_1 = \mu_0$), we see that

$$R_{\parallel} = \frac{\hat{N}_0^2 \cos \theta_1 - (\hat{N}_0^2 - \sin^2 \theta_1)^{1/2}}{\hat{N}_0^2 \cos \theta_1 + (\hat{N}_0^2 - \sin^2 \theta_1)^{1/2}} \qquad (4.153)$$

where θ_1 is the angle of incidence that $\hat{N}_0 = (\epsilon_0/\epsilon_1)^{1/2}$. Now, clearly, there is a situation where $\hat{N}_0^2 < \sin^2 \theta_1$. Then we would write

$$R_{\parallel} = \frac{\hat{N}_0^2 \cos \theta_1 + i(\sin^2 \theta_1 - \hat{N}_0^2)^{1/2}}{\hat{N}_0^2 \cos \theta_1 - i(\sin^2 \theta_1 - \hat{N}_0^2)^{1/2}} \qquad (4.154)$$

Here we see that $|R_{\parallel}| = 1$, and there is total reflection. This result occurs whenever $\theta_1 > \arcsin \hat{N}_0$.

The field in the less dense medium is easily shown to be

$$H_y = H_0(1 + R_{\parallel})e^{-u_0 z}e^{-i\lambda x} \qquad (4.155)$$

In the case where we are dealing with pure dielectrics,

$$u_0 = i\beta \left(1 - \frac{\sin^2 \theta_1}{\hat{N}_0^2}\right)^{1/2}$$

$$= \frac{i\beta}{\hat{N}_0}(\hat{N}_0^2 - \sin^2 \theta_1)^{1/2} \qquad (4.156)$$

is pure imaginary if $\sin^2 \theta_1 < \hat{N}_0^2$. In this case we have unattenuated plane wave propagation, and since $\lambda = \beta \sin \theta_0$ and $u_0 = i\beta \cos \theta_0$, the wave normal makes an angle θ_0 with the positive z axis. Critical reflection occurs when θ_0 exceeds $\pi/2$. In that case

$$u_0 = \frac{\beta}{\hat{N}_0}(\sin^2 \theta_1 - \hat{N}_0^2)^{1/2} \qquad (4.157)$$

which indicates that the field is evanescent (that is, it decays exponentially in the positive z direction). The decay rate is actually u_0 Np/m. The situation is illustrated in Figure 4.14. There is no mean power flow across the interface, so all the incident power flow is reflected. In the

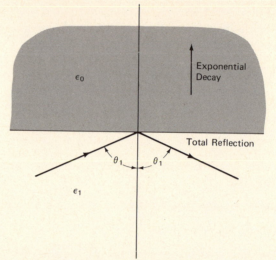

Figure 4.14   Totally reflected wave within relatively dense medium.

equivalent transmission-line circuit the medium on the incident side of the interface is represented by a semi-infinite line with characteristic impedance K_1, and it is terminated in a pure reactance, which is given by

$$K_0 = \frac{u_0}{i\epsilon_0\omega} = -i\eta_0 \left(\frac{\sin^2 \theta_1}{\hat{N}_0^2} - 1\right)^{1/2} \tag{4.158}$$

where $\eta_0 = (\mu_0/\epsilon_0)^{1/2}$.

Exercise: Show that the condition for critical or total reflection does not depend on the polarization of the incident wave. Contrast this result with the condition for the Brewster angle.

4.11 OBLIQUE INCIDENCE FOR LAYERED STRUCTURES

The generalization of the formulation to reflection at oblique incidence from any number of layers is straightforward. The situation is shown in Figure 4.15, where the structure is the same as that shown in Figure 4.9. We will choose the magnetic field to have only a y component. Clearly, the total field in the "incident medium" for $z < 0$ will have the form

$$H_y = H_0(e^{-u_0 z} + R_\parallel e^{u_0 z})e^{-i\lambda x} \tag{4.159}$$

where $u_0 = \gamma_0 \cos\theta$ and $i\lambda = \gamma_0 \sin\theta$. If we set $\sigma_0 = 0$, $\gamma_0 = i\beta$, where $\beta = (\epsilon_0\mu_0)^{1/2}\omega$, then θ can be regarded as the real angle of incidence and in this case $u_0 = i\beta \cos\theta$ and $\lambda = \beta \sin\theta$. In our discussion below we will assume that the incident medium is lossless, but obviously the formula-

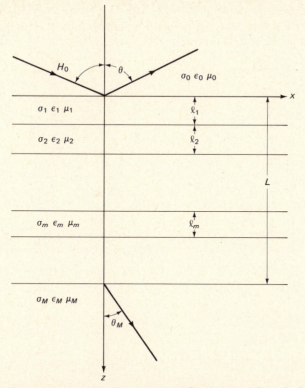

Figure 4.15   Illustrating transverse magnetic reflection at oblique incidence from layered structure. (Reverse direction of arrow at upper left.)

tion is not restricted to this case. The objective here is to determine the reflection coefficient $R_{\parallel}$ for the composite structure.

Now in each layer or region

$$(\nabla^2 - \gamma_m^2)H_y = 0 \tag{4.160}$$

where $m = 0, 1, 2, 3, \ldots, M$ and

$$\gamma_m = [i\mu_m\omega(\sigma_m + i\epsilon_m\omega)]^{1/2} \tag{4.161}$$

Here

$$\nabla^2 = \frac{\partial^2}{\partial x^2} + \frac{\partial^2}{\partial z^2} \tag{4.162}$$

The solution in this mth layer must have the form

$$H_y = (a_m e^{-u_m z} + b_m e^{u_m z})e^{-i\lambda x} \tag{4.163}$$

where

$$u_m = (\lambda^2 + \gamma_m^2)^{1/2} \qquad \mathrm{Re}\, u_m > 0$$

The presence of the common factor $\exp(-i\lambda x)$ is dictated by the requirement that the x variation of the field is imposed by the incident field. If we define an angle θ_m by

$$u_m = \gamma_m \cos \theta_m$$

then clearly,

$$i\lambda = \gamma_m \sin \theta_m$$

but

$$i\lambda = \gamma_0 \sin \theta$$

The identity

$$\gamma_m \sin \theta_m = \gamma_0 \sin \theta \tag{4.164}$$

is really a statement of a generalized form of Snell's law. In particular, if the bottom region (that is, for $m = M$) is lossless, we would have

$$N_r \sin \theta_M = \sin \theta \tag{4.165}$$

where

$$N_r = \left(\frac{\epsilon_M \mu_M}{\epsilon_0 \mu_0} \right)^{1/2}$$

The real angle θ_M is independent of the properties of the intervening layers.

Now from Maxwell's equation

$$E_x = -(\sigma_m + i\epsilon_m \omega)^{-1} \frac{\partial H_y}{\partial z} \tag{4.166}$$

For the mth layer we see that

$$E_x = K_m(a_m e^{-u_m z} - b_m e^{u_m z})e^{-i\lambda x} \tag{4.167}$$

where

$$K_m = \frac{u_m}{\sigma_m + i\epsilon_m \omega}$$

is the characteristic impedance within the layer.

In the bottom region we can see that the appropriate forms of the tangential fields are

$$H_y = H_0 T_\| e^{-u_M(z-L)} e^{-i\lambda x} \tag{4.168}$$

and

$$E_x = H_0 K_M T_\| e^{-u_M(z-L)} e^{-i\lambda x} \tag{4.169}$$

Here, as indicated, we have defined $T_\|$ as a transmission factor that relates the magnetic field as the "input" and "output" surfaces $z = 0$ and $z = L$, respectively.

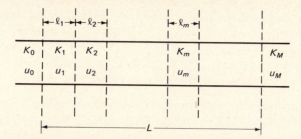

Figure 4.16 Equivalent transmission-line circuit for obliquely incident, transverse, magnetic plane wave onto *M*-layered structure.

The problem can be solved either by the archaic boundary-value approach, where we match tangential fields and solve simultaneous equations, or by using transmission-line concepts. We will choose the latter. The equivalent circuit is shown in Figure 4.16. The mth section of length ℓ_m has a characteristic impedance K_m and propagation constant u_m. The similarity with the model shown in Figure 4.10 is close. In the present case we easily deduce that the reflection coefficient is given by

$$R_{\parallel} = -\frac{Z_1 - K_0}{Z_1 + K_0} \tag{4.170}$$

where

$$Z_m = K_m \frac{Z_{m+1} + K_m \tanh u_m \ell_m}{K_m + Z_{m+1} \tanh u_m \ell_m} \tag{4.171}$$

where $m = 1, 2, \ldots, M-1$ and, of course, $Z_M = K_M$.

The corresponding transmission coefficient is given by

$$T_{\parallel} = \left(\frac{2K_0}{K_1 + K_0} \frac{2K_1}{K_2 + K_1} \cdots \frac{2K_{m-1}}{K_m + K_{m-1}} \cdots \frac{2K_{M-1}}{K_M + K_{M-1}} \right)$$
$$\times D^{-1} \exp(-u_1 \ell_1 - u_2 \ell_2 - \cdots - u_m \ell_m - u_{M-1} \ell_{M-1}) \tag{4.172}$$

where

$$D = \left[1 - \left(\frac{K_0 - K_1}{K_0 + K_1} \right) \left(\frac{Z_2 - K_1}{Z_2 + K_1} \right) e^{-2u_1 \ell_1} \right]$$
$$\times \left[1 - \left(\frac{K_1 - K_2}{K_1 + K_2} \right) \left(\frac{Z_3 - K_2}{Z_3 + K_2} \right) e^{-2u_2 \ell_2} \right] \cdots$$
$$\times \left[1 - \left(\frac{K_{m-1} - K_m}{K_{m-1} + K_m} \right) \left(\frac{Z_{m+1} - K_m}{Z_{m+1} + K_m} \right) e^{-2u_m \ell_m} \right] \cdots$$
$$\times \left[1 - \left(\frac{K_{M-2} - K_{M-1}}{K_{M-2} + K_{M-1}} \right) \left(\frac{Z_M - K_{M-1}}{Z_M + K_{M-1}} \right) e^{-2u_{M-1} \ell_{M-1}} \right] \tag{4.173}$$

where

$$Z_M = K_M$$

Clearly, a more compact definition of the denominator D is the finite product

$$D = \prod_{m=1}^{M-1} \left[1 - \left(\frac{K_{m-1} - K_m}{K_{m-1} + K_m} \right) \left(\frac{Z_{m+1} - K_m}{Z_{m+1} + K_m} \right) e^{-2u_m\ell_m} \right] \qquad (4.174)$$

The expression for the transmission coefficient has a lucid physical interpretation. Each of the factors $2K_{m-1}/(K_m + K_{m-1})$ is the transmission coefficient for the tangential magnetic field at the interface between the $(m-1)$th and the mth sections. The total transmission loss through the system is dominated by the exponential factor $\exp(\quad)$. On the other hand, D accounts for the multiple interactions within the structure and includes all orders of such interactions. If $|u_m\ell_m| \gg 1$ for $m = 1$, $2, \ldots, M - 1$, all internal interactions (that is, multiple reflections) can be ignored, and $D \to 1$.

It is useful to note that an alternative form for $T_\parallel$ is

$$T_\parallel = \frac{K_0}{K_M} T'_\parallel \qquad (4.175)$$

where

$$T'_\parallel = \left(\frac{2K_1}{K_1 + K_0} \frac{2K_2}{K_2 + K_1} \cdots \frac{2K_m}{K_m + K_{m-1}} \cdots \frac{2K_M}{K_M + K_{M-1}} \right)$$
$$\times \mathbf{D}^{-1} \exp(-u_1\ell_1 - u_2\ell_2 - \cdots - u_m\ell_m - u_{M-1}\ell_{M-1}) \qquad (4.176)$$

Thus we note that $T'_\parallel$ is the transmission factor that relates the tangential electric field at the "input" and "output" surfaces $z = 0$ and $z = L$, respectively. For example, if for the region $z < 0$,

$$E_x = E_0(e^{-u_0z} - R_\parallel e^{u_0z})e^{-i\lambda x} \qquad (4.177)$$

where $E_0 = K_0 H_0$, then for the region $z > L$

$$E_x = E_0 T'_\parallel e^{-u_M(z-L)}e^{-i\lambda x} \qquad (4.178)$$

In the limiting cases of normal incidence (that is, $\lambda \to 0$), the solution above reduces to the results given by (4.78) and (4.84) if we note that $T'_\parallel \to T$ and $R_\parallel \to -R$.

4.12 REFLECTION OF RADIO WAVES FROM A STRATIFIED GROUND

An interesting application of the preceding general formulation is to examine the wave characteristics at the earth's surface when the incident signal arrives at grazing angles. While this incident field is not strictly a plane wave, it can be approximated by such, at least for the following discussion.

The surface impedance Z_1 is given exactly by

$$Z_1 = \frac{E_x}{H_y}\bigg]_{z=0} \tag{4.179}$$

where Z_1 is defined by (4.171) with $m = 1$. If we are interested in the value of the surface impedance for grazing incidence (that is, $\theta \to \pi/2$), we see that $\lambda = -i\gamma_0 = \beta$. Then

$$u_m = (\gamma_m^2 - \gamma_0^2)^{1/2} \tag{4.180}$$

and

$$K_m = \frac{(\gamma_m^2 - \gamma_0^2)^{1/2}}{\sigma_m + i\epsilon_m \omega} \tag{4.181}$$

for $m = 1, 2, 3, \ldots, M$.

If the ground were homogeneous with properties $\sigma_1, \epsilon_1, \mu_1$, it is clear that for this case of grazing incidence the surface impedance is given by

$$Z_1 = \eta_1 \left(1 - \frac{\gamma_0^2}{\gamma_1^2}\right)^{1/2} = K_1 \tag{4.182}$$

where $\eta_1 = i\mu_1\omega/\gamma_1 = [i\mu_1\omega/(\sigma_1 + i\epsilon_1\omega)]^{1/2}$. When the ground or half-space model is layered, we would write

$$Z_1 = \eta_1 \left(1 - \frac{\gamma_0^2}{\gamma_1^2}\right)^{1/2} Q \tag{4.183}$$

where Q is a useful correction factor to account for the layer structure. For example, in the case of a two-layer model (that is, $\ell_2 \to \infty$), we see that

$$Q = \frac{K_2 + K_1 \tanh u_1\ell_1}{K_1 + K_2 \tanh u_1\ell_1} \tag{4.184}$$

where

$$K_1 = \eta_1 \left[1 - \left(\frac{\gamma_0}{\gamma_1}\right)^2\right]^{1/2}$$

$$u_1 = \gamma_1 \left[1 - \left(\frac{\gamma_0}{\gamma_1}\right)^2\right]^{1/2}$$

$$K_2 = \eta_2 \left[1 - \left(\frac{\gamma_0}{\gamma_2}\right)^2\right]^{1/2}$$

and

$$\eta_2 = \left(\frac{i\mu_2\omega}{\sigma_2 + i\epsilon_2\omega}\right)^{1/2}$$

It is rather evident that Q approaches one when $\text{Re}(u_1\ell_1) \gg 1$, which means that the subsurface layer is not detectable.

A related parameter that can also be measured at the earth's surface is the "wave tilt" W. This parameter is defined by the ratio of electric field components in the manner

$$W = \frac{E_x}{-E_z}\bigg]_{z \to -0} \tag{4.185}$$

where, as indicated above, the limit should be taken as z approaches zero from negative values. It is a simple matter to show that

$$W = \frac{i\epsilon_0 \omega}{\gamma_0 \sin \theta} Z_1 \tag{4.186}$$

which has no approximations. In the case of a homogeneous ground

$$W = \frac{i\epsilon_0 \omega}{\gamma_0 \sin \theta} K_1 = \frac{\eta_1}{\eta_0} \frac{[1 - (\gamma_0^2/\gamma_1^2)\sin^2 \theta]^{1/2}}{\sin \theta} \tag{4.187}$$

In the grazing-incidence limit this term becomes

$$W = \frac{\eta_1}{\eta_0}\left(1 - \frac{\gamma_0^2}{\gamma_1^2}\right)^{1/2} \tag{4.188}$$

or

$$W = \left(\frac{\mu_1}{\mu_0}\right)\frac{\gamma_0}{\gamma_1}\left(1 - \frac{\gamma_0^2}{\gamma_1^2}\right)^{1/2} \tag{4.189}$$

Again, if the half space is layered, we would have

$$W = \frac{\eta_1}{\eta_0}\left(1 - \frac{\gamma_0^2}{\gamma_1^2}\right)^{1/2}Q \tag{4.190}$$

where Q is the correction factor discussed above.

Exercise: Show that the wave tilt $W_h = -E_x/E_y$ at height $z = -h$ is given by

$$W_h = W\frac{1 - R_\parallel \exp(-2u_0 h)}{1 + R_\parallel \exp(-2u_0 h)} \tag{4.191}$$

While the formulas for the two-layer earth model are quite tractable, even further simplifications are possible when displacements in the conductive layers are neglected. For example, if $\epsilon_1 \omega/\sigma_1$ and $\epsilon_2 \omega/\sigma_2 \ll 1$, we see that (4.184) can be simplified to

$$Q \simeq \frac{\eta_2 + \eta_1 \tanh \gamma_1 \ell_1}{\eta_1 + \eta_2 \tanh \gamma_1 \ell_1} \tag{4.192}$$

where $\eta_1 = i\mu_1\omega/\gamma_1$, $\eta_2 = i\mu_2\omega/\gamma_2$, $\gamma_1 = (i\sigma_1\mu_1\omega)^{1/2}$, and $\gamma_2 = (i\sigma_2\mu_2\omega)^{1/2}$. If, in addition, we assume $\mu_2 = \mu_1 = \mu_0$, we see that

$$Q \simeq \frac{(\sigma_1/\sigma_2)^{1/2} + \tanh(\sqrt{i}V)}{1 + (\sigma_1/\sigma_2)^{1/2}\tanh(\sqrt{i}V)} \tag{4.193}$$

where

$$V = (\sigma_1\mu_0\omega)^{1/2}\ell_1$$

An equivalent form is

$$Q = \tanh[(\sqrt{i}V) + \operatorname{arctanh} C] \tag{4.194}$$

where $C = (\sigma_1/\sigma_2)^{1/2}$.

The two parameters of interest to the experimentalist are the surface impedance and the wave tilt. Under the simplification that displacement currents are negligible in the conductive layers, it is seen that

$$Z_1 = \eta_1 Q \tag{4.195}$$

and

$$W = \left(\frac{\gamma_0}{\gamma_1}\right)Q \tag{4.196}$$

Thus the quantity Q plays an important role. The amplitude $|Q|$ and the phase of Q, denoted q, are plotted in Figure 4.17(a) and 4.17(b) as a function of V, which is a normalized thickness of the upper layer. As indicated, when V is greater than one, the function Q is near one in amplitude, but it does differ in phase from zero up until V is about 2. Of particular interest is that q approaches $+45°$ when $\sigma_2 \gg \sigma_1$ and V tends to small values. In such a case the phase of the surface impedance Z_1 is tending toward $90°$, so the surface is actually reactive. This result has important consequences in radio wave transmission problems from both the geophysical and the telecommunications standpoint.

To give some idea of the order of magnitudes of the various quantities, we choose a specific example. The frequency $f = \omega/2\pi = 125$ kHz, $\sigma_1 = 10^{-3}$ S/m, $\epsilon_1 = 10\epsilon_0 = 8.85 \times 10^{-11}$ F/m, $\mu_0 = 4\pi \times 10^{-7}$ H/m. Then we find that the wave tilt W_0 for a homogeneous ground is given by

$$W_0 = 0.082\angle 41.1° \tag{4.197}$$

Now when the thickness of the upper layer is finite, we see that the wave tilt is modified to

$$W_0 = 0.082|Q|\angle(41.1° + q) \tag{4.198}$$

The lower layer with properties σ_2, ϵ_2, and μ_0 is contained in the correction factor $|Q|\angle q$. Since $\epsilon_1\omega/\sigma_1 = 6.9 \times 10^{-2}$, we can safely neglect displacement currents, so the curves in Figure 4.17 are directly applicable. Then, in fact, the parameter V can be replaced by $\ell_1/30$, where ℓ_1 is the thickness of the upper stratum, in meters.

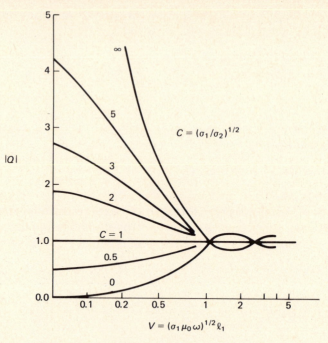

$$C = (\sigma_1/\sigma_2)^{1/2}$$

$C = 1$

$$V = (\sigma_1 \mu_0 \omega)^{1/2} \ell_1$$

(a)

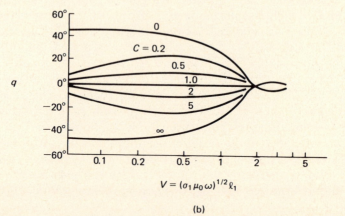

$$V = (\sigma_1 \mu_0 \omega)^{1/2} \ell_1$$

(b)

Figure 4.17(a) Amplitude $|Q|$ for two-layer earth model; (b) phase q for two-layer earth model.

4.13 EXTENSION TO TRANSVERSE ELECTRIC POLARIZATION FOR REFLECTION AT OBLIQUE INCIDENCE FROM LAYERED STRUCTURES

The situation illustrated in Figure 4.15 serves the purpose for describing transverse electric (TE) polarization in the context of reflection at oblique incidence. We merely replace H_0 there by E_0 so that the electric field, in the region $z < 0$, is now described by

$$E_y = E_0(e^{-u_0 z} + R_\perp e^{u_0 z})e^{-i\lambda x} \tag{4.199}$$

where $R_\perp$ is the corresponding reflection coefficient and the subscript $\perp$ designates that the electric vector is now perpendicular to the plane of incidence (that is, the xz plane). The electric field in the bottom region is written

$$E_y = E_0 T_\perp e^{-u_M(z-L)}e^{-i\lambda x} \tag{4.200}$$

where $T_\perp$ is the corresponding transmission coefficient.

The solution for the present problem is entirely analogous to the case for transverse magnetic (TM) incidence. In place of impedances K_m, we now have admittances N_m, where the latter is defined by

$$N_m = \frac{u_m}{i\mu_m\omega} \qquad m = 0, 1, 2, \ldots, M \tag{4.201}$$

and all other quantities have the same definitions.

In place of (4.170) and (4.171), we now write

$$R_\perp = \frac{N_0 - Y_1}{N_0 + Y_1} \tag{4.202}$$

where

$$Y_m = N_m \frac{Y_{m+1} + N_m \tanh u_m \ell_m}{N_m + Y_{m+1} \tanh u_m \ell_m} \tag{4.203}$$

for $m = 1, 2, \ldots, M - 1$, and

$$Y_M = N_M$$

Furthermore, in place of (4.172) and (4.174), we now have

$$T_\perp = \left(\frac{2N_0}{N_1 + N_0} \frac{2N_1}{N_2 + N_1} \cdots \frac{2N_{m-1}}{N_m + N_{m-1}} \cdots \frac{2N_{M-1}}{N_M + N_{M-1}} \right)$$
$$\times D^{-1} \exp(-u_1\ell_1 - u_2\ell_2 - \cdots$$
$$- u_m\ell_m \cdots - u_{M-1}\ell_{M-1}) \tag{4.204}$$

where

$$D = \prod_{m=1}^{M-1} \left[1 - \left(\frac{N_{m-1} - N_m}{N_{m-1} + N_m} \right) \left(\frac{Y_{m+1} - N_m}{Y_{m+1} + N_m} \right) e^{-2u_m\ell_m} \right] \tag{4.205}$$

As before,

$$L = \ell_1 + \ell_2 + \ell_m + \cdots + \ell_{M-1}$$

is the total thickness of the structure.

In the limiting case of normal incidence we immediately recover the earlier results given, for example, by (4.73) and (4.83). Here we should note that

$$\lim_{\lambda \to 0} \frac{2N_{m-1}}{N_m + N_{m-1}} = \frac{2\eta_{m-1}^{-1}}{\eta_m^{-1} + \eta_{m-1}^{-1}} = \frac{2\eta_m}{\eta_m + \eta_{m-1}} \tag{4.206}$$

and

$$R_\perp \to R \quad \text{and} \quad T_\perp \to T$$

Exercise: Show that for a two-layer model (that is, $\ell_2 \to \infty$),

$$R_\perp = \frac{N_0 - Y_1}{N_0 + Y_1} \tag{4.207}$$

where

$$Y_1 = N_1 \frac{N_2 + N_1 \tanh u_1 \ell_1}{N_1 + N_2 \tanh u_1 \ell_1} \tag{4.208}$$

and

$$T_\perp = \frac{[2N_0/(N_1 + N_0)][2N_1/(N_2 + N_1)]e^{-u_1\ell_1}}{1 - [(N_0 - N_1)/(N_0 + N_1)][(N_2 - N_1)/(N_2 + N_1)]e^{-2u_1\ell_1}} \tag{4.209}$$

Now consider the special case where $\sigma_1 \to \infty$ and $\ell_1 \to 0$ in such a fashion that $\sigma_1\ell_1$ is a bounded quantity. Show that

$$Y_1 \simeq N_2 + \sigma_1\ell_1 \tag{4.210}$$

and

$$T_\perp \simeq \frac{2N_0}{N_0 + N_2 + \sigma_1\ell_1} \tag{4.211}$$

Exercise: Show that (4.210) and (4.211) can be obtained by applying a jump boundary condition that could be written

$$\lim_{\delta \to 0}[H_x(z = \delta) - H_x(z = -\delta)] = \sigma_1\ell_1 E_y(0)$$

where also E_y is assumed to be continuous through the sheet.

4.14 LINE-SOURCE EXCITATION

It is useful to consider a specific source of the excitation of the fields other than a postulated incident plane wave. To this end, we will first consider a very basic problem in a two-dimensional geometry. The situation is illustrated in Figure 4.18. The surface $z = 0$ is a perfect conductor. At a height h above this surface we have a postulated current

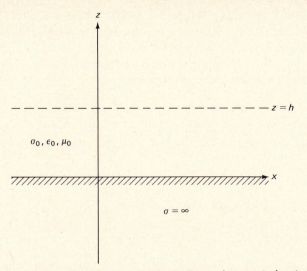

Figure 4.18 Two-dimensional geometry for y-directed current sheet at z = h over perfectly conducting plane surface at z = 0.

sheet $j_y(x)$ A/m that is directed parallel to the y axis. As indicated, the strength of this current density may be a function of the horizontal coordinate x.

For the problem as posed it is clear that the resulting fields exciting will have only E_y, H_x, and H_z components. Thus the field configuration is transverse electric (TE). If this result is not self-evident, one could show, in hindsight, that the transverse magnetic (TM) fields are not excited by this source specification.

According to the Maxwell-Ampere equation in integral form, we can easily deduce that the source condition is

$$\lim_{\delta \to 0}[H_x(z = h + \delta) - H_x(z = h - \delta)] = j_y(x) \tag{4.212}$$

Also, we assert that

$$\lim_{\delta \to 0}[E_y(z = h + \delta) - E_y(z = h- \delta)] = 0 \tag{4.213}$$

Now since

$$\left(\frac{\partial^2}{\partial x^2} + \frac{\partial^2}{\partial z^2} - \gamma_0^2\right)E_y = 0 \tag{4.214}$$

solutions will be of the form

$$\exp(\pm u_0 z)e^{-i\lambda x}$$

where

$$u_0 = (\lambda^2 + \gamma_0^2)^{1/2}$$

and λ can take any value. However, here we will restrict λ^2 to be a positive real parameter, and u_0 is defined such that Re $u_0 > 0$, bearing in mind that Re $\gamma_0 > 0$ if $\sigma_0 > 0$ can be vanishingly small.

Because of the assumed perfectly conducting surface at $z = 0$, our basic solution for E_y must be of the form

$$(e^{+u_0 z} - e^{-u_0 z})e^{-i\lambda x}$$

for the region $0 < z < h$ (that is, it must vanish at $z = 0$). Now for the semi-infinite region $z > h$ it is clear that the solution would be of the form

$$e^{-u_0 z}e^{-i\lambda x}$$

if it is to vanish or be bounded as $z \to \infty$.

The preceding discussion suggests that we express our current (source) density as an integral of the form

$$j_y(x) = \int_{-\infty}^{+\infty} J(\lambda)e^{-i\lambda x} \, d\lambda \tag{4.215}$$

This term is a Fourier integral, and the integration contour is taken to be along the real λ axis. The function $J(\lambda)$ is the Fourier transform of $j_y(x)$ and is given by the inverse

$$J(\lambda) = \frac{1}{2\pi} \int_{-\infty}^{+\infty} j_y(x)e^{i\lambda x} \, dx \tag{4.216}$$

where the integration is now along the (real) x coordinate.

The next step is to represent the fields as appropriate Fourier integrals in the manner

$$E_y = \int_{-\infty}^{+\infty} a(\lambda)(e^{u_0 z} - e^{-u_0 z})e^{-i\lambda x} \, d\lambda \tag{4.217}$$

for $0 < z < h$, and

$$E_y = \int_{-\infty}^{+\infty} b(\lambda)e^{-u_0 z}e^{-i\lambda x} \, d\lambda \tag{4.218}$$

for $z > h$. Then $a(\lambda)$ and $b(\lambda)$ are yet to be determined.

Now from Maxwell's equation

$$i\mu_0\omega H_x = \frac{\partial E_y}{\partial z} \tag{4.219}$$

we have

$$H_x = \int_{-\infty}^{+\infty} a(\lambda)N_0(e^{u_0 z} + e^{-u_0 z})e^{-i\lambda x} \, d\lambda \tag{4.220}$$

for $0 < z < h$, and

$$H_x = - \int_{-\infty}^{+\infty} b(\lambda) N_0 e^{-u_0 z} e^{-i\lambda x} \, d\lambda \tag{4.221}$$

where $N_0 = u_0/i\mu_0\omega$. On imposing the source conditions (4.212) and (4.213), we see that

$$a(e^{u_0 h} + e^{-u_0 h}) + be^{-u_0 h} = -\frac{J(\lambda)}{N_0} \tag{4.222}$$

and

$$a(e^{u_0 h} - e^{-u_0 h}) - be^{-u_0 h} = 0 \tag{4.223}$$

Thus we determine that

$$a(\lambda) = -\frac{J(\lambda)}{2N_0} e^{-u_0 h} \tag{4.224}$$

and

$$b(\lambda) = -\frac{J(\lambda)}{2N_0} (e^{u_0 h} - e^{-u_0 h}) \tag{4.225}$$

On inserting these expressions into (4.217) and (4.218), we can write the unified result

$$E_y = -\frac{1}{2} \int_{-\infty}^{+\infty} \frac{J(\lambda)}{N_0} (e^{\pm u_0(z-h)} - e^{-u_0(z+h)}) e^{-i\lambda x} \, d\lambda \tag{4.226}$$

where we use plus in the exponent for $0 < z < h$ and minus for $z > h$. This result is the exact formal solution for the problem as posed. Here the appropriate transform $J(\lambda)$ must be inserted and the integration over λ must be performed.

It is useful to interpret (4.226) as a spectrum of plane waves. If $\lambda = \beta \sin \theta$ and $u_0 = i\beta \cos \theta$, where $\beta = -i\gamma_0 \approx (\epsilon_0\mu_0)^{1/2}\omega$ for negligible conductivity σ_0, then θ is the angle of incidence in the usual sense. But since λ varies over real values from $-\infty$ to $+\infty$, we see that $\sin \theta$ must also vary from $-\infty$ to $+\infty$. Thus we must admit imaginary angles when $|\sin \theta| > 1$. Therefore the corresponding range of θ is from $-i\infty - (\pi/2)$, to $-\pi/2$, to $+\pi/2$, to $(+\pi/2) + i\infty$.

To deal with the present problem in a more specific manner, we choose a line-source excitation at $x = 0$, which means

$$j_y(x) = I_0 \, \delta(x) \tag{4.227}$$

where $\delta(x)$ is the unit-impulse function and I_0 is the current, in amperes. Now on making use of (4.216), we see that for this case

$$J(\lambda) = \frac{I_0}{2\pi} \tag{4.228}$$

The integrals in (4.226), with this substitution, are of a standard form, and we can immediately write for all $z > 0$

$$E_y = -\frac{I_0 i \mu_0 \omega}{2\pi}(K_0\{\gamma_0[x^2 + (z-h)^2]^{1/2}\}$$
$$- K_0\{\gamma_0[x^2 + (z+h)^2]^{1/2}\}) \tag{4.229}$$

where K_0 is the modified Bessel function of the second type [3]. It is useful to rewrite (4.229) as

$$E_y = -\frac{I_0 i \mu_0 \omega}{2\pi}[K_0(\gamma_0 R_0) - K_0(\gamma_0 R_1)] \tag{4.230}$$

where

$$R_0 = [x^2 + (z-h)^2]^{1/2}$$

and

$$R_1 = [x^2 + (z+h)^2]^{1/2}$$

Clearly, R_0 is the linear distance from the observer at (x, z) to the line current source at $(0, h)$, while R_1 is the linear distance from (x, z) to an image line source (with opposite sign) at $(0, -h)$. The situation is illustrated in Figure 4.19.

When $|\gamma_0 R_0|$ and $|\gamma_0 R_1| \gg 1$, we can use the large-argument asymptotic approximations for the modified Bessel functions to write

$$E_y \simeq -\frac{I_0 i \mu_0 \omega}{2\pi}\left(\frac{\pi}{2\gamma_0}\right)^{1/2}\left(\frac{e^{-\gamma_0 R_0}}{R_0^{1/2}} - \frac{e^{-\gamma_0 R_1}}{R_1^{1/2}}\right) \tag{4.231}$$

Furthermore, if $R_0 \gg 2h$, this expression is further simplified to

$$E_y \simeq -\frac{i I_0 \mu_0 \omega e^{-\gamma_0 R_0}}{2(\pi 2\gamma_0 R_0)^{1/2}}(1 - e^{-2\gamma_0 h \cos\theta}) \tag{4.232}$$

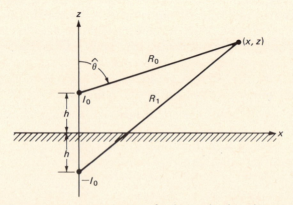

Figure 4.19   Line current source over perfectly conducting plane surface showing image location.

where $\sin \hat{\theta} = x/R_0$. Or if we specifically deal with the lossless case $\gamma_0 = i\beta$,

$$E_y \simeq -\frac{I_0\mu_0\omega e^{i\pi/4}}{2(2\pi\beta R_0)^{1/2}}\, e^{-i\beta R_0}(1 - e^{-2i\beta h\cos\theta}) \tag{4.233}$$

and thus

$$|E_y| \simeq \frac{I_0\mu_0\omega}{(2\pi\beta R_0)^{1/2}}\, |\sin(\beta h \cos \hat{\theta})| \tag{4.234}$$

The function $\sin(\beta h \cos \hat{\theta})$ is the radiation pattern, where $\hat{\theta}$ is the angle measured from the vertical. Of course, it vanishes as $\hat{\theta} \to \pi/2$. Depending on the magnitude of βh, it has lobes for values of $\hat{\theta}$ where there is constructive interference between the source line current and it's image [that is, where $2\beta h \cos \hat{\theta} = (2m + 1)(\pi/2)$, where $m = 0, 1, 2, \ldots$].

4.15 LINE SOURCE OVER *M*-LAYERED STRUCTURE

We are now in the position to generalize the results in the previous section to the case where the line source is located over an M-layered half space. Again, we may begin with the case where the sheet-current density is specified to be $j_y(x)$ at the level $z = h$. The situation is illustrated in Figure 4.20.

We can immediately write the desired form for the y-directed electric field in upper half space $z > 0$:

$$E_y = -\frac{1}{2}\int_{-\infty}^{\infty} \frac{J(\lambda)}{N_0}[e^{\pm u_0(z-h)} + R_\perp(\lambda)e^{-u_0(z+h)}]e^{-i\lambda x}\, d\lambda \tag{4.235}$$

where the $\pm$ sign is used for $z \lessgtr h$. The coefficient $R_\perp \to -1$ in the case where the conductivity $\sigma_1 \to \infty$ (that is, perfectly conducting surface), as

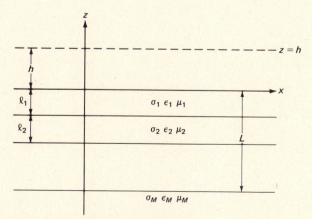

Figure 4.20   Geometry for problem of y-directed current sheet at $z = h$ over M-layered structure occupying region $z < 0$.

can be confirmed by (4.226). In fact, the solution can be carried through in the same manner as the previous development for oblique incidence of a TE plane wave. Thus in the bottom region

$$E_y = -\frac{1}{2} \int_{-\infty}^{\infty} \frac{J(\lambda)e^{-u_0 h}}{N_0} T_\perp(\lambda) e^{u_M(z+L)} e^{-i\lambda x} \, d\lambda \tag{4.236}$$

The functional form and definition of $R_\perp(\lambda)$ are precisely the same as given by (4.202) and (4.204), respectively. In the case where we have a line current source I_0 at $z = h$ and $x = 0$, it is required only to replace $J(\lambda)$ by $I_0/(2\pi)$.

The evaluation of the integrals such as in (4.235) and (4.236) requires considerable effort in the general case. A detailed discussion is out of the question here. However, we examine the problem in a heuristic (that is, nonrigorous) fashion that permits us to achieve a tractable form in the far field. In the case where we have a line-source excitation, the field in the upper half space can be written without approximation in the form, for $z > 0$,

$$E_y = -\frac{i\mu_0\omega I_0}{2\pi}\left[K_0(\gamma_0 R_0) + \frac{1}{2}\int_{-\infty}^{+\infty} \frac{1}{u_0} e^{-u_0(z+h)} R_\perp(\lambda) e^{-i\lambda x} \, d\lambda \right] \tag{4.237}$$

where $R_0 = [(z-h)^2 + x^2]^{1/2}$. We now assume that $R_\perp(\lambda)$ can be taken outside the integral and replaced by its value at $\lambda = \bar\lambda$, where $\bar\lambda$ is the "predominant" value of λ. Then we find that

$$E_y \simeq -\frac{i\mu_0\omega I_0}{2\pi}[K_0(\gamma_0 R_0) + R_\perp(\bar\lambda) \, K_0(\gamma_0 R_1)] \tag{4.238}$$

where

$$R_1 = [x^2 + (z+h)^2]^{1/2}$$

As indicated in Figure 4.21, the angle $\bar\theta$ is the local angle of incidence for

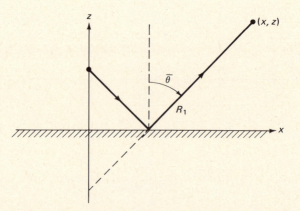

Figure 4.21 Geometry for calculating reflected field from line source over layered structure.

the reflected or secondary field. If we note that $\bar{\lambda} = -i\lambda_0 \sin\bar{\theta} = \beta \sin\bar{\theta}$ then $R_\perp(\bar{\lambda})$ is the local reflection coefficient. The geometry of the situation indicates that

$$\sin\bar{\theta} = \frac{x}{R_1} \tag{4.239}$$

A more rigorous development, using the saddle-point method [3], indicates that (4.238) is indeed a valid approximation in an asymptotic sense where $|\gamma_0 R_1| \gg 1$ and $\bar{\theta}$ is not too near grazing (that is, not near $\pi/2$). Thus in place of (4.231), we can write

$$E_y \simeq -\frac{I_0 i\mu_0\omega}{2\pi}\left(\frac{\pi}{2\gamma_0}\right)^{1/2}\left[\frac{e^{-\gamma_0 R_0}}{R_0^{1/2}} + R_\perp(\bar{\lambda})\frac{e^{-\gamma_0 R_1}}{R_1^{1/2}}\right] \tag{4.240}$$

where

$$i\bar{\lambda} = \gamma_0 \sin\bar{\theta} \tag{4.241}$$

If, in addition, $R_0 \gg 2h$, and dealing specifically with the lossless case (that is, $\gamma_0 = i\beta$), we have

$$E_y \simeq -\frac{I_0\mu_0\omega e^{i\pi/4}}{2(2\pi\beta R_0)^{1/2}} e^{-i\beta R_0}P(\hat{\theta}) \tag{4.242}$$

where

$$P(\hat{\theta}) = 1 + R_\perp(\beta\sin\hat{\theta})e^{-2i\beta h\cos\theta} \tag{4.243}$$

In this case we note that $\bar{\lambda} = \beta\sin\hat{\theta}$, where $\hat{\theta}$ is the angle subtended by R_0 and the z axis (that is, as shown in Figure 4.19).

The function $|P(\hat{\theta})|$ can be regarded as the radiation pattern of the line source at a height h over the layered structure. The value of the reflection coefficient $R_\perp(\beta\sin\hat{\theta})$ needed for the pattern calculation is given explicitly by (4.202), with λ replaced by $\beta\sin\hat{\theta}$.

The foregoing asymptotic development can be applied to the transmitted field. To illustrate this, we will simplify the situation by assuming that the top and bottom regions are both free space, whence $\sigma_0 = 0$ and $\sigma_M = 0$, $\epsilon_M = \epsilon_0$, and $\mu_M = \mu_0$. Also, we assume that the observer at (x, z) is at a large distance from the layer stack of thickness L. Specifically, we assume that

$$\beta R_0 \gg 1 \quad \text{and} \quad R_0 \gg h + L$$

The situation is illustrated in Figure 4.22. Now we are able to approximate the integral representation in the following manner:

$$E_y = -\frac{i\mu_0\omega I_0}{4\pi}\int_{-\infty}^{+\infty}[T_\perp(\lambda)e^{u_0 L}]\frac{1}{u_0}e^{-\mu_0(h-z)}e^{-i\lambda x}\,d\lambda \tag{4.244}$$

or

$$E_y \simeq -\frac{i\mu_0\omega I_0}{2\pi}T_\perp(\beta\sin\theta_0)e^{i\beta L\cos\theta_0}\left(\frac{\pi}{2i\beta R_0}\right)^{1/2}e^{-i\beta R_0} \tag{4.245}$$

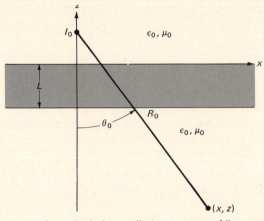

Figure 4.22 Geometry for calculating radiation pattern of line source on far side of layered structure.

where $R_0 = [(h - z)^2 + r^2]^{1/2}$ and $\sin \theta_0 = x/R_0$. Again, a more thorough analytical development of the integral shows this result is a valid approximation for the far field in the sense that terms containing $(\beta R_0)^{-3/2}$, $(\beta R_0)^{-5/2}$, and so on are neglected [3]. Also, the observer should not be near the layer stack (that is, θ_0 not near $\pi/2$). The pattern is now characterized by the function $|T_\perp(\beta \sin \theta_0)|$, which can be computed from (4.204) with λ replaced by $\beta \sin(\pi - \theta_0) = \beta \sin \theta_0$ and u_0 by $i\beta \cos \theta_0$.

4.16 FIELDS OF LONG-SLIT OR -SLOT ANTENNA

An interesting and important problem is to determine the influence of the immediate environment of an antenna on it's radiation pattern. An idealized representation is an infinitely long slit cut in a perfectly conducting sheet that is covered with a layered medium. The slit is excited on the bottom side by a voltage V_0, which is assumed here not to vary along the slit.

The situation is illustrated in Figure 4.23 where the perfectly conducting ground plane of infinite extent is the surface $z = 0$. The slit is located at $x = x_0$ and is parallel to the y axis. The region $0 < z < \ell_0$ is taken to be free space with properties ϵ_0 and μ_0. A conductive slab of conductivity σ, permittivity ϵ, and permeability μ is bounded by $z = \ell_0$ and $z = \ell + \ell_0$. In the external region (that is, $z > \ell + \ell_0$), the free-space conditions prevail.

The fields excited by such a slot source will obviously have $E_y = 0$, and the only nonvanishing field components are E_x, E_z, and H_y. Maxwell's equations for such a transverse magnetic (TM) configuration, in

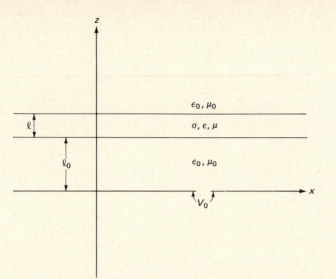

Figure 4.23 Slot excited by voltage V_0 in ground plane with overhead conductive slab.

the slab region, show readily that

$$\left(\frac{\partial^2}{\partial x^2} + \frac{\partial^2}{\partial z^2} - \gamma^2\right)H_y = 0 \tag{4.246}$$

where

$$\gamma = [i\mu\omega(\sigma + i\epsilon\omega)]^{1/2}$$

and

$$E_x = -\frac{1}{\sigma + i\epsilon\omega}\frac{\partial H_y}{\partial z} \tag{4.247}$$

and

$$E_z = \frac{1}{\sigma + i\epsilon\omega}\frac{\partial H_y}{\partial x} \tag{4.248}$$

The same equations apply in the free-space regions if $\sigma + i\epsilon\omega$ is replaced by $i\epsilon_0\omega$ and $i\mu\omega$ is replaced by $i\mu_0\omega$.

Now for the region $0 < z < \ell_0$

$$\left(\frac{\partial^2}{\partial x^2} + \frac{\partial^2}{\partial z^2} - \gamma_0^2\right)H_y = 0 \tag{4.249}$$

where $\gamma_0 = i\beta = i(\epsilon_0\mu_0)^{1/2}\omega$. Thus solutions are of the form $\exp(\pm u_0 z)$, where $u_0 = (\lambda^2 + \gamma_0^2)^{1/2}$. In view of our earlier experience we are thus led to write

$$H_y = \int_{-\infty}^{+\infty} [a(\lambda)e^{u_0 z} + b(\lambda)e^{-u_0 z}]e^{-i\lambda x} \, d\lambda \tag{4.250}$$

as a suitable integral representation for the field in the region $0 < z < \ell_0$. As yet, $a(\lambda)$ and $b(\lambda)$ are not known. But our condition at the ground plane tells us that

$$E_x \bigg]_{z=0} = V_0 \, \delta(x - x_0) = \frac{V_0}{2\pi} \int_{-\infty}^{+\infty} e^{-i\lambda(x - x_0)} \, d\lambda \tag{4.251}$$

But also,

$$E_x \bigg]_{z=0} = -(i\epsilon_0 \omega)^{-1} \frac{\partial H_y}{\partial z} \bigg]_{z=0}$$
$$= -\int_{-\infty}^{+\infty} [a(\lambda) - b(\lambda)]K_0 e^{-i\lambda x} \, d\lambda \tag{4.252}$$

where $K_0 = u_0/(i\epsilon_0 \omega)$. Thus

$$-a + b = \frac{V_0}{2\pi K_0} e^{i\lambda x_0} \tag{4.253}$$

Another relationship between a and b is found by imposing the impedance condition at $z = \ell_0$. This amounts to saying that in λ-transform space

$$\left[\frac{\hat{E}_x(\lambda)}{\hat{H}_y(\lambda)} \right]_{z=\ell_0} = Z(\lambda) \tag{4.254}$$

where

$$Z = K \frac{K_0 + K \tanh u\ell}{K + K_0 \tanh u\ell} \tag{4.255}$$

and

$$K = \frac{u}{\sigma + i\epsilon\omega}$$

and

$$u = (\lambda^2 + \gamma^2)^{1/2}$$

Here we note that

$$E_x(x) = \int_{-\infty}^{+\infty} \hat{E}_x(\lambda)e^{-i\lambda x} d\lambda \tag{4.256}$$

and

$$H_y(x) = \int_{-\infty}^{+\infty} \hat{H}_y(\lambda)e^{-i\lambda x} \, d\lambda \tag{4.257}$$

define the transforms $\hat{E}_x$ and $\hat{H}_y$. At $z = \ell_0$ we have

$$\hat{E}_x(\lambda) = -K_0(ae^{u_0\ell_0} - be^{-u_0\ell_0}) \tag{4.258}$$

and

$$\hat{H}_y(\lambda) = a^{u_0\ell_0} + be^{-u_0\ell_0} \tag{4.259}$$

and

$$-K_0 \frac{a - be^{-2u_0\ell_0}}{a + be^{-2u_0\ell_0}} = Z \tag{4.260}$$

or

$$\frac{a}{b} = \frac{K_0 - Z}{K_0 + Z} e^{-2u_0\ell_0} \tag{4.261}$$

Combining this with (4.253), we get

$$a = \frac{V_0}{2\pi K_0} \frac{\hat{R}e^{-2u_0\ell_0}e^{i\lambda x_0}}{1 - \hat{R}e^{-2u_0\ell_0}} \tag{4.262}$$

and

$$b = \frac{V_0}{2\pi K_0} \frac{e^{i\lambda x_0}}{1 - \hat{R}e^{-2u_0\ell_0}} \tag{4.263}$$

where

$$\hat{R} = \frac{K_0 - Z}{K_0 + Z} \tag{4.264}$$

It is also useful to note that

$$a + b = \frac{V_0e^{i\lambda x_0}}{2\pi K_0} \frac{1 + \hat{R}e^{-2u_0\ell_0}}{1 - \hat{R}e^{-2u_0\ell_0}} = \frac{V_0}{2\pi} \frac{e^{i\lambda x_0}}{Z_0(\lambda)} \tag{4.265}$$

where $Z_0(\lambda)$ can be identified as the upward-looking impedance at the surface $z = 0$. Thus we can write

$$H_y\Big]_{z=0} = \frac{V_0}{2\pi} \int_{-\infty}^{+\infty} \frac{1}{Z_0(\lambda)} e^{-i\lambda(x-x_0)} \, d\lambda \tag{4.266}$$

This result is certainly consistent with (4.251) for the tangential electric field at this surface. In this case, where $0 < z < \ell_0$, we have

$$H_y = \frac{V_0}{2\pi} \int_{-\infty}^{+\infty} \frac{1}{K_0} \frac{e^{-u_0z} + \hat{R}e^{u_0(z-2\ell_0)}}{1 - \hat{R}e^{-2u_0\ell_0}} e^{-i\lambda(x-x_0)} \, d\lambda \tag{4.267}$$

In a straightforward manner we also find that for $z > \ell_0 + \ell$

$$H_y = \frac{V_0}{2\pi} \int_{-\infty}^{+\infty} \frac{1}{K_0} \frac{e^{+u_0\ell}T_0}{1 - \hat{R}e^{-2u_0\ell_0}} e^{-u_0z}e^{-i\lambda(x-x_0)} \, d\lambda \tag{4.268}$$

where

$$T_0 = \frac{4KK_0}{(K + K_0)^2} e^{-u_0\ell} \left[1 - \left(\frac{K - K_0}{K + K_0} \right)^2 e^{-2u\ell} \right]^{-1} \tag{4.269}$$

is the transmission coefficient across the slab of thickness ℓ. Here we see that H_y can be written in the form

$$H_y = \frac{i\epsilon_0 \omega V_0}{2\pi} \int_{-\infty}^{+\infty} f(\lambda) \frac{1}{u_0} e^{-u_0 z} e^{-i\lambda x} \, d\lambda \tag{4.270}$$

where

$$f(\lambda) = \frac{T_0(\lambda) e^{u_0\ell}}{1 - \hat{R}e^{-2u_0\ell_0}} e^{i\lambda x_0} \tag{4.271}$$

The far-field asymptotic approximation is achieved by taking $f(\lambda)$ outside the integral and replacing it by $f(\bar{\lambda})$, where $\bar{\lambda} = -i\gamma \sin \theta_s = \beta \sin \theta_s$. Here $\sin \theta_s = x/R$, where $R = [(x - x_0^2) + z^2]^{1/2}$. Thus our final approximation is

$$H_y \simeq \frac{i\epsilon_0 \omega V_0}{2\pi} f(\beta \sin \theta_s) \left(\frac{\pi}{2i\beta R} \right)^{1/2} e^{-i\beta R} e^{i\beta \sin\theta_s x_0} \tag{4.272}$$

The result is valid only if $\beta R \gg 1$ and θ_s is not near $\pi/2$.

The radiation pattern showing how H_y varies with θ_s in the far field is characterized by the function $|f(\beta \sin \theta_s)|$.

4.17 FIELD BETWEEN SLAB AND GROUND PLANE

We now devote some attention to the field behavior in the region between the ground plane and the overhead slab. The relevant expression is (4.267). First of all, we expand the denominator in the fashion

$$\frac{1}{1 - \hat{R}e^{-2u_0\ell}} = \sum_{n=0}^{\infty} \hat{R}^n e^{-2nu_0\ell} \tag{4.273}$$

which is valid if $|\hat{R}e^{-2u_0\ell}| < 1$. Then (4.267) can be written

$$H_y = \frac{i\epsilon_0 \omega V_0}{2\pi} \sum_{n=0}^{\infty} (H_n^u + H_n^d) \tag{4.274}$$

where

$$H_n^u = \int_{-\infty}^{+\infty} \frac{1}{u_0} \hat{R}^n e^{-u_0(z + 2n\ell_0)} e^{-i\lambda(x - x_0)} \, d\lambda \tag{4.275}$$

and

$$H_n^d = \int_{-\infty}^{+\infty} \frac{1}{u_0} \hat{R}^{n+1} e^{u_0[z - 2(n+1)\ell_0]} e^{-i\lambda(x - x_0)} d\lambda \tag{4.276}$$

To interpret the terms H_n^u and H_n^d, we assume that $R(\lambda)$ can be approximated by $R(\lambda_n)$, where λ_n is the predominant value of λ. The value for H_n^u will be designated λ_n^u, and the one for H_n^d will be designated λ_n^d. Then the integrals are approximated by

$$H_n^u \simeq \hat{R}^n(\lambda_n^u)\left(\frac{\pi}{2i\beta r_n^u}\right)^{1/2} e^{-i\beta r_n^u} \tag{4.277}$$

and

$$H_n^d \simeq \hat{R}^{n+1}(\lambda_n^d)\left(\frac{\pi}{2i\beta r_n^d}\right)^{1/2} e^{-i\beta r_n^d} \tag{4.278}$$

These are valid provided βr_n^u and βr_n^d are both much greater than 1. The distances r_n^u and r_n^d are defined by

$$r_n^u = [(x - x_0)^2 + (z + 2n\ell_0)^2]^{1/2} \tag{4.279}$$

and

$$r_n^d = \{(x - x_0)^2 + [z - 2(n + 1)\ell_0]^2\}^{1/2} \tag{4.280}$$

We now can interpret our result in a geometrical-optical sense. The terms in H_n^u can be identified as up-going rays that emanate from the source and image locations at $(x_0, 0)$, $(x_0, -2\ell_0)$, $(x_0, -4\ell_0)$, ... On the other hand, the terms in H_n^d can be identified as down-going rays that emanate from image locations at $(x_0, 2\ell_0)$, $(x_0, 4\ell_0)$,. . . . The situation is illustrated in Figure 4.24(a). Actually, we can regard these rays as discrete reflections from the horizontal interfaces at $z = 0$ and $z = \ell_0$. This viewpoint is illustrated in Figure 4.24(b), where only the H_1^u and H_0^d rays are shown. In this case of H_1^u the local angle of incidence θ_1^u, at upper boundary $z = \ell_0$, is defined by

$$\sin \theta_1^u = \frac{x - x_0}{r_1^u}$$

and the corresponding λ value is

$$\lambda_1^u = \beta \sin \theta_1^u$$

In fact, in general, we can assert that

$$\lambda_n^u = \beta \sin \theta_n^u \tag{4.281}$$

where

$$\sin \theta_n^u = \frac{x - x_0}{r_n^u}$$

and

$$\lambda_n^d = \beta \sin \theta_n^d$$

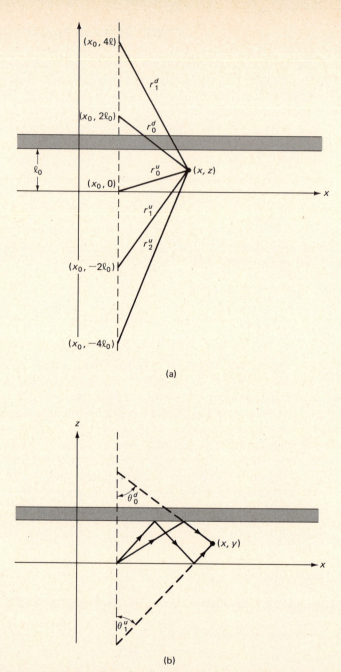

Figure 4.24(a) Image locations for source on lower boundary; (b) up-going ray H_1^u and down-going ray H_0^d that have both suffered one reflection at upper boundary, showing local angles of incidence.

where

$$\sin \theta_n^d = \frac{x - x_0}{r_n^u} \tag{4.282}$$

It should be stressed that the asymptotic forms for H_n^u and H_n^d given by (4.277) and (4.278) are valid only if the energy does not penetrate appreciably into the conductive slab. Ideally, the conductivity σ should be large compared with $\epsilon_0 \omega$. In the limiting case where $\sigma \to \infty$, the reflection coefficient R becomes 1 and the geometrical-optical series is formally correct. However, the series is poorly convergent except at "short distances" where $x - x_0$ is comparable with $2\ell_0$. In radio propagation investigations the geometrical rays are called wave hops [3] since they are reflected successively by the ionosphere and the ground surface. Actually, the validity and usefulness of the wave hop concept can be extended if the integral representations for H_n^u and H_n^d given by (4.275) and (4.276), respectively, are employed in their exact form. In such a case it is necessary to insert the required functional form for $R(\lambda)$ before evaluating the integrals.

4.18 WAVEGUIDE REPRESENTATION

Another representation for the fields in the region $0 < z < \ell_0$ is possible. To illustrate the method that leads to waveguide modes, we assume that the conductivity σ of the bounding slab is sufficiently large that R can be replaced by 1. Thus the field H_y is given by

$$H_y = \frac{i\epsilon_0 \omega V_0}{2\pi} \int_{-\infty}^{+\infty} \frac{1}{u_0} \frac{e^{-u_0 z} + e^{+u_0(z - 2\ell_0)}}{1 - e^{-2u_0 \ell_0}} e^{-i\lambda(x - x_0)} \, d\lambda \tag{4.283}$$

The integrand in this case has poles in the complex λ plane, where

$$1 - e^{-2u_0 \ell_0} = 0 \tag{4.284}$$

or where

$$2u_0 \ell_0 = i2m\pi$$

where $m = 0, 1, 2, \ldots$. This result is equivalent to

$$(\lambda^2 - \beta^2)\ell_0^2 = -m^2 \pi^2 \tag{4.285}$$

These values of λ, designated λ_m, locate the poles. Explicitly, we write

$$\lambda_m = \pm \left(\beta^2 - \frac{m^2 \pi^2}{\ell_0^2} \right)^{1/2} \tag{4.286}$$

for

$$\beta^2 > \frac{m^2 \pi^2}{\ell_0^2}$$

and

$$\lambda_m = \mp i \left(\frac{m^2 \pi^2}{\ell_0^2} - \beta^2 \right)^{1/2} \tag{4.287}$$

for

$$\beta^2 < \frac{m^2 \pi^2}{\ell_0^2}$$

In locating the poles in the complex λ plane, we regard $\beta = -i\gamma_0$ to have a vanishingly small but nonzero, positive imaginary part. Then it is not difficult to show that λ_m are either in the second or the fourth quadrant. The complex λ plane is shown in Figure 4.25, where the integration contour is along the entire real axis. The pole locations, which are symmetrical about the origin, are also shown. A finite number of these are very near the real axis. An infinite number are very near the imaginary axis, and these are symmetrical about the real axis. It can be shown for the present problem that no other singularities occur in the λ plane in spite of the factor u_0 (but this is not true, in general).

We now apply residue calculus to obtain an exact series formula for the field H_y at any point (x, z) within the interval $0 < z < \ell_0$. Initially, we assume $x - x_0 > 0$. The principal step is to close the integration contour along the real axis by a semicircle of radius R_ℓ in the lower half plane, as indicated in Figure 4.25. Because of the factor $\exp[i\lambda(x - x_0)]$ in the integrand, the contribution along this contour vanishes as $R_\ell \to \infty$. This result is Jordan's lemma, and it can be proved rigorously for a problem of

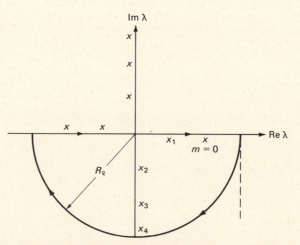

Figure 4.25 Complex λ plane showing pole singularities, original contour along real axis, and closing semicircle of infinite radius.

this type. The consequence is that the integration contour, which initially was along the real axis of λ only, can be replaced by a *closed* contour that is made up of the real axis and the infinite semicircle in the lower half plane. Such a *closed* contour encloses or "captures" all the poles that are given by

$$\lambda_m = + \left(\beta^2 - \frac{\pi^2 m^2}{\ell_0^2} \right)^{1/2}$$

when

$$\beta > \frac{\pi m}{\ell_0}$$

and

$$\lambda_m = -i \left(-\beta^2 + \frac{\pi^2 m^2}{\ell_0^2} \right)^{1/2}$$

when

$$\beta < \frac{\pi m}{\ell_0}$$

These are actually the poles shown in Figure 4.25 that lie in the fourth quadrant.

The next step is to invoke Cauchy's theorem, which tells us that the integral around the closed contour, in the clockwise sense, is $-2\pi i$ times the sum of the residues of the enclosed poles. Thus

$$H_y = \epsilon_0 \omega V_0 \sum_{m=0}^{\infty} \frac{(e^{-\bar{u}_0 z} + e^{\bar{u}_0(z - 2\ell_0)}) e^{-i\lambda_m(x - x_0)}}{(\partial/\partial\lambda)(1 - e^{-2u_0\ell_0}) u_0|_{\lambda = \lambda_m}} \tag{4.288}$$

where $\bar{u}_0 = im\pi/\ell_0$ is the value of u_0 at $\lambda = \lambda_m$. Now we note that

$$-\frac{\partial}{\partial\lambda} e^{-2u_0\ell_0} = 2\ell_0 e^{-2u_0\ell_0} \frac{\partial u_0}{\partial\lambda} = \frac{2\ell_0\lambda}{u_0} e^{-2u_0\ell_0} \tag{4.289}$$

Then we find that

$$H_y = \frac{\epsilon_0 \omega V_0}{\ell_0} \sum_{m=0}^{\infty} \frac{\delta_m}{\lambda_m} \cos \frac{m\pi z}{\ell_0} e^{-i\lambda_m(x - x_0)} \tag{4.290}$$

where $\delta_0 = \frac{1}{2}$ and $\delta_m = 1$ *for* $m = 1, 2, 3, \ldots$. This equation is a waveguide-mode representation for the field in the region $0 < z < \ell_0$, which is valid for $x > x_0$. For the region $z < x_0$ we merely change the exponential term to $\exp[+i\lambda_m(x - x_0)]$.

It is useful to note that the above result is consistent with a normal-mode formulation. We outline this approach briefly. Here we would postulate an expansion of the form

$$H_y = \sum_{m=0}^{\infty} A_m \cos \frac{m\pi z}{\ell_0} e^{-i\lambda_m(x-x_0)} \tag{4.291}$$

for $x > x_0$. Then

$$E_z = \frac{1}{i\epsilon_0 \omega} \frac{\partial H_y}{\partial x} = \frac{-1}{\epsilon_0 \omega} \sum_{m=0}^{\infty} A_m \cos \frac{m\pi z}{\ell_0} \lambda_m e^{-i\lambda_m(x-x_0)} \tag{4.292}$$

An equivalent statement of the source condition is that

$$E_z]_{x=x_0} = -\frac{V_0}{2} \delta(z) \tag{4.293}$$

Thus A_m is to be determined from

$$\frac{V_0}{2} \delta(z) = \frac{1}{\epsilon_0 \omega} \sum_{m=0}^{\infty} A_m \lambda_m \cos \frac{m\pi z}{\ell_0} \tag{4.294}$$

We now multiply both sides by $\cos (m'\pi z/\ell_0)$ and integrate with respect to z over the interval 0 to ℓ_0. Then on noting the orthogonality property, we have

$$\int_0^{\ell_0} \cos \frac{m\pi z}{\ell_0} \cos \frac{m'\pi z}{\ell_0} \, dz = \begin{cases} \ell_0 & \text{if } m = m' = 0 \\ \dfrac{\ell_0}{2} & \text{if } m = m' \neq 0 \\ 0 & \text{if } m \neq m' \end{cases} \tag{4.295}$$

We readily deduce that

$$A_m = \frac{\epsilon_0 \omega V_0}{\ell_0} \delta_m \tag{4.296}$$

which is consistent with (4.290).

4.19 FURTHER WORKED EXAMPLES

Exercise: Deduce the fields of a narrow slot or slit of infinite length in a perfectly conducting ground plane. The voltage V_0 across the slot is specified to be a constant. The region above the slot is assumed to be free space (see Figure 4.26).

SOLUTION: Method I — Choose rectangular coordinates (x, y, z) with the ground plane defined by $y = 0$ and the slot coincident with the z axis. The objective is to determine the fields for $y > 0$. Now, obviously, the

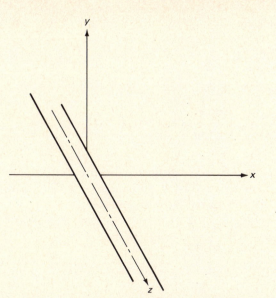

Figure 4.26   Slot aperture in perfectly conducting ground plane at $y = 0$.

magnetic field has only the component H_z, which satisfies

$$\left(\frac{\partial^2}{\partial x^2} + \frac{\partial^2}{\partial y^2} - \gamma_0^2\right) H_z = 0 \tag{4.297}$$

where $\gamma_0 = i(\epsilon_0\mu_0)^{1/2}\omega = i\omega/c$. A suitable form of the solution is

$$H_z = \int_{-\infty}^{+\infty} b(\lambda)e^{-u_0 y}e^{-i\lambda x}\, d\lambda \tag{4.298}$$

where $u_0 = (\lambda^2 + \gamma_0^2)^{1/2}$ and $b(\lambda)$ is to be determined. The corresponding electric field components are obtained from

$$E_x = +(i\epsilon_0\omega)^{-1}\frac{\partial H_z}{\partial y} \tag{4.299}$$

and

$$E_y = -(i\epsilon_0\omega)^{-1}\frac{\partial H_z}{\partial x} \tag{4.300}$$

The source condition is

$$E_x\big|_{y=0} = -V_0\,\delta(x) = -\frac{V_0}{2\pi}\int_{-\infty}^{+\infty} e^{-i\lambda x}\, d\lambda \tag{4.301}$$

But note that

$$E_x\big|_{y=0} = \frac{-1}{i\epsilon_0\omega}\int_{-\infty}^{+\infty} u_0 b(\lambda)e^{-i\lambda x}\, d\lambda \tag{4.302}$$

Then we see that

$$b(\lambda) = \frac{i\epsilon_0 \omega V_0}{2\pi u_0}$$

Thus the desired solution for $y > 0$ is

$$H_z = \frac{i\epsilon_0 \omega}{2\pi} V_0 \int_{-\infty}^{+\infty} \frac{1}{u_0} e^{-u_0 y} e^{-i\lambda x} \, d\lambda \tag{4.303}$$

SOLUTION: Method II — We choose cylindrical coordinates (ρ, ϕ, z) with the ground plane defined by $\phi = 0$ and $\phi = \pi$. The slot is located to be coincident with the z axis. Now H_z satisfies

$$\left[\frac{1}{\rho} \frac{\partial}{\partial \rho} \left(\rho \frac{\partial}{\partial \rho} \right) + \frac{1}{\rho^2} \frac{\partial^2}{\partial \phi^2} - \gamma_0^2 \right] H_z = 0 \tag{4.304}$$

In this system a suitable form of the solution is

$$H_z = \sum_{m=0}^{\infty} A_m K_m(\gamma_0 \rho) \cos m\phi \tag{4.305}$$

where $K_m(\gamma_0 \rho)$ is the modified Bessel function of order m and argument $\gamma_0 \rho$. The electric field components are

$$E_\rho = (i\epsilon_0 \omega)^{-1} \rho^{-1} \frac{\partial H_z}{\partial \phi} \tag{4.306}$$

and

$$E_\phi = -(i\epsilon_0 \omega)^{-1} \frac{\partial H_z}{\partial \rho} \tag{4.307}$$

We see that $E_\rho = 0$ for all $\rho > 0$ when $\phi = 0$ or π. Now

$$E_\phi = -\frac{\gamma_0}{i\epsilon_0 \omega} \sum_{m=0}^{\infty} A_m K_m'(\gamma_0 \rho) \cos m\phi \tag{4.308}$$

The source condition can now be written

$$\lim_{\rho \to 0} \left\{ \int_0^\pi E_\phi \rho \, d\phi \right\} = V_0 \tag{4.309}$$

Only the $m = 0$ term survives:

$$\lim_{\rho \to 0} \left[\frac{-\gamma_0}{i\epsilon_0 \omega} A_0 K_0'(\gamma_0 \rho) \pi \rho \right] = V_0 \tag{4.310}$$

or

$$\frac{\gamma_0 A_0 \pi}{i\epsilon_0 \omega \gamma_0} = V_0$$

Thus we find that

$$H_z = \frac{i\epsilon_0 \omega}{\pi} V_0 K_0(\gamma_0 \rho) \tag{4.311}$$

On equating (4.303) and (4.311), we establish the integral identity:

$$\frac{1}{2} \int_{-\infty}^{+\infty} \frac{1}{u_0} e^{-u_0 y} e^{-i\lambda x} \, d\lambda = K_0[\gamma_0(x^2 + y^2)^{1/2}] \tag{4.312}$$

which is known in the theory of Bessel functions.

Exercise: Extend the above solution to the case where the slit is of finite width, using method I.

SOLUTION: Now we assume our boundary condition has the form

$$E_x]_{y=0} = \begin{cases} -\dfrac{V_0}{2w} & \text{for } -w < x < w \\ 0 & \text{otherwise} \end{cases} \tag{4.313}$$

or

$$E_x]_{y=0} = \int_{-\infty}^{+\infty} f(\lambda) e^{-i\lambda x} \, d\lambda \qquad \text{for all } x \tag{4.314}$$

Thus

$$f(\lambda) = -\frac{1}{2\pi} \int_{-w}^{w} \frac{V_0}{2w} e^{+i\lambda x} \, dx = -\frac{V_0}{2\pi} \frac{\sin \lambda w}{\lambda w} \tag{4.315}$$

Then we easily deduce that for $y > 0$

$$H_z = \frac{i\epsilon_0 w}{2\pi} V_0 \int_{-\infty}^{+\infty} \frac{1}{u_0} \frac{\sin \lambda w}{\lambda w} e^{-u_0 y} e^{-i\lambda x} d\lambda \tag{4.316}$$

As $w \to 0$, we recover the appropriate expression for the slot of negligible width. (That is, $(\lambda w)^{-1} \sin \lambda w \to 1$ as $w \to 0$.)

Exercise: Given the expression for the fields of an infinitely long slot antenna in the ground plane, deduce the resultant fields of two such slots separated by a distance $2s$. Assume that the two slots are excited by constant voltages V_1 and V_2. (See Figure 4.27.)

SOLUTION: We choose rectangular coordinates (x, y, z) with the ground plane defined by $y = 0$ and the slots located at $x = \pm s$. By simple

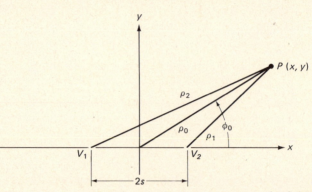

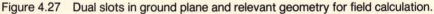

Figure 4.27   Dual slots in ground plane and relevant geometry for field calculation.

superposition we immediately write, for the region $y > 0$, that

$$H_z = \frac{i\epsilon_0\omega}{\pi} \left(V_1 K_0(\gamma_0\rho_1) + V_2 K_0(\gamma_0\rho_2)\right] \tag{4.317}$$

where

$$\rho_1 = [(x-s)^2 + y^2]^{1/2} \qquad \rho_2 = [(x+s)^2 + y^2]^{1/2}$$

Exercise: For the geometry of the previous problem, ascertain the radiation pattern of the two-slot array and discuss the limiting case when the observer recedes to infinity.

SOLUTION: Set $\gamma_0 = i\beta$ where β is real. Then consider $\beta\rho_1$ and $\beta\rho_2 \gg 1$. The leading term in the asymptotic form for the modified Bessel function is then adequate. Thus, we are led to write

$$H_z = \frac{i\epsilon_0\omega}{\pi} \left(\frac{\pi}{2i\beta}\right)^{1/2} \frac{e^{-i\beta\rho_0}}{\rho_0^{1/2}} \left[V_1 e^{-i\beta(\rho_1-\rho_0)}\left(\frac{\rho_0}{\rho_1}\right)^{1/2}\right.$$
$$\left. + V_2 e^{-i\beta(\rho_2-\rho_0)}\left(\frac{\rho_0}{\rho_2}\right)^{1/2}\right] \tag{4.318}$$

where

$$\rho_0 = (x^2 + y^2)^{1/2}$$

If an angle ϕ_0 is defined by $\cos\phi_0 = x/\rho_0$ we see that the ϕ_0 dependence of the square bracket term in the previous equation is by definition the radiation pattern. It also depends on the range ρ_0 in general. But there is further simplification if $\rho_0 \gg s$ whence

$$\left(\frac{\rho_0}{\rho_1}\right)^{1/2} = (\rho_0/\rho_1)^{1/2} \cong 1$$

and

$$\rho_1 - \rho_0 \cong -s \cos\phi_0, \, \rho_2 - \rho_0 \cong +s \cos\phi_0$$

Then it follows that

$$H_z \cong i\epsilon_0\omega \left(\frac{i}{2\pi\beta}\right)^{1/2} \frac{e^{-i\beta\rho_0}}{\rho_0^{1/2}} V_1 \, F\left(\frac{V_2}{V_1}, \phi_0\right) \tag{4.319}$$

where now the radiation pattern

$$F\left(\frac{V_2}{V_1}, \phi_0\right) = e^{+\beta s i \cos\phi_0} + \frac{V_2}{V_1} e^{-i\beta s \cos\phi_0} \tag{4.320}$$

does not depend on range. In the limiting case where $V_1 = V_2$ we have simply that

$$F(1, \phi_0) = 2 \cos(\beta s \cos \phi_0) \tag{4.321}$$

which exhibits the expected maximum in the end-fire direction ($\phi_0 = 0°$) when $\beta s = m\pi$, where m is an integer (that is, when the spacing $2s = m$ times the free-space wavelength). The broadside radiation (at $\phi_0 = 90°$) clearly always has a maximum for this case.

Exercise: Consider an array of $M + 1$ such slots located successively at x_0, $x_1, x_2, \ldots, x_M$ in the ground plane $y = 0$ and fed by voltages $V_0, V_1, V_2, \ldots, V_M$. Show that the far-field radiation pattern is given by

$$F(\phi_0) = \sum_{m=0}^{M} \frac{V_m}{V_0} e^{-\beta s i \cos\phi_0} \tag{4.322}$$

where we have normalized everywhere by the reference-slot voltage at $x = x_0 = 0$.

Exercise: As a further extension, consider the following problem. A voltage-excited slot is located in a vertical, perfectly conducting sheet of assumed infinite height, and a horizontal ground plane is located beneath. The situation is shown in Figure 4.28 where rectangular coordinates are chosen such that the slot is located at $x = 0$ and $z = h$ with the perfectly conducting ground plane now defined by $z = 0$. The objective is to determine the fields everywhere within the right-angled sector defined by $x > 0$, $z > 0$, bearing in mind that the problem is invariant in the y direction (that is, $\partial/\partial y = 0$).

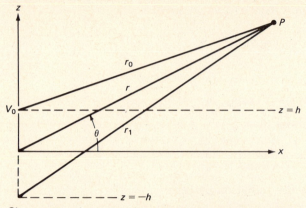

Figure 4.28 Slot aperture in vertical face of right-angle-sector region.

SOLUTION: We are led to write the solution as the sum of the primary field of the slot and an image slot located at $x = 0$ and $z = -h$. Thus our "guess" is

$$H_y = V_0 \frac{i\epsilon_0\omega}{\pi} \left[K_0(i\beta r_0) + K_0(i\beta r_1) \right] \tag{4.323}$$

where

$$r_0 = [x^2 + (z - h)^2]^{1/2}$$

and

$$r_1 = [x^2 + (z + h)^2]^{1/2}$$

We readily confirm this expression is indeed the right form by noting that

$$E_x = -(i\epsilon_0\omega)^{-1} \frac{\partial H_y}{\partial z} \to 0 \qquad \text{at } z = 0 \tag{4.324}$$

We now examine the far-field radiation field, where βr_0 and $\beta r_1 \gg 1$ and $r_0, r_1 \gg h$. Then we find without difficulty that

$$H_y = V_0\epsilon_0\omega \left(\frac{i\pi}{2\beta} \right)^{1/2} \frac{e^{-i\beta r}}{r^{1/2}} \left(e^{i\beta h\sin\theta} + e^{-i\beta h\sin\theta} \right) \tag{4.325}$$

or if we are just interested in the magnitude of the field, we see that

$$|H_y| = |V_0|\epsilon_0\omega \left(\frac{\pi}{2\beta} \right)^{1/2} \frac{2}{r^{1/2}} |\cos(\beta h \sin\theta)| \tag{4.326}$$

This expression always has a maximum along the ground plane at $\theta_0 \to 0$.

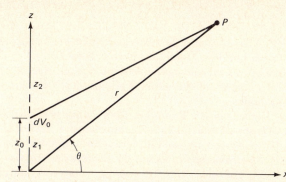

Figure 4.29 Aperture of finite vertical extent in vertical face of right-angle-sector region.

Exercise: Now consider a vertical aperture extending from $z = z_1$ to z_2 in the vertical sheet rather than just a slot. The situation is depicted in Figure 4.29. We wish to determine the field at $P(r, \theta)$ for a specified electric field $\hat{E}_z(z_0)$ within the aperture.

SOLUTION: We consider that the source is the integrated effect of an infinite number of narrow slots with incremental voltage dV_0. Now $dV_0 = -\hat{E}_z(z_0) \, dz_0$ for a slot of width dz_0 located at $z = z_0$. Thus we are led to write for the far field

$$H_y = -\epsilon_0 \omega \left(\frac{i\pi}{2\beta} \right)^{1/2} \frac{e^{-i\beta r}}{r^{1/2}} \int_{z_1}^{z_2} E(z_0)(e^{+i\beta z_0 \sin\theta} + e^{-i\beta z_0 \sin\theta}) \, dz_0 \qquad (4.327)$$

An alternative form is

$$H_y = -\epsilon_0 \omega \left(\frac{2i\pi}{\beta} \right)^{1/2} \frac{e^{-i\beta r}}{r^{1/2}} \int_0^{\infty} e(z_0) \cos(\bar{\lambda} z_0) \, dz_0 \qquad (4.328)$$

where $\bar{\lambda} = \beta \sin \theta$ and where

$$e(z_0) = \begin{cases} \hat{E}(z_0) & \text{for } z_1 < z < z_2 \\ 0 & \text{for } 0 < z < z_1 \text{ and } z > z_2 \end{cases}$$

Now the problem of synthesis is to determine the excitation $e(z_0)$ that gives the radiation pattern function

$$P(\lambda) = \int_0^{\infty} e(z_0) \cos \lambda z_0 \, dz_0 \qquad (4.329)$$

In principle, the desired form is obtained by noting that $e(z_0)$ is the cosine Fourier transform of $P(\lambda)$. Thus

$$e(z_0) = \frac{2}{\pi} \int_0^{\infty} P(\lambda) \cos \lambda z_0 \, d\lambda \qquad (4.330)$$

References

1. J. C. Maxwell: *A Treatise on Electricity and Magnetism,* unabridged 3d ed., Clarendon Press, London, 1891 (also Dover, New York).
2. S. A. Schelkunoff: *Electromagnetic Fields,* Blaisdell, New York, 1963. Includes a good summary of the general telegraphist's equation for dealing with coupled modes.
3. J. R. Wait: *Electromagnetic Radiation from Cylindrical Structures,* Pergamon Press, Elmsford, N.Y., 1959. Comprehensive review and developments of radiation scattering and diffraction for two-dimensional problems.
4. W. L. Weeks: *Electromagnetic Theory for Engineering Applications,* Wiley, New York, 1964. This older text still contains some very useful methods of solving separable boundary-value problems with an engineering slant.
5. M. Kline, ed.: *The Theory of Electromagnetic Waves,* Dover, New York, 1965. Contains a collection of electromagnetic classics from a 1950 conference held at the Courant Institute at New York University.
6. L. B. Felsen and N. Marcuvitz: *Radiation and Scattering of Waves,* Prentice-Hall, Englewood Cliffs, N.J., 1973. This is a unified formalism for "systematic eigenmode and transmission line analyses of linear field problems" with large doses of demanding function-theoretic methods, but the effort is rewarding.
7. J. A. Kong: *Theory of Electromagnetic Waves,* Wiley, New York, 1975. A formidable and admirable treatise on macroscopic electrodynamics developed from relativistic principles. This is not for the fainthearted.
8. P. L. E. Uslenghi, ed.: *Electromagnetic Scattering,* Academic Press, New York, 1978. Tightly edited collection of authorative papers covering recent advances in diffraction and scattering from material media.
9. J. R. Wait: *Wave Propagation Theory,* Pergamon Press, Elmsford, N.Y., 1981. Mostly collected reprints with revisions and supplemental material.
10. C. A. Balanis: *Antenna Theory — Analysis and Design,* Harper & Row, New York, 1982. This lavishly illustrated volume is the most comprehensive and the most clearly presented of the latest crop of antenna texts.
11. L. C. Shen and J. A. Kong: *Applied Electromagnetism,* Brooks/Cole, Monterey, Calif., 1983. This is the latest and maybe the best of the current crop of junior-level electromagnetic texts with many applications to the real world.

Appendix A. Cylindrical Electromagnetic Waves

In the foregoing chapter we have restricted attention to problems in planar geometry. Also, the constitutive media have been assumed to be isotropic.

A.1 STATEMENT OF PROBLEM

Our purpose here is to set forth a general analysis for the electromagnetic fields that can be induced in an anisotropic cylindrical structure of infinite length with a circular cross section. The procedure to be employed bears some similarity to the analyses of isotropic cylindrical structures [1]. We simplify the problem to some extent by choosing uniaxial forms for the conductivity and permeability tensors with their principal axes to be taken parallel to the axis of the cylinder.

In our discussion below we will call attention to the relevance of these results to the nondestructive testing of wire ropes and cables that have a cylindrical geometry, but the internal structure may be anisotropic [1 – 6].

A.2 FORMULATION

To be specific, we chose a cylindrical coordinate system (ρ, ϕ, z) such that the surface of the cylinder is $\rho = a$, where a is the radius. The external region $\rho > a$ will contain the sources, but for the time being we will restrict our attention to the internal region $\rho < a$.

Maxwell's equations for the internal region, for a time factor $\exp(i\omega t)$, are

$$(\sigma)\mathbf{E} = \text{curl } \mathbf{H} \tag{A.1}$$

$$-i\omega(\mu)\mathbf{H} = \text{curl } \mathbf{E} \tag{A.2}$$

where $\mathbf{E}$ and $\mathbf{H}$ are the vector electric and magnetic fields, respectively, and (σ) and (μ) are the tensor conductivity and permeability, respectively. In view of our stated assumptions we may write

$$(\sigma) = \begin{pmatrix} \sigma_\rho & 0 & 0 \\ 0 & \sigma_\phi & 0 \\ 0 & 0 & \sigma_z \end{pmatrix} \tag{A.3}$$

and

$$(\mu) = \begin{pmatrix} \mu_\rho & 0 & 0 \\ 0 & \mu_\phi & 0 \\ 0 & 0 & \mu_z \end{pmatrix} \tag{A.4}$$

Here it should be noted that σ_ρ, σ_ϕ, and σ_z are scalar complex conductivities; thus, for example, $\sigma_\rho = g_\rho + i\omega\epsilon_\rho$, where g_ρ and ϵ_ρ are the real conductivity and the real permittivity, respectively, in the ρ direction. Similarly, we could write $\mu_\rho = \mu_\rho' - i\mu_\rho''$, where μ_ρ' and μ_ρ'' are both positive real. However, for virtually all applications in Nondestructive Testing, the elements of (σ) and (μ) can be regarded as real since the imaginary parts are entirely negligible. A more important limitation is that we restrict attention to time-harmonic fields of sufficiently small magnitude that the elements of (σ) and (μ) do not vary with the field magnitude (that is, we are within the linear regime).

Equations (A.1) and (A.2), expressed in cylindrical coordinates, are

$$\frac{1}{\rho}\frac{\partial E_z}{\partial \phi} - \frac{\partial E_\phi}{\partial z} = -i\omega\mu_\rho H_\rho \tag{A.5}$$

$$\frac{\partial E_\rho}{\partial z} - \frac{\partial E_z}{\partial \rho} = -i\omega\mu_\phi H_\phi \tag{A.6}$$

$$\frac{1}{\rho}\frac{\partial}{\partial \rho}(\rho E_\phi) - \frac{1}{\rho}\frac{\partial E_\rho}{\partial \phi} = -i\omega\mu_z H_z \tag{A.7}$$

$$\frac{1}{\rho}\frac{\partial H_z}{\partial \phi} - \frac{\partial H_\phi}{\partial z} = \sigma_\rho E_\rho \tag{A.8}$$

$$\frac{\partial H_\rho}{\partial z} - \frac{\partial H_z}{\partial \rho} = \sigma_\phi E_\phi \tag{A.9}$$

$$\frac{1}{\rho} \frac{\partial}{\partial \rho} (\rho H_\phi) - \frac{1}{\rho} \frac{\partial H_\rho}{\partial \phi} = \sigma_z E_z \tag{A.10}$$

Leading up to later developments, we now assume that the field components vary according to $\exp(-im\phi)\exp(-i\lambda z)$. For single-valuedness m is restricted to positive or negative integers, including zero. The parameter λ, as yet, is unrestricted. Maxwell's equations are now simplified to

$$-\frac{im}{\rho} E_z + i\lambda E_\rho = -i\omega\mu_\rho H_\phi \tag{A.11}$$

$$-i\lambda E_\rho - \frac{\partial E_z}{\partial \rho} = -i\omega\mu_\phi H_\phi \tag{A.12}$$

$$\frac{1}{\rho} \frac{\partial}{\partial \rho} (\rho E_\phi) + \frac{im}{\rho} E_\rho = -i\omega\mu_z H_z \tag{A.13}$$

$$-\frac{im}{\rho} H_z + i\lambda H_\phi = \sigma_\rho E_\rho \tag{A.14}$$

$$-i\lambda H_\rho - \frac{\partial H_z}{\partial \rho} = \sigma_\phi E_\phi \tag{A.15}$$

$$\frac{1}{\rho} \frac{\partial}{\partial \rho} (\rho H_\phi) + \frac{im}{\rho} H_\rho = \sigma_z E_z \tag{A.16}$$

To proceed further, we can now decompose or decouple the above set of equations into two cases. In the first case we set $H_z = 0$, whence it is a simple matter to deduce that E_z satisfies

$$\frac{1}{\rho} \frac{\partial}{\partial \rho} \rho \frac{\partial}{\partial \rho} E_z - \frac{\sigma_\phi}{\sigma_\rho} \frac{m^2}{\rho^2} \frac{\lambda^2 + i\sigma_\rho\mu_\phi\omega}{\lambda^2 + i\sigma_\phi\mu_\rho\omega} E_z$$
$$- (\lambda^2 + i\sigma_\rho\mu_\phi\omega) \frac{\sigma_z}{\sigma_\rho} E_z = 0 \tag{A.17}$$

The other case is when we set $E_z = 0$, whence H_z is found to satisfy

$$\frac{1}{\rho} \frac{\partial}{\partial \rho} \rho \frac{\partial}{\partial \rho} H_z - \frac{\mu_\phi}{\mu_\rho} \frac{m^2}{\rho^2} \frac{\lambda^2 + i\sigma_\phi\mu_\rho\omega}{\lambda^2 + i\sigma_\rho\mu_\phi\omega} H_z$$
$$- (\lambda^2 + i\sigma_\phi\mu_\rho\omega) \frac{\mu_z}{\mu_\rho} H_z = 0 \tag{A.18}$$

Now, clearly, the total field is the superposition of these two cases. In fact, it is not difficult to show that

$$E_\rho = \frac{1}{(\lambda^2 + i\sigma_\rho\mu_\phi\omega)} \left(i\lambda \frac{\partial E_z}{\partial \rho} + \frac{m\omega\mu_\phi}{\rho} H_z \right) \tag{A.19}$$

$$E_\phi = \frac{1}{(\lambda^2 + i\sigma_\phi\mu_\rho\omega)} \left(\frac{\lambda m}{\rho} E_z - i\omega\mu_\rho \frac{\partial H_z}{\partial \rho} \right) \qquad (A.20)$$

$$H_\rho = \frac{1}{(\lambda^2 + i\sigma_\phi\mu_\rho\omega)} \left(im\sigma_\phi E_z/\rho + i\lambda \frac{\partial H_z}{\partial \rho} \right) \qquad (A.21)$$

$$H_\phi = \frac{1}{(\lambda^2 + i\sigma_\rho\mu_\phi\omega)} \left(\sigma_\rho \frac{\partial E_z}{\partial \rho} + \frac{\lambda m}{\rho} H_z \right) \qquad (A.22)$$

A.3 GENERAL SOLUTION AND ITS PROPERTIES

Solutions of (A.17) and (A.18) are modified Bessel functions. In fact, for fields that are finite at $\rho = 0$, we can verify that

$$E_z = f_{e,m}(\lambda) I_{\alpha_m}(u\rho) e^{-im\phi} e^{-i\lambda z} \qquad (A.23)$$

$$H_z = f_{h,m}(\lambda) I_{\beta_m}(v\rho) e^{-im\phi} e^{-i\lambda z} \qquad (A.24)$$

where

$$u^2 = \frac{(\lambda^2 + i\sigma_\rho\mu_\phi\omega)\sigma_\rho}{\sigma_\rho} \qquad (A.25)$$

$$v^2 = \frac{(\lambda^2 + i\mu_\rho\sigma_\phi\omega)\mu_z}{\mu_\rho} \qquad (A.26)$$

$$\alpha_m^2 = m^2 \frac{\sigma_\phi}{\sigma_\rho} \frac{\lambda^2 + i\sigma_\rho\mu_\phi\omega}{\lambda^2 + i\sigma_\phi\mu_\rho\omega} \qquad (A.27)$$

$$\beta_m^2 = m^2 \frac{\mu_\phi}{\mu_\rho} \frac{\lambda^2 + i\sigma_\phi\mu_\rho\omega}{\lambda^2 + i\sigma_\rho\mu_\phi\omega} \qquad (A.28)$$

and where $f_{e,m}(\lambda)$ and $f_{h,m}(\lambda)$ are unspecified functions of λ and m.

Without actually solving anything, we can at this stage draw a number of important inferences about the eddy-current testing of wire ropes and cables. When we impress an axial electric current, we expect, of course, that the axial electric field in the cable will be dominant. In fact, if the axial variation of the field is small (that is, λ is small), the forms of the solutions simplify. Then

$$E_z = f_{e,m}(\lambda) I_{\hat{m}}(\hat{u}\rho) e^{-im\phi} \qquad (A.29)$$

where $\hat{u} = (i\sigma_z\mu_\phi\omega)^{1/2}$ and $\hat{m} = m(\mu_\phi/\mu_\rho)^{1/2}$. Furthermore, since H_z is negligible in such cases, E_ρ and E_ϕ are also small. The induced currents and the secondary external fields are thus predominantly a function of the axial electrical conductivity σ_z and the azimuthal magnetic permeability μ_ϕ. However, the higher-order harmonics (that is, $m > 0$) are influenced to some extent by the radial magnetic permeability μ_ρ.

Another comparable situation is when the impressed magnetic field is axial and uniform. Then

$$H_z = f_{h,m}(\lambda)I_{\overline{m}}(\overline{v}\rho)e^{-im\phi} \tag{A.30}$$

where $\overline{v} = (i\mu_z\sigma_\phi\omega)^{1/2}$ and $\overline{m} = m(\sigma_\phi/\sigma_\rho)^{1/2}$. Now the significant induced currents are azimuthal and driven by the azimuthal electric field

$$E_\phi = -\frac{1}{\sigma_\phi}\frac{\partial H_z}{\partial\rho} \tag{A.31}$$

Clearly, in this case the secondary fields are mainly a function of the axial magnetic permeability μ_z and the azimuthal electric conductivity σ_ϕ. Analogously, the higher-order harmonics are influenced by the radial electric conductivity σ_ρ.

The two cases considered above can be described as the E-field or the H-field method of excitation. Actually, it is the latter H-field configuration that is the basis for nearly all existing methods of eddy-current testing of cables and wire ropes. In the case of the so-called dc technique the external excitation is a solenoid that carries a large azimuthal current. The secondary H field is then heavily influenced by the axial magnetic permeability. On the other hand, in the so-called ac method the secondary field results from the induced eddy currents in the cable. These are primarily a function of the azimuthal conductivity σ_ϕ in addition to the axial magnetic permeability μ_z.

The E-field configuration would arise if the cable were excited by a toroidal or doughnut-shaped coil [4]. The resulting induced currents now flow principally in the axial direction. In general, in this case the finite value of λ needs to be considered. Nevertheless, the secondary fields are now mainly a function of the axial conductivity σ_z and the azimuthal permeability μ_ϕ.

The situation generally becomes quite complicated when the effects of the nonzero values of λ and m are considered. The resulting fields then become hybrid [3], being neither purely E or H. In such cases the secondary fields will depend, in general, on all tensor elements of the conductivity and permeability. Then it appears that quantitative calculations must be resorted to in order to gather further insight.

A.4 CASE OF AXIAL SYMMETRY

In the following discussion we will restrict attention to azimuthally uniform excitation such that all the harmonics for $m \neq 0$ may be discarded. We will consider, however, fields that have a significant variation in the axial direction. Now the field components are decomposed into two sets. For the E-field type

$$E_z = f_e(\lambda)I_0(u\rho)e^{-i\lambda z} \tag{A.32}$$

$$H_\phi = \frac{\sigma_\rho u f_e(\lambda)}{\lambda^2 + i\omega\sigma_\rho\mu_\phi} I_1(u\rho)e^{-i\lambda z} \tag{A.33}$$

and

$$E_\rho = \frac{i\lambda}{\sigma_\rho} H_\phi \tag{A.34}$$

For the H-field type we would have

$$H_z = f_h(\lambda)I_0(v\rho)e^{-i\lambda z} \tag{A.35}$$

$$E_\phi = \frac{-i\omega\mu_\rho v f_h(\lambda)}{\lambda^2 + i\omega\mu_\rho\sigma_\phi}I_1(v\rho)e^{-i\lambda z} \tag{A.36}$$

and

$$H_\rho = \frac{-\lambda}{\mu_\rho\omega} E_\phi \tag{A.37}$$

Two important physical parameters are the axial series impedance $Z_s(\lambda)$ and the axial series admittance $Y_s(\lambda)$ of the cable. These are defined and given as follows:

$$Z_s(\lambda) = \frac{E_z}{2\pi a H_\phi}\bigg]_{\rho=a} = \frac{(\lambda^2 + i\omega\sigma_\rho\mu_\phi)I_0(ua)}{2\pi a\sigma_\rho u I_1(ua)} \tag{A.38}$$

and

$$Y_s(\lambda) = -\frac{H_z}{2\pi a E_\phi}\bigg]_{\rho=a} = \frac{(\lambda^2 + i\omega\mu_\rho\sigma_\phi)I_0(va)}{2\pi a i\omega\mu_\rho v I_1(va)} \tag{A.39}$$

Two limiting cases follow immediately: (1) $|ua|$ and $|va| \ll 1$, and then we obtain the quasi-static forms

$$Z_s(\lambda) \simeq \frac{1}{\pi a^2 \sigma_z} \tag{A.40}$$

and

$$Y_s(\lambda) \simeq \frac{1}{\pi a^2 i\omega\mu_z} \tag{A.41}$$

and (2) $|ua|$ and $|va| \gg 1$, and then we obtain the asymptotic high-frequency forms

$$Z_s(\lambda) \simeq \left(\frac{i\omega\mu_\phi}{\sigma_z}\right)^{1/2}\frac{1}{2\pi a} \tag{A.42}$$

and

$$Y_s(\lambda) \simeq \left(\frac{\sigma_\phi}{i\omega\mu_z}\right)^{1/2}\frac{1}{2\pi a} \tag{A.43}$$

These limiting cases have the expected dependencies on the electric properties. Not surprisingly, the static impedance (that is, resistance) depends only on the axial conductivity. The analog admittance depends in the same way on the axial permeability.

A.5 THE EXCITATION PROBLEM

We now consider the excitation problem. The external source is taken to be a solenoid of radius b of finite axial extent that is concentric with the cable. The situation is illustrated in Figure 4.A.1. The surface electric current density in the solenoid, in the present idealization, has both an axial and an azimuthal current density defined by

$$j_z(z) = j_0 \sin \psi \qquad \text{for } -\frac{\ell}{2} < z < \frac{\ell}{2} \tag{A.44}$$

and

$$j_\phi(z) = j_0 \cos \psi \qquad \text{for } -\frac{\ell}{2} < z < \frac{\ell}{2} \tag{A.45}$$

where ψ is a pitch angle of the current flow. Here one should note that $\psi = 0$ corresponds to purely azimuthal current flow, which is the condition approached in a solenoid of many turns. In general, for a pitched winding the axial component of the current flow should also be

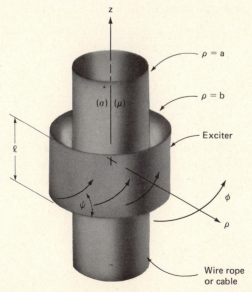

Figure 4.A.1 Anisotropic wire rope or cable excited by concentric current sheet of finite axial extent.

accounted for. The corresponding boundary or initial conditions for the tangential magnetic fields, according to Ampere's law, must be

$$H_z(b^-, z) - H_z(b^+, z) = j_\phi(z) \tag{A.46}$$

and

$$H_\phi(b^-, z) - H_\phi(b^+, z) = -j_z(z) \tag{A.47}$$

where $b^\pm = $ limit of $b \pm \Delta$ when the positive quantity Δ tends to zero.

For the same excitation we also have continuity of the tangential electric field. Thus

$$E_z(b^-, z) - E_z(b^+, z) = 0 \tag{A.48}$$

$$E_\phi(b^-, z) - E_\phi(b^+, z) = 0 \tag{A.49}$$

We now use the Fourier integral representation

$$j_\phi(z) = \int_{-\infty}^{+\infty} f(\lambda) e^{-i\lambda z} \, d\lambda \tag{A.50}$$

where it follows that

$$f(\lambda) = \frac{1}{2\pi} \int_{-\infty}^{+\infty} j_\phi(z) e^{i\lambda z} \, dz = \frac{1}{2\pi} (j_0 \cos \psi) \int_{-\ell/2}^{\ell/2} e^{i\lambda z} \, dz$$
$$= (j_0 \cos \psi)(\pi\lambda)^{-1} \sin \frac{\lambda \ell}{2} \tag{A.51}$$

The desired representations are thus

$$j_\phi(z) = \frac{j_0 \cos \psi}{\pi} \int_{-\infty}^{+\infty} \frac{\sin(\lambda\ell/2)}{\lambda} e^{-i\lambda z} \, d\lambda \tag{A.52}$$

and

$$j_z(z) = \frac{j_0 \sin \psi}{\pi} \int_{-\infty}^{+\infty} \frac{\sin(\lambda\ell/2)}{\lambda} e^{-i\lambda z} \, d\lambda \tag{A.53}$$

In the free-space region $\rho > a$, it is convenient to derive the fields from electric and magnetic Hertz vectors [3] that have only z components U and V, respectively. The fields are thus obtained from

$$E_\rho = \frac{\partial^2 U}{\partial \rho \, \partial z} \tag{A.54}$$

$$H_\rho = \frac{\partial^2 V}{\partial \rho \, \partial z} \tag{A.55}$$

$$E_z = \left(k_0^2 + \frac{\partial^2}{\partial z^2} \right) U \tag{A.56}$$

$$H_z = \left(k_0^2 + \frac{\partial^2}{\partial z^2}\right)V \tag{A.57}$$

$$H_\phi = -i\epsilon_0\omega\,\frac{\partial U}{\partial \rho} \tag{A.58}$$

$$E_\phi = i\mu_0\omega\,\frac{\partial V}{\partial \rho} \tag{A.59}$$

where $k_0^2 = \epsilon_0\mu_0\omega^2$.

To be compatible with the selected source fields, we assume the respective forms, for $a < \rho < b$,

$$U = \int_{-\infty}^{+\infty} A^e(\lambda)[I_0(\beta\rho) + \delta_e(\lambda)K_0(\beta\rho)]e^{-i\lambda z}\,d\lambda \tag{A.60}$$

and

$$V = \int_{-\infty}^{+\infty} A^h(\lambda)[I_0(\beta\rho) + \delta_h(\lambda)K_0(\beta\rho)]e^{-i\lambda z}\,d\lambda \tag{A.61}$$

where $\beta^2 = \lambda^2 - k_0^2$ and A^e and A^h are functions of λ yet to be determined. Now the functions δ_e and δ_h can be determined immediately by imposing the impedance and admittance conditions at $\rho = a$ that are specified by (A.38) and (A.39). Then using (A.56), (A.57), (A.58), and (A.59), it is easy to show that

$$\delta_e(\lambda) = -\left[\frac{\beta I_0(\beta a) - 2\pi a Z_s(\lambda)i\epsilon_0\omega I_1(\beta a)}{\beta K_0(\beta a) + 2\pi a Z_s(\lambda)i\epsilon_0\omega K_1(\beta a)}\right] \tag{A.62}$$

and

$$\delta_h(\lambda) = -\left[\frac{\beta I_0(\beta a) - 2\pi a Y_s(\lambda)i\mu_0\omega I_1(\beta a)}{\beta K_0(\beta a) + 2\pi a Y_s(\lambda)i\mu_0\omega K_1(\beta a)}\right] \tag{A.63}$$

Now in the outer region $\rho > b$ the appropriate forms for the Hertz potentials are

$$U = \int_{-\infty}^{+\infty} C^e(\lambda)K_0(\beta\rho)e^{-i\lambda z}\,d\lambda \tag{A.64}$$

and

$$V = \int_{-\infty}^{+\infty} C^h(\lambda)K_0(\beta\rho)e^{-i\lambda z}\,d\lambda \tag{A.65}$$

because the fields must be noninfinite as $\rho \to \infty$.

The source conditions specified by (A.46) and (A.47) can now be applied to yield

$$A^e(\lambda) = \frac{+(j_0/\pi\lambda)\sin\psi\,\sin(\lambda\ell/2)}{i\epsilon_0\omega\beta[I_1 - (\delta_e + R_e)K_1]} \tag{A.66}$$

and

$$A^h(\lambda) = \frac{-(j_0/\pi\lambda)\cos\psi\,\sin(\lambda\ell/2)}{\beta^2[I_0 + (\delta_h - R_h)K_0]} \tag{A.67}$$

where

$$R_e = \frac{C^e}{A^e} = \frac{I_0 + \delta_e K_0}{K_0} \tag{A.68}$$

and

$$R_h = \frac{C^h}{A^h} = \frac{-(I_1 - \delta_h K_1)}{K_1} \tag{A.69}$$

In the above four equations the indicated modified Bessel functions are all functions of βh.

In principle, we have now solved the problem, since the resultant fields are now given in terms of the source current. The next step would be to deduce the pickup voltage in a sensor such as a small loop or an electric probe. This process is straightforward once we have specified the source and sensor configuration.

References

1. J. R. Wait: "Review of Electromagnetic Methods in Nondestructive Testing of Wire Ropes," *Proceedings of the IEEE,* vol. 47, 1979, pp. 892–903.
2. J. R. Wait and R. L. Gardner: "Electromagnetic Testing of Cylindrically Layered Conductors," *IEEE Transactions,* vol. IM–28, 1979, pp. 159–161.
3. D. A. Hill and J. R. Wait: "Electromagnetic Field Perturbation by an Internal Void in a Conducting Cylinder Excited by a Wire Loop," *Applied Physics* (Springer-Verlag), vol. 18, 1979, vol. 18, pp. 141–147.
4. J. R. Wait: "Electromagnetic Response of an Anisotropic Conducting Cylinder to an External Source," *Radio Science,* vol. 13, 1978, pp. 789–792.
5. J. R. Wait and D. A. Hill: "Electromagnetic Interaction Between a Conducting Cylinder and a Solenoid in Relative Motion," *Journal of Applied Physics,* vol. 50, 1979, pp. 5115–5119.
6. J. R. Wait and D. A. Hill: "Dynamic Electromagnetic Response of a Homogeneous Cylinder for Symmetric Excitation," *Applied Physics* (Springer-Verlag), vol. 20, 1979, pp. 89–96.

Appendix B. Power Flow and Radiation Resistance

Associated with time-varying electromagnetic fields is a flow of energy. This fact can be demonstrated as follows. We again write Maxwell's equations for vector fields $\mathbf{E}$ and $\mathbf{H}$ that vary harmonically with time according to $\exp{(i\omega t)}$:

$$\text{curl } \mathbf{E} = -i\mu\omega\mathbf{H} \tag{B.1}$$

and

$$\text{curl } \mathbf{H} = (\sigma + i\epsilon\omega)\mathbf{E} + \mathbf{J} \tag{B.2}$$

where σ, ϵ, and μ are the electrical properties of the medium and here $\mathbf{J}$ (without any subscript) denotes the impressed current density in the medium.

We now multiply the first Maxwell equation by the complex conjugate $\mathbf{H}^*$ and the complex conjugate of the second Maxwell equation by $\mathbf{E}$ to get

$$\mathbf{H}^* \cdot \text{curl } \mathbf{E} - \mathbf{E} \cdot \text{curl } \mathbf{H}^* = -\mathbf{E} \cdot \mathbf{J}^* - \sigma E \cdot \mathbf{E}^*$$
$$+ i\omega(\epsilon \mathbf{E} \cdot \mathbf{E}^* - \mu\mathbf{H} \cdot \mathbf{H}^*) \tag{B.3}$$

The left-hand side of this equation is $\text{div}(E \times H^*)$, and if this quantity is integrated over a volume V, Green's theorem can be applied to give

$$\int_V (\mathbf{H}^* \cdot \text{curl } \mathbf{E} - \mathbf{E} \cdot \text{curl } \mathbf{H}^*) \, dv = \int_S (\mathbf{E} \times \mathbf{H}^*)_n \, da \qquad \text{(B.4)}$$

where S is a surface enclosing V and $(\mathbf{E} \times \mathbf{H}^*)_n$ is the outward normal component of the vector $\mathbf{E} \times \mathbf{H}^*$ at the surface element da.

Rearranging terms, we can cast (B.3) into the form

$$-\frac{1}{2} \int_V \mathbf{E} \cdot \mathbf{J}^* \, dv = \frac{1}{2} \int_V \sigma \mathbf{E} \cdot \mathbf{E}^* \, dv$$

$$+ \frac{i\omega}{2} \int_V (\mu \mathbf{H} \cdot \mathbf{H}^* - \epsilon \mathbf{E} \cdot \mathbf{E}^*) \, dv$$

$$+ \frac{1}{2} \int_S (\mathbf{E} \times \mathbf{H}^*)_n \, da \qquad \text{(B.5)}$$

where we have divided through by 2. The real part of the left-hand side is the average power spent by the impressed forces to sustain the field. The first term on the right, which is real, is the rate of energy dissipation in the volume V. The second term is pure imaginary since ϵ and μ are assumed real. It can be identified as the stored or reactive mean energy inside V. The third and last term must then be the complex power crossing the surface S. Its real part is the power P transferred across S:

$$P = \frac{1}{2} \text{Re} \int_S (\mathbf{E} \times \mathbf{H}^*)_n \, da \qquad \text{(B.6)}$$

In the case of an antenna located in free space, the power supplied to the terminals must all be transmitted across an enclosing surface S if ohmic losses in the antenna can be neglected. In fact, the radiation resistance R is defined by

$$R = \frac{2P}{|I_0|^2} = \frac{2P}{I_0 I_0^*} \qquad \text{(B.7)}$$

where I_0 is the (complex) current at the terminals. Therefore

$$|I_0|^2 R = \text{Re} \int_S (\mathbf{E} \times \mathbf{H}^*)_n \, da \qquad \text{(B.8)}$$

We can also define a radiation conductance G by $G = 2PV_0 V_0^*$, where V_0 is the (complex) voltage at the antenna terminals. Therefore for this analogous situation

$$|V_0|^2 G = \text{Re} \int_S (\mathbf{E} \times \mathbf{H}^*)_n \, dz \qquad \text{(B.9)}$$

Chapter 5
Guided Cylindrical Waves

5.1 INTRODUCTION

There are a broad class of problems where we seek solutions for electromagnetic waves that propagate along or within cylindrical structures. Some examples are coaxial cables, tunnels, bore holes, hollow circular waveguides, and dielectric rods or fibers, just to name a few. The common objective is to determine the axial propagation constant Γ that characterizes the transmission of the mode in question.

While we can certainly formulate the problem in a general context for any cylindrical geometry, it is preferable to deal at the outset with specific cases. This is the approach that we will adopt.

5.2 THE IDEAL COAXIAL CABLE

Probably the best known conveyor of electromagnetic waves is the coaxial cable. The theory of wave transmission in the ideal case is very simple, as we will indicate.

With reference to a cylindrical coordinate system (ρ, ϕ, z), we define two reference concentric boundaries by $\rho = a$ and $\rho = b$, where

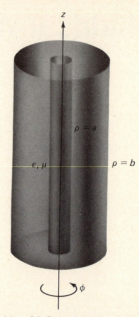

Figure 5.1   Ideal coaxial cable of infinite length with perfectly conducting walls.

$b > a$. The dielectric region $a < \rho < b$, for $0 < \phi < 2\pi$ and $-\infty < z < \infty$, is taken to be homogeneous with electrical properties (ϵ, μ). Initially, we will assume that the boundary surfaces are perfectly conducting so that the tangential electric fields E_ϕ and E_z vanish at $\rho = a$ and at $\rho = b$. The situation is illustrated in Figure 5.1, where we show only a section of the concentric cable that is of infinite length.

We will make a further simplifying assumption at this stage. Specifically, we say that the fields within the cable do not vary in the azimuthal or ϕ direction. Thus Maxwell's equations for this case are

$$-i\mu\omega H_\rho = -\frac{\partial E_\phi}{\partial z} \tag{5.1}$$

$$-i\mu\omega H_\phi = \frac{\partial E_\rho}{\partial z} - \frac{\partial E_z}{\partial \rho} \tag{5.2}$$

$$-i\mu\omega\rho H_z = \frac{\partial}{\partial \rho}(\rho E_\phi) \tag{5.3}$$

$$i\epsilon\omega E_\rho = -\frac{\partial H_\phi}{\partial z} \tag{5.4}$$

$$i\epsilon\omega E_\phi = \frac{\partial H_\rho}{\partial z} - \frac{\partial H_z}{\partial \rho} \tag{5.5}$$

$$i\epsilon\omega\rho E_z = \frac{\partial}{\partial \rho}(\rho H_\phi) \tag{5.6}$$

By inspection, we see that the fields break into two independent sets. In one case using (5.1), (5.3), and (5.5), we find that H_ρ and H_z are derivable from E_ϕ, while in the other case using (5.2), (5.4), and (5.6), E_ρ and E_z are derivable from H_ϕ. Also, it is a simple matter to show that both E_ϕ and H_ϕ satisfy the same equation:

$$i\epsilon\omega E_\phi = \frac{1}{i\mu\omega}\left\{\frac{\partial^2 E_\phi}{\partial z^2} + \frac{\partial}{\partial\rho}\left[\frac{1}{\rho}\frac{\partial}{\partial\rho}(\rho E_\phi)\right]\right\} \tag{5.7}$$

We now introduce Debye potentials U and V such that

$$H_\phi = -i\epsilon\omega\frac{\partial U}{\partial\rho} \tag{5.8}$$

and

$$E_\phi = i\mu\omega\frac{\partial V}{\partial\rho} \tag{5.9}$$

It is found that

$$(\nabla^2 - \gamma^2){}^U_V = 0 \tag{5.10}$$

where

$$\nabla^2 = \frac{1}{\rho}\frac{\partial}{\partial\rho}\left(\rho\frac{\partial}{\partial\rho}\right) + \frac{\partial^2}{\partial z^2} \tag{5.11}$$

and

$$\gamma^2 = (i\mu\omega)(i\epsilon\omega) = -\mu\epsilon\omega^2 = -k^2$$

The U and V functions are now operated on to give

$$E_\rho = \frac{\partial^2 U}{\partial\rho\,\partial z} \tag{5.12}$$

$$E_z = \left(-\gamma^2 + \frac{\partial^2}{\partial z^2}\right)U \tag{5.13}$$

and

$$H_\rho = \frac{\partial^2 V}{\partial\rho\,\partial z} \tag{5.14}$$

$$H_z = \left(-\gamma^2 + \frac{\partial^2}{\partial z^2}\right)V \tag{5.15}$$

The field components E_ρ, E_z, and H_ϕ, derivable from U, are called transverse magnetic (TM), because $\mathbf{H}$ is transverse to the direction of propagation. Correspondingly, the field components H_ρ, H_z, and E_ϕ, derivable from V, are called transverse electric (TE), because $\mathbf{E}$ is transverse to the direction of propagation.

In the present situation where $\partial/\partial\phi = 0$ (that is, azimuthal symmetry), it is clear that the TE and TM fields are uncoupled, so they may both exist without reference to each other.

We will deal with the TM configuration in an explicit fashion. Solutions of (5.10) have the form

$$U = f(\rho)\exp(\mp\Gamma z) \tag{5.16}$$

where $f(\rho)$ satisfies

$$\left[\frac{1}{\rho}\frac{\partial}{\partial\rho}\left(\rho\frac{\partial}{\partial\rho}\right) + u^2\right]f(\rho) = 0 \tag{5.17}$$

where

$$u^2 = k^2 + \Gamma^2$$

Now

$$f(\rho) = BJ_0(u\rho) + CY_0(u\rho) \qquad \text{for } a < \rho < b \tag{5.18}$$

where B and C are constants. We identify Γ as the resultant propagation in the z direction. Without loss of generality we may just deal with the form $\exp(-\Gamma z)$ if attention is to be focused on transmission in the positive z direction. Also, we require that Im $\Gamma > 0$.

We can now write the corresponding explicit forms of the field components:

$$E_z = u^2[BJ_0(u\rho) + CY_0(u\rho)] \tag{5.19}$$

$$H_\phi = -\frac{ik^2}{\mu\omega}\left[B\frac{\partial}{\partial\rho}J_0(u\rho) + C\frac{\partial}{\partial\rho}Y_0(u\rho)\right] \tag{5.20}$$

and

$$E_\rho = -\Gamma\left[B\frac{\partial}{\partial\rho}J_0(u\rho) + C\frac{\partial}{\partial\rho}Y_0(u\rho)\right] \tag{5.21}$$

where, for sake of brevity, the common factor $\exp(-\Gamma z + i\omega t)$ has been omitted.

The wall-boundary conditions are that E_z vanishes at both $\rho = a$ and $\rho = b$. But, in fact, $E_z = 0$ for all ρ in the range $a \leq \rho \leq b$ if $u^2 = 0$, whence $\Gamma = ik$. To deal with (5.20) and (5.21) in this limiting situation, we need to evaluate

$$\lim_{u\to 0}\left[B\frac{\partial}{\partial\rho}J_0(u\rho) + C\frac{\partial}{\partial\rho}Y_0(u\rho)\right]$$

$$= \lim_{u\to 0}[-uBJ_1(u\rho) - uCY_1(y\rho)]$$

$$= \lim_{u\to 0}\left[-uB\left(\frac{u\rho}{2}\right) + uC\left(\frac{2}{\pi u\rho}\right)\right] = \frac{2C}{\pi\rho} \tag{5.22}$$

Thus we find that

$$H_\phi = \frac{-2i\epsilon\omega C}{\pi\rho} \tag{5.23}$$

and

$$E_\rho = \frac{-2ikC}{\pi\rho} = \eta H_\phi \tag{5.24}$$

where

$$\eta = \left(\frac{\mu}{\epsilon}\right)^{1/2}$$

This particular configuration is called a transverse electromagnetic (TEM) mode because both the electric field E_ρ and the magnetic field H_ϕ are transverse to the direction of propagation [1].

The current I carried by the center conductor is obtained from Ampere's law, which can be applied to any closed path at $\rho =$ constant:

$$I = \oint \mathbf{H} \cdot \mathbf{ds} = 2\pi\rho H_\phi = -4i\epsilon\omega C \tag{5.25}$$

Thus

$$E_\rho = \frac{\eta I}{2\pi\rho} \tag{5.26}$$

We can also deduce the voltage between the inner and outer conductor by the equation

$$V = \int_a^b E_\rho \, d\rho = \frac{\eta}{2\pi} I \int_a^b \frac{1}{\rho} \, d\rho = \frac{\eta}{2\pi} I \ln \frac{b}{a} \tag{5.27}$$

The characteristic impedance K of the TEM mode is thus given by

$$K = \frac{V}{I} = \frac{\eta}{2\pi} \ln \frac{b}{a} \tag{5.28}$$

We can also write

$$K = \left(\frac{\hat{Z}}{\hat{Y}}\right)^{1/2} \quad \text{and} \quad \Gamma = (\hat{Z}\hat{Y})^{1/2}$$

if

$$\hat{Z} = \frac{i\mu\omega}{2\pi} \ln \frac{b}{a}$$

and

$$\hat{Y} = \frac{2\pi i\epsilon\omega}{\ln b/a}$$

where $\hat{Z}$ and $\hat{Y}$ can be interpreted as the series impedance and series admittance, respectively, per unit length.

Now there are other solutions of the equation $E_z = 0$ at $\rho = a$ and b. In fact, from (5.19) we see that the wall condition is satisfied if both

$$BJ_0(ua) + CY_0(ua) = 0 \tag{5.29}$$

and

$$BJ_0(ub) + CY_0(ub) = 0 \tag{5.30}$$

In order for both equations to yield the same value of B/C, clearly, we must have,

$$\frac{Y_0(ua)}{J_0(ua)} = \frac{Y_0(ub)}{J_0(ub)} \tag{5.31}$$

When the ratio is fixed, there are an infinite number of real solutions for u, which we designate u_n for $n = 1, 2, 3, \ldots$. The corresponding propagation constants are obtained from

$$\Gamma_n = (u_n^2 - k^2)^{1/2} = i(k^2 - u_n^2)^{1/2} \tag{5.32}$$

These are purely imaginary if $k > u_n$ and, as such, the associated fields are freely propagating waveguide modes. In the case where $u_n > k$, Γ_n are purely real, and the modes are heavily damped. When the radius b is much less than a wavelength (that is, $kb \ll 1$), the modes are still heavily attenuated except the **TEM** mode, which always propagates freely.

Exercise: Show that the TE modes of the above coaxial structure satisfy

$$\frac{Y_1(ua)}{J_1(ua)} = \frac{Y_1(ub)}{J_1(ub)} \tag{5.33}$$

These modes are always heavily attenuated if $kb \ll 1$.

5.3 GENERAL CABLE STRUCTURE

We now consider a more interesting cable structure. The situation is illustrated in Figure 5.2, where the cross section is shown. The inner conductor has radius a and it has electrical properties σ_w, ϵ_w, and μ_w. This conductor is covered by a concentric layer, of outer radius b, of a perfect dielectric with permittivity ϵ and permeability μ_0. To allow for the presence of a metal-braided shield, we locate a thin uniform sheath of radius b that has a designated transfer impedance Z_T (defined below). Surrounding this sheath we have a coating whose permittivity is ϵ_c; it is also assumed lossless and has a free-space permeability μ_0. Finally to account for the environmental influences, we have a thin outer layer of conductive material with a transfer impedance Z_L.

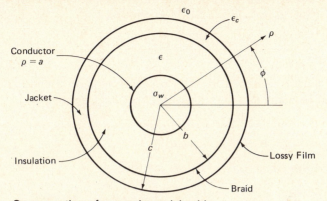

Figure 5.2   Cross section of general coaxial cable.

In spite of the complexity of the problem, the analysis is quite straightforward. We solve the problem in successive steps, beginning with the inner core. The objective is to obtain a viable expression for the effective series impedance of the cable structure as seen by an external observer. We also wish to determine the propagation constant of the cable.

As indicated, we deal with the homogeneous inner conductor first. Also, we again restrict attention to azimuthal symmetry, TM field configuration, and axial variations of the form $\exp(-\Gamma z)$. Then as a minor extension of (5.8), (5.12), and (5.14), we can write for the region $0 < \rho < a$

$$H_\phi = -(\sigma_w + i\epsilon_w\omega)\frac{\partial U_w}{\partial \rho} \tag{5.34}$$

$$E_\rho = -\Gamma\frac{\partial U_w}{\partial \rho} \tag{5.35}$$

$$E_z = (k_w^2 + \Gamma^2)U_w \tag{5.36}$$

where U_w satisfies

$$(\nabla^2 - \gamma_w^2)U_w = 0 \tag{5.37}$$

with

$$\gamma_w = [(i\mu_w\omega)(\sigma_w + i\epsilon_w\omega)]^{1/2} = ik_w$$

The appropriate form of the solution of (5.37) is $I_0[i(k_w^2 + \Gamma^2)^{1/2}\rho]$ for the region $0 < \rho < a$, where A is a constant. A useful definition is the axial impedance defined by

$$Z_i = \frac{E_z}{2\pi\rho H_\phi}\bigg]_{\rho=a} \tag{5.38}$$

Then on using (5.34) and (5.36), we see that

$$Z_i = \frac{i(k_w^2 + \Gamma^2)^{1/2}}{2\pi(\sigma_w + i\epsilon_w\omega)a} \frac{I_0[i(k_w^2 + \Gamma^2)^{1/2}a]}{I_1[i(k_w^2 + \Gamma^2)^{1/2}a]} \tag{5.39}$$

where we have used the identity $dI_0(x)/dx = I_1(x)$ for modified Bessel functions. In many cases of interest $|\Gamma^2| \ll |k_w^2|$, so that

$$Z_i \simeq \frac{(i\mu_w\omega)^{1/2}}{2\pi(\sigma_w + i\epsilon_w\omega)^{1/2}a} \frac{I_0(ik_wa)}{I_1(ik_wa)} \tag{5.40}$$

In the dc limit (that is, $\omega \to 0$) we see that

$$Z_i = (\pi a^2\sigma_w)^{-1} \tag{5.41}$$

which is the expected form. Another useful limit is when $|(k_w^2 + \Gamma^2)^{1/2}a| \gg 1$, in which case (5.39) simplifies to

$$Z_i \simeq \frac{i(k_w^2 + \Gamma^2)^{1/2}}{2\pi(\sigma_w + i\epsilon_w\omega)a} \tag{5.42}$$

and if, in addition $|\Gamma|^2 \ll |k_w^2|^2$,

$$Z_i \simeq \frac{\eta_w}{2\pi a} \tag{5.43}$$

where

$$\eta_w = \left(\frac{i\mu_w\omega}{\sigma_w + i\epsilon_w\omega}\right)^{1/2}$$

These limiting, high-frequency forms are valid for communication cables when the central conductor is copper or aluminum. But, in general, Z_i depends on the axial propagation constant Γ of the mode or wave. In such a case we refer to Z_i as being spatially dispersive.

We now wish to deal with the fields in the concentric regions external to the core. In each case the fields will vary as $\exp(i\omega t - \Gamma z)$, and azimuthal symmetry is again assumed. Now we deduce that

$$\left.\begin{aligned} E_z &= (k^2 + \Gamma^2)U = u^2U \\ H_\phi &= -i\epsilon\omega \frac{\partial U}{\partial \rho} \end{aligned}\right\} \text{ for } a < \rho < b \tag{5.44}$$

$$\left.\begin{aligned} E_z &= (k_c^2 + \Gamma^2)U_c = u_c^2U_c \\ H_\phi &= -i\epsilon_c\omega \frac{\partial U_c}{\partial \rho} \end{aligned}\right\} \text{ for } b < \rho < c \tag{5.45}$$

where U and U_c satisfy

$$(\nabla^2 + k^2)U = 0 \qquad (\nabla^2 + k_c^2)U_c = 0 \tag{5.46}$$

in the respective regions. Thus

$$\left.\begin{array}{l} E_z = u^2[PJ_0(u\rho) + QY_0(u\rho)] \\ H_\phi = i\epsilon\omega u[PJ_1(u\rho) + QY_1(u\rho)] \end{array}\right\} \text{ for } a < \rho < b \qquad (5.47)$$

and

$$\left.\begin{array}{l} E_{cz} = u_c^2[MJ_0(u_c\rho) + NY_0(u_c\rho)] \\ H_{c\phi} = i\epsilon_c\omega u_c[MJ_1(u_c\rho) + NY_1(u_c\rho)] \end{array}\right\} \text{ for } b < \rho < c \qquad (5.48)$$

Here P, Q, M, and N are constants yet to be determined. We also employ Bessel functions J and Y of arguments $u\rho$ and $u_c\rho$.

The boundary conditions of the problem may now be applied. With reference to (5.47) and (5.48) these can be written

$$E_z = 2\pi a Z_i H_\phi \qquad \text{at } \rho = a \qquad (5.49)$$

$$E_z = E_{cz} \qquad \text{at } \rho = b \qquad (5.50)$$

$$H_\phi - H_{c\phi} = -(2\pi b Z_T)^{-1} E_z \qquad \text{at } \rho = b \qquad (5.51)$$

$$E_{cz} = E_{0z} \qquad \text{at } \rho = c \qquad (5.52)$$

$$H_{c\phi} - H_{0\phi} = -(2\pi c Z_L)^{-1} E_z \qquad \text{at } \rho = c \qquad (5.53)$$

The tangential fields E_{0z} and $H_{0\phi}$ are the appropriate values at the outer surface of the cable. They are related by

$$Z(\Gamma) = \frac{E_{0z}}{2\pi c H_{0\phi}}\bigg|_{\rho=c} \qquad (5.54)$$

where $Z(\Gamma)$ is the effective surface impedance of the outer surface of the cable and it is a function of the axial propagation constant.

We see that (5.49) is equivalent to (5.38) where Z_i is defined by (5.39) without approximation. Equations (5.50) and (5.52) specify that the tangential electric fields are continuous at $\rho = b$ and $\rho = c$, respectively, as appropriate for thin sheaths. On the other hand, (5.51) and (5.53) indicate that tangential magnetic fields are discontinuous at $\rho = b$ and $\rho = c$ by the amount of current in the braid and conductive coating, respectively.

Now (5.49)–(5.53) may be applied to (5.47) and (5.48) to obtain a set of linear algebraic equations in the unknown coefficients. In particular, we may deduce an explicit result for the ratio N/M. The process can be somewhat simplified if we utilize the appropriate quasi-static forms for the fields within the dielectric and coating regions. To this end, we utilize the small argument approximations $J_0(x) \simeq 1$, $J_1(x) \simeq x/2$, $Y_0(x) \simeq (2/\pi)\ln 0.89x$, and $Y_1(x) \simeq -2/(\pi x)$. Thus

$$E_z \simeq u^2 \left(P + \frac{2}{\pi} Q \ln 0.89 \, u\rho \right)$$

$$H_\phi \simeq \frac{-2}{\pi} \frac{i\epsilon\omega Q}{\rho}$$

$$\left. \right\} \quad \text{for } a < \rho < b \qquad (5.55)$$

$$E_{cz} \simeq u_c^2 \left(M + \frac{2}{\pi} N \ln 0.89 \, u_c\rho \right)$$

$$H_{c\phi} \simeq -\frac{2}{\pi} \frac{i\epsilon_c\omega N}{\rho}$$

$$\left. \right\} \quad \text{for } b < \rho < c \qquad (5.56)$$

Certainly, these forms are justified because $kb \ll 1$ and $k_c c \ll 1$, but note that we do not invoke the small-argument approximation in the expression for Z_i that characterizes the central metal core.

When the boundary conditions are applied to (5.55) and (5.56), we obtain the following explicit result for the effective series impedance of the cable:

$$Z(\Gamma) = \frac{Z_L(Z_c + Z_b)}{Z_L + Z_c + Z_b} \qquad (5.57)$$

where

$$Z_b = \frac{Z_T(Z' + Z_i)}{Z_T + Z' + Z_i} \qquad (5.58)$$

and

$$Z' = -\frac{k^2 + \Gamma^2}{2\pi i\epsilon\omega} \ln \frac{b}{a} \qquad (5.59)$$

and

$$Z_c = \frac{k_c^2 + \Gamma^2}{2\pi i\epsilon_c\omega} \ln \frac{c}{b} \qquad (5.60)$$

The equivalent circuit for this situation is the ladder network shown in Figure 5.3. The terminating element Z_i is the impedance of the central-core conductor, while the shunt elements Z_T and Z_L are the transfer impedance of the braid and the lossy coating, respectively. The series elements Z' and Z_c can be identified as short uniform sections of transmission lines. Then, of course, $Z(\Gamma)$ is the input impedance of the network, as indicated in Figure 5.3.

This quasi-static analysis is amazingly versatile. In particular, we can employ (5.54) as an effective boundary condition when dealing with a cable placed in a more complicated environment. This point will be illustrated below in connection with radio transmission in tunnels. But we can also formulate the appropriate mode equation for guided waves in a direct fashion by using the ladder network representation described

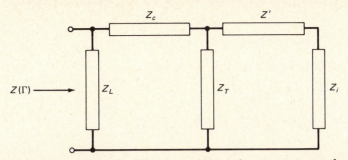

Figure 5.3 Equivalent circuit for effective axial impedance at outer surface of general coaxial cable structure.

above. For example, in the rather extreme case treated earlier, where the coaxial cable had perfectly conducting walls, we are essentially saying that Z_T and Z_i are zero. Then the appropriate mode equation must be $Z' = 0$, where Z' is given by (5.59). Clearly, this result yields the expected solution $\Gamma = ik$ for the TEM mode. Not surprisingly, the possible waveguide modes in the region are excluded because we had invoked the condition $kb \ll 1$ in writing the quasi-static forms.

5.4 SURFACE-WAVE TRANSMISSION LINE

If we wish to obtain a mode equation to predict the guided waves that will propagate along or within the composite cable, we need to say something about the external region $\rho > c$. In the simplest case we have a free-space environment with electrical properties (ϵ_0, μ_0). Then again restricting attention to azimuthal symmetry, the external solutions are written as follows:

$$E_{e\rho} = \frac{\partial^2 U_0}{\partial \rho \, \partial z} = -\Gamma \frac{\partial U_0}{\partial \rho} \tag{5.61}$$

$$E_{ez} = \left(k_0^2 + \frac{\partial^2}{\partial z^2} \right) U_0 = (k_0^2 + \Gamma^2) U_0 \tag{5.62}$$

$$H_{e\phi} = -i\epsilon_0 \omega \frac{\partial U_0}{\partial \rho} \tag{5.63}$$

where U_0 satisfies

$$(\nabla^2 + k_0^2) U_0 = 0 \tag{5.64}$$

Clearly, a solution of (5.64) that applies in the region $c < \rho < \infty$ is

$$H_0^{(2)}[(k_0^2 + \Gamma^2)^{1/2} \rho] \exp(-\Gamma z)$$

An equivalent form, more convenient for the present application, is

$$K_0[v\rho] \exp(-\Gamma z)$$

where

$$v = i(\Gamma^2 + k_0^2)^{1/2} = [(-\Gamma^2) - k_0^2]^{1/2} \tag{5.65}$$

Thus we write, for $\rho > c$,

$$E_{ez} = -v^2 A K_0(v\rho) \tag{5.66}$$

$$H_{ez} = -i\epsilon_0 \omega A \frac{\partial}{\partial \rho} K_0(v\rho) = i\epsilon_0 \omega v A K_1(v\rho) \tag{5.67}$$

where we have omitted the usual $\exp(-\Gamma z + i\omega t)$. The external impedance for this case is now obtained:

$$Z_{ex}(\Gamma) = -\frac{E_{ez}}{2\pi c H_{e\phi}}\bigg|_{\rho=c} = \frac{v}{2\pi i \epsilon_0 \omega c} \frac{K_0(vc)}{K_1(vc)} \tag{5.68}$$

The corresponding mode equation is then

$$Z_{ex}(\Gamma) + Z(\Gamma) = 0 \tag{5.69}$$

where $Z(\Gamma)$ is defined by (5.54). In general, surface-wave solutions yield
large values of v with small imaginary parts. Then for large $v\rho$ we see that
$K_0(v\rho) \simeq (\pi/2v\rho)^{1/2}\exp(-v\rho)$ is highly damped in the radial direction.

Exercise: Consider a dielectric-coated wire conductor of radius a located
in free space. Show that the mode equation for the propagation constant
$\Gamma = i\beta$ of the guided TM mode is given by

$$\frac{K_1(vb)}{vK_0(vb)} = \frac{(\epsilon/\epsilon_0)J_0(ua)Y_1(ub) - J_1(ub)Y_0(ua)}{uJ_0(ub)Y_0(ua) - J_0(ua)Y_0(ub)} \tag{5.70}$$

where $v^2 = \beta^2 - k_0^2$ and $u^2 = k^2 - \beta^2$. The outer radius of the coating is b,
and the relative permittivity of the coating is $\epsilon/\epsilon_0 = k^2/k_0^2$. Simplify the
above by using $K_0(x) \simeq -\ln 0.89x$ and $K_1(x) \simeq 1/x$ in addition to the
thin-coating approximation for $J_0(ua)$ and $Y_0(ua)$ in the manner

$$J_0(ua) \simeq J_0(ub) - (b-a)\frac{dJ_0(u\rho)}{d\rho}\bigg|_{\rho=b} \simeq J_0(ub) + utJ_1(ub) \tag{5.71}$$

where $t = b - a$. Then confirm that

$$\frac{\epsilon}{\epsilon_0}v^2 \ln(0.89vb) \simeq -\left(\frac{\epsilon}{\epsilon_0} - 1\right)k_0^2 t + v^2 t \tag{5.72}$$

Assume the following parameters: $b = 1$ mm, $t = 0.1$ mm, $\epsilon/\epsilon_0 = 2.56$,
and wavelength $2\pi/k_0 = 3.14 \times 10^{-2}$ m. Then confirm that $v = 25.8$ Np/m, $\beta = 202$ rad/m, or $(\beta/k_0 - 1) = 0.01$. In this example $v^2 t$ is
negligible compared with other terms in (5.72). This example was given
by Collin [1]. Note that it is a special case of the equivalent circuit in
Figure 5.3 when $Z_i = 0$, $Z_c = 0$, $Z_L = Z_T = \infty$, and $(b - a)/a \ll 1$.

5.5 ASYMMETRICAL FORMULATION

When azimuthal symmetry no longer prevails, the situation is somewhat more complicated even for idealized circular cylinders. But, again, it is desirable to express the fields in terms of two auxiliary functions U and V (that is, the Debye potentials). In terms of the electric and magnetic Hertz vectors, we are saying that

$$\Pi = (0, 0, U) \tag{5.73}$$

and

$$\Pi^\circ = (0, 0, V) \tag{5.74}$$

Thus for homogeneous regions we can write

$$E_\rho = \frac{\partial^2 U}{\partial\rho\,\partial z} - \frac{i\mu\omega}{\rho}\frac{\partial V}{\partial\phi} \tag{5.75}$$

$$E_\phi = \frac{1}{\rho}\frac{\partial^2 U}{\partial\phi\,\partial z} + i\mu\omega\frac{\partial V}{\partial\rho} \tag{5.76}$$

$$E_z = \left(-\gamma^2 + \frac{\partial^2}{\partial z^2}\right)U \tag{5.77}$$

and

$$H_\rho = \frac{\partial^2 V}{\partial\rho\,\partial z} + \frac{i\epsilon\omega}{\rho}\frac{\partial V}{\partial\phi} \tag{5.78}$$

$$H_\phi = \frac{1}{\rho}\frac{\partial^2 V}{\partial\phi\,\partial z} - i\epsilon\omega\frac{\partial U}{\partial\rho} \tag{5.79}$$

$$H_z = \left(-\gamma^2 + \frac{\partial^2}{\partial z^2}\right)V \tag{5.80}$$

where $\gamma^2 = i\mu\omega(\sigma + i\epsilon\omega)$. Equations (5.75)–(5.80), of course, reduce to (5.8), (5.9), and (5.12)–(5.15) when $\partial/\partial\phi = 0$. Now, however, as we will see, the TM modes associated with U and the TE modes associated with V are coupled by the boundary conditions.

In the homogeneous region under consideration U and V satisfy

$$(\nabla^2 - \gamma^2)_V^U = 0 \tag{5.81}$$

where

$$\nabla^2 = \frac{1}{\rho}\frac{\partial}{\partial\rho}\left(\rho\frac{\partial}{\partial\rho}\right) + \frac{1}{\rho^2}\frac{\partial^2}{\partial\phi^2} + \frac{\partial^2}{\partial z^2} \tag{5.82}$$

is the laplacian operator in cylindrical coordinates. Solutions of (5.81) have the form

$$\begin{matrix} I_m(v\rho) \\ K_m(v\rho) \end{matrix} e^{-im\phi}e^{-\Gamma z}$$

where $v = (\gamma^2 - \Gamma^2)^{1/2}$ and where m is an integer. The order of the modified Bessel function is m, which can range from $-\infty$ to $+\infty$. Thus (5.75)–(5.80) show, for any value of m, that

$$E_{m,\rho} = -\Gamma \frac{\partial U_m}{\partial \rho} - \frac{\mu\omega m}{\rho} V_m \qquad (5.83)$$

$$E_{m,\phi} = \frac{im\Gamma}{\rho} U_m + i\mu\omega \frac{\partial V_m}{\partial \rho} \qquad (5.84)$$

$$E_{m,z} = -v^2 U_m \qquad (5.85)$$

and

$$H_{m,\rho} = -\Gamma \frac{\partial V_m}{\partial \rho} + \frac{\epsilon\omega m}{\rho} U_m \qquad (5.86)$$

$$H_{m,\phi} = \frac{im\Gamma}{\rho} - i\epsilon\omega \frac{\partial U_m}{\partial \rho} \qquad (5.87)$$

$$H_{m,z} = -v^2 V_m \qquad (5.88)$$

5.6 WAVES ON A DIELECTRIC ROD

We now deal with a very simple structure that is illustrated in Figure 5.4. A homogeneous cylinder of radius a with electrical properties σ_1, ϵ_1, and μ_1 is considered. The surrounding region is unbounded and has electrical properties σ_2, ϵ_2, and μ_2. The appropriate form for the Debye potentials are

$$U_m = A_m I_m(v_1\rho) \qquad (5.89)$$
$$V_m = B_m I_m(v_1\rho) \qquad \left.\right\} 0 < \rho < a \qquad (5.90)$$
$$U_m = C_m K_m(v_2\rho) \qquad (5.91)$$
$$V_m = D_m K_m(v_2\rho) \qquad \left.\right\} a < \rho < \infty \qquad (5.92)$$

where the common factor $\exp(-im\phi - \Gamma z + i\omega t)$ has been omitted in each case. The coefficients A_m, B_m, C_m, and D_m are not known, but they do not depend on ρ, ϕ, or z. Also, we note that $v_1 = (\gamma_1^2 - \Gamma^2)^{1/2}$ and $v_2 = (\gamma_2^2 - \Gamma^2)^{1/2}$ The corresponding field components are obtained by using (5.83)–(5.88) along with (5.89)–(5.92). Subscripts 1 or 2 are added as appropriate.

A modal equation can now be obtained by insisting that the tangen-

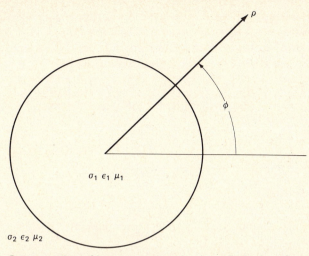

Figure 5.4   Cross section of homogeneous circular cylinder in unbounded medium with different electrical properties.

tial field components $E_{m\phi}$, E_{mz}, $H_{m\phi}$, and H_{mz} are continuous across the boundary at $\rho = a$. This assumption leads to

$$0 = -\Gamma v_1 I_m' A_m - \frac{\mu_1 \omega m}{a} I_m B_m + \Gamma v_2 K_m' C_m + \frac{\mu_2 \omega m}{a} K_m D_m \quad (5.93)$$

$$0 = -\frac{im\Gamma}{a} I_m A_m + i\mu_1 \omega v_1 I_m' B_m + \frac{im\Gamma}{a} K_m C_m - i\mu_2 \omega v_2 K_m' D_m \quad (5.94)$$

$$0 = -v_1^2 I_m A_m + 0 + v_2^2 K_m C_m + 0 \quad (5.95)$$

$$0 = 0 - v_1^2 I_m B_m + 0 + v_2^2 K_m D_m \quad (5.96)$$

where

$$I_m = I_m(v_1 a) \qquad I_m'(x) = \frac{dI_m(x)}{dx}$$

and

$$K_m = K_m(v_2 a) \qquad K_m'(x) = \frac{dK_m(x)}{dx}$$

In order for (5.93)–(5.96) to yield a nonzero field, the determinant of the coefficients of A_m, B_m, C_m, and D_m must vanish. The resulting equation

$$D(\Gamma) = 0 \quad (5.97)$$

yields the appropriate value of the axial propagation constants.

An important special case is a lossless dielectric rod where we set

$\epsilon_1 = \epsilon$, $\mu_1 = \mu_0$, and $\sigma_1 = 0$ and where the surrounding medium is free space so that $\epsilon_2 = \epsilon_0$, $\mu_2 = \mu_0$, $\sigma_2 = 0$. Then if we set $\Gamma = i\beta$, change the Bessel function° I_m to the unmodified form J_m, and designate $dJ_m(Z)/dZ$ by $J'_m(Z)$, we see that (5.97) takes the explicit form

$$\left[\frac{(\epsilon/\epsilon_0)J'_m(p)}{pJ_m(p)} + \frac{K'_m(q)}{qK_m(q)}\right]\left[\frac{J'_m(p)}{pJ_m(p)} + \frac{K'_m(q)}{qK_m(q)}\right] = \left[\frac{m\beta}{k_0}\left(\frac{p^2 + q^2}{p^2 q^2}\right)\right]^2$$

(5.98)

where

$$p = (k^2 - \beta^2)^{1/2}a \qquad q = (\beta^2 - k_0^2)^{1/2}a$$

We also note that

$$k^2 = \frac{\epsilon}{\epsilon_0}k_0^2 \qquad \text{and} \qquad p^2 + q^2 = \left(\frac{\epsilon}{\epsilon_0} - 1\right)k_0^2 a^2$$

In the special case of azimuthal symmetry (5.98) splits into the two independent equations

$$\frac{(\epsilon/\epsilon_0)J_1(p)}{pJ_0(p)} + \frac{K_1(q)}{qK_0(q)} = 0$$

(5.99)

and

$$\frac{J_1(p)}{pJ_0(p)} + \frac{K_1(q)}{qK_0(q)} = 0$$

(5.100)

where we have used $J'_0(p) = -J_1(p)$ and $K'_0(q) = -K_1(q)$. Here we note that (5.99) and (5.100) are the appropriate mode equations for the purely TM and TE modes, respectively.

Exercise: Derive (5.99) and (5.100) directly for the homogeneous dielectric cylinder working with (5.8), (5.12), and (5.13) for the TM modes and with (5.9), (5.14), and (5.15) for the TE modes.

When $m \neq 0$, the modes are hybrid in the sense that the two factors on the left-hand side of (5.98) are not identically zero although they be small. The solution for (5.98) for the propagation constants $\Gamma = i\beta$ is quite difficult, although graphical methods were used by earlier workers with some success. Using numerical data from Elsasser [2], which was also abstracted by Collin [1], we plot β/k_0 as a function of the cylinder radius a in Figure 5.5. For this example $\epsilon/\epsilon_0 = 2.56$, corresponding to polystyrene. The TE_1 and TM_1 designate the first-order modes that begin to propagate when $2a/\lambda_0$ becomes larger than 0.613. Higher-order TE and TM modes are also possible at even larger values of $2a/\lambda_0$. These are not shown on Figure 5.5. The lowest-order hybrid mode,

---

° Note that $I_m(iZ) = \exp(im\pi/2)J_m(Z)$.

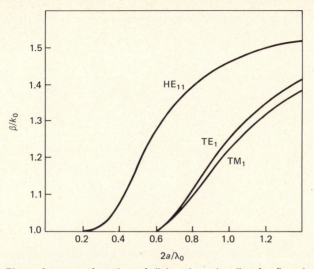

Figure 5.5 Phase factor as function of dielectric rod radius for first three dominant modes of propagation on structure.

designated HE_{11}, is sometimes called the dipole mode [3]. Actually, it has no low-frequency cutoff, so it plays a role in conveying useful signals along very slender dielectric fibers. Further discussion of the optical fiber applications is found in the excellent books by Sodha and Ghatak [4] and by Marcuse [5].

5.7 WAVE TRANSMISSION IN TUNNELS

The transmission of electromagnetic waves through tunnels can also be studied as a guided-wave problem [6–13]. If we assume the tunnel is air-filled and bounded by a homogeneous surround, the formulation is identical to that for cylindrical rod discussed above. Now, of course, the dense medium is on the outside. Furthermore, to make the problem more relevant to the real world, we allow for the presence of an axial conductor or cable that is located within and runs parallel to the axis of the tunnel. The situation is illustrated in Figure 5.6, where we designate the tunnel wall by $\rho = a_0$ with respect to a cylindrical coordinate system (ρ, ϕ, z). As indicated, the cable is located at $\rho = \rho_0$ and $\phi = \phi_0$. The external region (that is, $\rho > a_0$) has electrical properties σ_e, ϵ_e, and μ_e.

   The objective of the analysis is to deduce the propagation constants of the guided waves in the system as described. The principal restriction is that the outer radius c of the cable will be much smaller than the tunnel radius a_0. Also, $k_0 c \ll 1$, where $k_0 = 2\pi/\lambda_0 = (\epsilon_0 \mu_0)^{1/2} \omega$, can be assumed for any radio frequency of interest.

   The current on the cable for a particular propagation constant can be designated $I \exp(-\Gamma z)$. The primary fields within the tunnel pro-

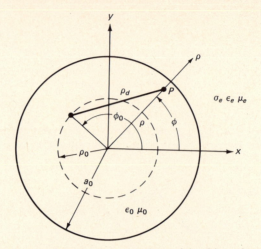

Figure 5.6   Circular-tunnel model that contains axial cable at (ρ_0, ϕ_0).

duced by this filamental current can be derived from a Debye potential U^p given by

$$U^p = \frac{I}{2\pi i \epsilon_0 \omega}\, e^{-\Gamma z} K_0(v\rho_d) \tag{5.101}$$

where

$$\gamma_0 = ik = i(\epsilon_0\mu_0)^{1/2}\omega \qquad v = (\gamma_0^2 - \Gamma^2)^{1/2}$$

and

$$\rho_d = [\rho^2 + \rho_0^2 - 2\rho\rho_0 \cos(\phi - \phi_0)]^{1/2}$$

This result can be verified by noting that the primary magnetic field, with reference to cylindrical coordinates (ρ_d, ϕ_d, z), centered about the cable, is obtained from

$$H^p_{\phi_d} = -i\epsilon_0\omega\, \frac{\partial U^p}{\partial \rho_d} = \frac{Iv}{2\pi}\, e^{-\Gamma z} K_1(v\rho_d) \tag{5.102}$$

Then we see that

$$2\pi c H^p_{\phi_d}\Big]_{\rho_d = 0} = Ie^{-\Gamma z} vc K_1(vc) \simeq Ie^{-\Gamma z} \qquad \text{if } |vc| \ll 1 \tag{5.103}$$

We note here that $I \exp(-\Gamma z)$ is the total or net current carried by the cable as seen by the observer at P. At this stage we do not have to specify the cable currents in more concrete terms.

Now the resultant Debye potential U within the tunnel (that is, $\rho < a_0$) can be written

$$U = U^p + U^s \tag{5.104}$$

where U^s, by definition, is the secondary potential. Now

$$(\nabla^2 - \gamma_0^2)U^s = 0 \tag{5.105}$$

and the appropriate form of the solutions, which remain finite at $\rho = 0$, is $I_m(v\rho)\exp(-im\phi - \Gamma z)$, where m is an integer. The general solution for U^s is a linear superposition of these elementary forms.

To deal with U^p, we need to employ an addition theorem from the theory of Bessel functions to express $K_0(v\rho_d)$ in terms of wave functions that are referred to the origin $\rho = 0$. The required form [6] is

$$K_0(v\rho_d) = \sum_{m=-\infty}^{+\infty} e^{-im(\phi-\phi_0)} \begin{cases} I_m(v\rho)K_m(v\rho_0) & \rho < \rho_0 \\ K_m(v\rho)I_m(v\rho_0) & \rho > \rho_0 \end{cases} \tag{5.106}$$

where, as indicated, we need to distinguish between the two cases where $\rho < \rho_0$ and $\rho > \rho_0$.

We are now in the position to write the desired form of the Debye potential U in the region $\rho_0 < \rho < a_0$:

$$U = \sum_{m=-\infty}^{+\infty} U_m e^{-im(\phi-\phi_0)} \tag{5.107}$$

where

$$U_m = \frac{I}{2\pi i \epsilon_0 \omega} e^{-\Gamma z} I_m(v\rho_0) \left[K_m(v\rho) - R_m \frac{K_m(va_0)}{I_m(va_0)} I_m(v\rho) \right] \tag{5.108}$$

The fields associated with U are TM. Also, in writing (5.108) in the form indicated, we have introduced R_m in such a manner that it would be $+1$ for the tunnel with perfectly conducting walls (in which case U_m and thus E_z would vanish at $\rho = a_0$).

In the general case we also need to allow for the presence of TE fields. Thus for the region $0 < \rho < a_0$, we write

$$V = \sum_{m=-\infty}^{+\infty} V_m e^{-im(\phi-\phi_0)} \tag{5.109}$$

where

$$V_m = \frac{I}{2\pi i \epsilon_0 \omega} e^{-\Gamma z} \Delta_m I_m(v\rho_0) I_m(v\rho) \tag{5.110}$$

where Δ_m is a coefficient yet to be determined. Of course, V_m does not have any singularity at the cable because $E_z = 0$ for TE fields.

To express the field components in terms of U_m and V_m, we find it convenient to employ the same Fourier series representation. Thus, for example, the field components are written

$$E_z = \sum_{m=-\infty}^{+\infty} E_{zm} e^{-im(\phi-\phi_0)} \tag{5.111}$$

and

$$H_z = \sum_{m=-\infty}^{+\infty} H_{zm} e^{-im(\phi-\phi_0)} \tag{5.112}$$

where

$$\begin{aligned} E_{z,m} &= -v^2 U_m \\ H_{z,m} &= -v^2 V_m \end{aligned} \quad \text{for } 0 < \rho < a_0 \tag{5.113}$$

and where

$$\begin{aligned} E_{z,m} &= -v_e^2 U_m \\ H_{z,m} &= -v_e^2 V_m \end{aligned} \quad \text{for } \rho > a_0 \tag{5.114}$$

and

$$v_e^2 = (\gamma_e^2 - \Gamma^2)^{1/2} \qquad \gamma_e = [i\mu_e \omega(\sigma_e + i\epsilon_e \omega)]^{1/2}$$

In a similar fashion, the azimuthal field components are written

$$E_\phi = \sum_{m=-\infty}^{+\infty} E_{\phi,m} e^{-im(\phi-\phi_1)} \tag{5.115}$$

and

$$H_\phi = \sum_{m=-\infty}^{+\infty} H_{\phi,m} e^{-im(\phi-\phi_0)} \tag{5.116}$$

where

$$\left. \begin{aligned} E_{\phi,m} &= \frac{im\Gamma}{\rho} U_m + i\mu_0 \omega \frac{\partial V_m}{\partial \rho} \\ H_{\phi,m} &= \frac{im\Gamma}{\rho} V_m - i\epsilon_0 \omega \frac{\partial U_m}{\partial \rho} \end{aligned} \right\} \quad \text{for } 0 < \rho < a_0 \tag{5.117}$$

and where

$$\left. \begin{aligned} E_{\phi,m} &= \frac{im\Gamma}{\rho} U_m + i\mu_e \omega \frac{\partial V_m}{\partial \rho} \\ H_{\phi,m} &= \frac{im\Gamma}{\rho} V_m - (\sigma_e + i\epsilon_e \omega) \frac{\partial U_m}{\partial \rho} \end{aligned} \right\} \quad \text{for } \rho > a_0 \tag{5.118}$$

Now in the external region we could use the same Fourier representation for U and V as given by (5.107) and (5.109), respectively. Thus for $\rho > a_0$ the Debye potentials must satisfy

$$(\nabla^2 - \gamma_e^2)_V^U = 0 \tag{5.119}$$

and solutions would be of the form $K_m(v_e\rho)\exp(-im\phi - \Gamma z)$, where again m is an integer. The boundary conditions at the tunnel wall $\rho = a_0$

are met if $E_{z,m}$, $H_{z,m}$, $E_{\phi,m}$, and $H_{\phi,m}$ are continuous. Using (5.117) and (5.118), which are valid in the limit $\rho = a_0$, we find that it is not difficult to show that an equivalent statement of the boundary conditions at the wall is

$$\left. \begin{array}{l} E_{\phi,m} = \alpha_m E_{z,m} + Z_m H_{z,m} \\ H_{\phi,m} = - Y_m E_{z,m} + \alpha_m H_{z,m} \end{array} \right]_{\rho=a_0} \tag{5.120}$$

$$\alpha_m = \frac{-im\Gamma}{v_e^2 a_0} \tag{5.121}$$

$$Z_m = -\left(\frac{i\mu_e\omega}{v_e}\right)\frac{K_m'(v_e a_0)}{K_m(v_e a_0)} \tag{5.122}$$

$$Y_m = -\left(\frac{\sigma_e + i\epsilon_e\omega}{v_e}\right)\frac{K_m'(v_e a_0)}{K_m(v_e a_0)} \tag{5.123}$$

Thus to obtain a modal equation for the system, we insert the forms given by (5.113) and (5.117), evaluated at $\rho = a_0$, into (5.120). This leads to the algebraic pair

$$\Delta_m(i\mu_0\omega v I_m' + Z_m v^2 I_m) + R_m\left(-\frac{im\Gamma K_m}{a_0} - \alpha_m v^2 K_m\right)$$

$$= -\frac{im\Gamma}{a_0} K_m - \alpha_m v^2 K_m \tag{5.124}$$

$$\Delta_m\left(\frac{im\Gamma}{a_0} I_m + \alpha_m v^2 I_m\right) + R_m\left(\frac{i\epsilon_0\omega v K_m I_m'}{I_m} + Y_m v^2 K_m\right)$$

$$= i\epsilon_0\omega v K_m' + Y_m v^2 K_m \tag{5.125}$$

where the arguments of the Bessel functions and their derivatives are va_0. Eliminating Δ_m from (5.124) and (5.125) yields

$$R_m = \frac{[(\gamma_0/v)K_m'(va_0)/K_m(va_0)] + Y_m\eta_0 + \delta_m\eta_0}{[(\gamma_0/v)I_m'(va_0)/I_m(va_0)] + Y_m\eta_0 + \delta_m\eta_0} \tag{5.126}$$

where

$$\delta_m\eta_0 = \frac{(im\Gamma/a_0)(v^{-2} - v_e^{-2})^2}{[(\gamma_0/v)I_m'(va)/I_m(va)] + (Z_m/\eta)} \tag{5.127}$$

and

$$\eta = \frac{i\mu_0\omega}{\gamma_0} \simeq 120\pi$$

We now express the axial electric field E_z anywhere within the tunnel guide (that is, $0 < \rho < a_0$) in terms of the still-unknown current on the cable. Using (5.101), (5.107), (5.111), and (5.113), we find that

$$
E_z = -\frac{Iv^2}{2\pi i \epsilon_0 \omega} e^{-\Gamma z} \Bigg[K_0(v\rho_d)
$$

$$
- \sum_{m=-\infty}^{+\infty} R_m \frac{K_m(va_0)}{I_m(va_0)} I_m(v\rho_0) I_m(v\rho) e^{-im(\phi-\phi_0)} \Bigg] \tag{5.128}
$$

A key step in our analysis is to apply the appropriate boundary condition at the surface of the cable. This can be stated in the form

$$
E_z|_{\rho_d=c} = IZ(\Gamma)\exp(-\Gamma z) \tag{5.129}
$$

which is to hold for all values of z where $Z(\Gamma)$ is the effective series impedance of the cable. Because of the assumed smallness of the cable radius, we may apply the condition at any point around the circumference. For convenience we apply (5.129) at $\rho = \rho_0 + c$ and at $\phi = \phi_0$. This leads to the mode equation

$$
-\frac{v^2}{2\pi i \epsilon_0 \omega} \Bigg[K_0(vc) - \sum_{m=-\infty}^{\infty} R_m \frac{K_m(va_0)}{I_m(va_0)} I_m(v\rho_0) I_m(v(\rho_0 + c)) \Bigg] = Z(\Gamma) \tag{5.130}
$$

where R_m is defined by (5.126).

To solve (5.130) for a given case, we need to specify $Z(\Gamma)$. Such an expression was given by (5.57) for a concentrically layered cylindrical structure. An interesting special case, which we select for discussion here, is a dielectric-coated conductor of outer radius c and inner radius a. The insulation of thickness $c - a$ has a permittivity ϵ, while the central conductor of radius a has a conductivity σ_w, permittivity ϵ_w, and permeability μ_w. Specializing (5.57), we see that

$$
Z(\Gamma) = Z + Z_i \tag{5.131}
$$

where

$$
Z = -\frac{\Gamma^2 + k^2}{2\pi i \epsilon \omega} \ln \frac{c}{a} \tag{5.132}
$$

and where $k^2 = \epsilon \mu_0 \omega^2$ and Z_i is given by (5.38). Now because $|\gamma_w^2| \gg |\Gamma^2|$, we can employ the simplified form of Z_i given by (5.40). Also, displacement currents in the conductor can be neglected since $\epsilon_w \omega / \sigma_w \ll 1$ for metals such as copper, where $\sigma_w = 5.7 \times 10^7$. Also, in such cases we can set $\mu_w = \mu_0$.

If we examine (5.130) in the case where $R_m = 0$, the resulting mode equation is seen to be identical to that for a dielectric-coated conductor located in free space. Thus the summation term in (5.130) represents the effect of the tunnel wall.

Using numerical techniques for solving (5.130), the attenuation rate (that is, Re Γ) for the dominant mode has been obtained for several

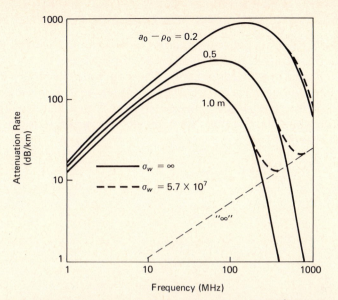

Figure 5.7   Attenuation rate of mode that behaves as axial surface wave at high frequencies for dielectric-coated conductor located in circular tunnel. Case indicated by "∞" corresponds to having conductor located in free space. For example shown, $a = 1$ mm, $c = 5$ mm, $a_0 = 2$ m, $\epsilon/\epsilon_0 = 2.56$, $\mu_e = \mu_w = \mu_0 = 4 \times 10^{-7}$, $\sigma_e/\sigma_0 = 10^{-3} = 1$ mmho/m, $\epsilon_e/\epsilon_0 = 5$, and $\epsilon_0 = 8.85 \times 10^{-12}$.

tunnel geometries [10, 14]. Here we show a particular example in Figure 5.7, where the attenuation rate, expressed in decibels per kilometer, is plotted as a function of the frequency. The assumed values of the parameters are given in the caption to Figure 5.7. The parameter of special interest is the distance of the cable to the tunnel wall (that is, $a_0 - \rho_0$). Not surprisingly, the attenuation rate is higher as the cable approaches the wall. Also, we note that for very high frequencies the attenuation rate becomes relatively small; in this case the cable is acting more in the manner of a surface-wave transmission line in free space. For comparison purposes we show the attenuation rate for the same cable located in free space [that is, $R_m = 0$ in (5.130)]. This is clearly a high-frequency asymptote for the curves in Figure 5.7. We also show the effect of the finite conductivity of the center conductor by choosing $\sigma_w = \infty$ and 5.7×10^7 S/m. The curves are only distinguishable at the higher frequencies.

Actually, the low-frequency behavior of the curves in Figure 5.7 is also important. As we see, the attenuation rate varies approximately as the first power of the frequency. Also, as other calculations show, the results depend only slightly on the conductivity of the surrounding rock in this frequency range [10, 14]. This seemingly strange behavior is explained by the fact that the return current for this mode is mainly through the adjacent medium.

5.8 QUASI-STATIC ANALYSIS FOR CABLE IN TUNNEL

Actually, (5.130) is a general equation to determine the axial propagation constants of any of the discrete modes for the tunnel waveguide with a contained axial conductor or cable. These include the modes that have a transmission-line characteristic at low frequencies. Here we would like to identify a low-frequency approximation [17] that leads neatly to the transmission-line description. First of all, we recast (5.130) into the equivalent form

$$\frac{v^2}{2\pi i \epsilon_0 \omega}\left(\Lambda + \frac{\gamma_0^2 \Omega}{v^2}\right) + Z(\Gamma) = 0 \tag{5.133}$$

where

$$\Omega = -\frac{v^2}{\gamma_0^2} \sum_{m=-\infty}^{+\infty} (R_m - 1)\frac{K_m(va_0)}{I_m(va_0)}[I_m(v\rho_0)]^2 \tag{5.134}$$

and

$$\Lambda = K_0(vc) - \sum_{m=-\infty}^{+\infty} \frac{K_m(va_0)}{I_m(va_0)}[I_m(v\rho_0)]^2 \tag{5.135}$$

Then we deduce from (5.133) that

$$\Gamma^2 = \gamma_0^2\left[1 + \frac{\Omega}{\Lambda} + \frac{2\pi Z(\Gamma)}{i\mu_0 \omega \Lambda}\right] \tag{5.136}$$

or alternatively,

$$\Gamma^2 = \hat{Z}\hat{Y} \tag{5.137}$$

where

$$\hat{Z} = \frac{i\mu_0 \omega}{2\pi}(\Lambda + \Omega) + Z(\Gamma) \tag{5.138}$$

and

$$\hat{Y} = \frac{2\pi i \epsilon_0 \omega}{\Lambda} \tag{5.139}$$

In view of the form of (5.137) we may interpret $\hat{Z}$ and $\hat{Y}$ as the general series impedance and series admittance, respectively, of the cable. However, we stress that $\hat{Z}$ and $\hat{Y}$ are spatially dispersive in the sense that they are functions of the axial propagation constant Γ. However, at sufficiently low frequencies we can use the transmission-line form to obtain explicit results. This development is illustrated below. Specifically, we invoke the conditions that $|va_0| \ll 1$, $(\gamma_e/\gamma_0)^2 \gg 1$, and $\mu_e = \mu_0$.

Then on using the small-argument approximation for the modified Bessel functions, we see that

$$\Lambda \simeq \ln \frac{a_0}{c} - \ln \frac{a_0^2}{a_0^2 - \rho_0^2} \tag{5.140}$$

Furthermore, if, in addition, $d = a_0 - \rho_0 \ll a_0$,

$$\Lambda \simeq \ln \frac{2d}{c} \tag{5.141}$$

To reduce Ω, we first note that (5.127) is simplified to

$$\delta_m \eta_0 \simeq \frac{-\Gamma^2 m}{\gamma_0^2 v^2 a_0} \tag{5.142}$$

when m is a positive integer. Furthermore, (5.123) is approximated by

$$Y_m \eta_0 \simeq -\frac{\gamma_e}{\eta_0} \left[\frac{m}{\gamma_e a_0} - \frac{K_{m+1}(\gamma_e a_0)}{K_m(\gamma_e a_0)} \right] \tag{5.143}$$

where we have utilized the derivative relation

$$K'_m(x) = \frac{m}{x} K_m(x) - K_{m+1}(x) \tag{5.144}$$

and also we have noted that

$$v a_0 = (\gamma_e^2 - \Gamma^2)^{1/2} a_0 \simeq \gamma_e a_0 \tag{5.145}$$

Then (5.126) becomes

$$(R_m - 1) \frac{K_m(v a_0)}{I_m(v a_0)} [I_m(v \rho_0)]^2 \simeq -\frac{\gamma_0}{v^2 a_0} \left(\frac{\rho_0}{a_0} \right)^{2m} \left[\frac{m}{\gamma_0 a_0} + Y_m \eta_0 \right]^{-1}$$

$$\simeq -\frac{\gamma_0^2}{v^2 \gamma_e a_0} \left(\frac{\rho_0}{a_0} \right)^{2m} \frac{K_m(\gamma_e a)}{K_{m+1}(\gamma_e a)} \tag{5.146}$$

Thus (5.134) is approximated by

$$\Omega = \frac{1}{\gamma_e a_0} \left[\frac{K_0(\gamma_e a)}{K_1(\gamma_e a)} + 2 \sum_{m=1}^{\infty} \left(\frac{\rho_0}{a} \right)^{2m} \frac{K_m(\gamma_e a_0)}{K_{m+1}(\gamma_e a_0)} \right] \tag{5.147}$$

which is independent of Γ. This form for Ω may not converge well if $|\gamma_e a_0| \ll 1$. Thus following a suggestion of Wu, Shen, and King [18], we find it expedient for numerical work to rewrite (5.147) in the equivalent form

$$\Omega = \frac{1}{\gamma_e a_0} \left\{ \frac{K_0(\gamma_e a_0)}{K_1(\gamma_e a_0)} + 2 \sum_{m=1}^{\infty} \left[\frac{K_m(\gamma_e a_0)}{K_{m+1}(\gamma_e a_0)} - \frac{\gamma_e a_0}{2m} \right] \left(\frac{\rho_0}{a_0} \right)^{2m} \right\}$$

$$+ \ln \frac{a^2}{a^2 - \rho_0^2} \tag{5.148}$$

where we have merely added and subtracted a series of the form

$$\sum_{m=1}^{\infty} \frac{1}{my^m} = \ln \frac{y}{y-1} \qquad \text{for } y > 1 \tag{5.149}$$

When we insert the Γ-independent forms for Λ and Ω into (5.138) and (5.139) or substitute directly into (5.136), we have a quasi-static form of the mode equation. But note that it is not an explicit formula to deduce the propagation constant Γ because $Z(\Gamma)$, the effective series impedance of the cable, may be strongly dependent on Γ. This is particularly the case when we are dealing with a braided coaxial cable or structures such as the Goubau surface-wave line [16] discussed above. Nevertheless, $Z(\Gamma)$ can be expressed adequately in terms of an algebraic formula involving powers of Γ^2. Then the resulting mode equation ends up being a quadratic or cubic equation to solve for the desired propagation modes of the system [11]. In the case of the braided coaxial cable the two principal solutions of the mode equation correspond to the so-called monofilar and bifilar modes. In the limit of very low frequencies the propagation constant Γ of the monofilar mode is approaching $\Gamma \simeq \gamma_0 \simeq ik_0$, while the propagation constant of the bifilar mode is approaching $\Gamma \simeq \gamma \simeq ik$. Clearly, the monofilar mode in this context is formed by a transmission line consisting of the cable and the surrounding (rock) medium. The attenuation for such a mode is strongly dependent on the distance of the cable to the tunnel wall. On the other hand, the bilfilar mode is confined mainly within the coaxial cable, so it is relatively insensitive to the location of the cable relative to the wall of the tunnel.

5.9 REMARKS CONCERNING MODE CONVERSION

The question of mode conversion between the propagating modes in cylindrical structures is a subject in its own right. Such conversion will always take place when there is any longitudinal variation of the structure such as a junction or tapered section. In the mine communication problem the mode conversion phenomena have been clearly recognized and, in some cases, exploited to improve the performance of two-way communication systems. Many of the technological advances in this field have been made by Delogne and his colleagues in Belgium [13, 19] and by Gabillard, Degauque, and their associates in France [8].

The interested reader is advised to consult the references cited and other related papers [20–27] for more details. Professor Paul Delogne's excellent book also has many useful nuggets of information [13].

References

1. R. E. Collin: *Field Theory of Guided Waves*, McGraw-Hill, New York, 1960. *Note:* In this book Eq. (56) is misprinted; the first factor should be the sum,

not the difference, of the J and K terms. Also, Eq. (58) should read $(K - 1)(k_0a)^2$ on the right-hand side.

2. W. M. Elasser: "Attenuation in a Dielectric Rod," *Journal of Applied Physics*, vol. 20, 1949, pp. 1193–1196.

3. C. H. Chandler: "Investigation of Dielectric Rod as a Waveguide," *Journal of Applied Physics*, vol. 20, 1949, pp. 1188–1192.

4. M. S. Sodha and A. K. Ghatak: *Inhomogeneous Optical Waveguides*, Plenum, New York, 1977.

5. D. Marcuse: *Theory of Dielectric Optical Waveguides*, Academic Press, New York, 1974.

6. J. R. Wait and D. A. Hill: "Guided Electromagnetic Waves Along an Axial Conductor in a Circular Tunnel," *IEEE Transactions*, vol. AP–22, 1974, pp. 627–630.

7. P. Delogne: "Basic Mechanisms of Tunnel Propagation," *Radio Science*, vol. 11, 1976, pp. 299–303.

8. P. Degauque, B. Demoulin, J. Fontaine, and R. Gabillard: "Theory and Experiment of a Mobile Communication in Tunnels by Means of a Leaky Braided Coaxial Cable, *Radio Science*, vol. 11, 1976, pp. 305–314.

9. D. J. R. Martin: "The Biaxial Leaky Line and Its Application to Underground Radio Communication, *Radio Science*, vol. 11, 1976, pp. 779–785.

10. J. R. Wait and D. A. Hill: "Electromagnetic Fields of a Dipole Source in a Circular Tunnel Containing a Surface Wave Line," *International Journal of Electronics*, vol. 42, 1977, pp. 377–391.

11. D. B. Seidel and J. R. Wait: "Transmission Modes in a Braided Coaxial Cable and Coupling to a Tunnel Environment," *IEEE Transactions*, vol. MTT–26, 1978, pp. 494–499.

12. S. F. Mahmoud: "Characteristics of EM Guided Waves for Communication in Coal Mine Tunnels," *IEEE Transactions*, vol. COM–22, 1974, pp. 1547–1554.

13. P. Delogne: *Leaky Feeders and Subsurface Radio Communications*, Peter Peregrinus Ltd. (Institution of Electrical Engineers), London and New York, 1982.

14. J. R. Wait and D. A. Hill: "Attenuation on a Surface Wave Line Suspended Within a Circular Tunnel," *Journal of Applied Physics*, vol. 47, 1976, pp. 5472–5473.

15. A. G. Emslie, R. L. Lagace, and P. Strong: "Theory of the Propagation of UHF Radio Waves in Coal Mine Tunnels," *IEEE Transactions*, vol. AP–23, 1975, pp. 182–205.

16. G. Goubau: "Single Conductor Surface Wave Transmission Line," *Proceedings of the IRE*, vol. 39, 1951, pp. 619–624.

17. J. R. Wait: "Quasi-Static Limit for the Propagating Mode Along a Thin Wire in a Circular Tunnel," *IEEE Transactions*, vol. AP–25, 1977, pp. 441–443.

18. T. T. Wu, L. C. Shen, and R. W. P. King: "The Dipole Antenna with Eccentric Coating in a Relatively Dense Medium," *IEEE Transactions*, vol. AP–23, 1975, pp. 57–62.

19. L. Deryck: "Control of Mode Conversions on Bifilar Lines in Tunnels," *Radio and Electronic Engineering*, 1975, pp. 241–247.

20. D. A. Hill and J. R. Wait: "Bandwidth of a Leaky Coaxial Cable in a Circular Tunnel," *IEEE Transactions*, vol. COM–26, 1978, pp. 1765–1771.

21. J. R. Wait: "EM Theory of the Loosely Braided Coaxial Cable, Part I," *IEEE Transactions*, vol. MTT–24, 1976, pp. 262–265.

22. D. A. Hill and J. R. Wait: "Propagation Along a Coaxial Cable with a Helical Shield," *IEEE Transactions*, vol. MTT–28, 1980, pp. 84–89.

23. D. A. Hill and J. R. Wait: "EM Theory of the Loosely Braided Coaxial Cable — Numerical Results," *IEEE Transactions*, vol. MTT–28, 1981, pp. 326–331.

24. K. F. Casey: "On the Effective Transfer Impedance of Thin Coaxial Cable Shields," *IEEE Transactions*, vol. EMC–18, 1976, pp. 110–117.

25. E. F. Kuester and D. B. Seidel: "Low Frequency Behavior of the Propagation Along a Thin Wire in an Arbitrarily Shaped Mine Tunnel," *IEEE Transactions*, vol. MTT–27, 1979, pp. 736–741.

26. S. F. Mahmoud and J. R. Wait: "Calculated Channel Characteristics of a Braided Coaxial Cable in a Mine Tunnel," *IEEE Transactions*, vol. COM–24, 1976, pp. 82–87.

27. R. J. Pogorzelski: "EM Propagation Along a Wire in a Tunnel-Approximate Analysis," *IEEE Transactions*, vol. AP–27, 1979, pp. 814–819.

Chapter 6
Basic Radiation Fields

6.1 RADIATION FROM A CURRENT ELEMENT

A basic building block in the analysis of radiating systems is the current element or electric dipole. We will present a direct derivation of the electromagnetic fields of such a dipole that begins with the static field behavior.

With respect to conventional spherical coordinates (r, θ, ϕ), the dipole is located at the origin and is oriented in the polar or z direction, as illustrated in Figure 6.1. The surrounding medium is assumed to be homogeneous with electrical constants σ, ϵ, and μ. The "dipole" is to be regarded as a current element or filament of length ℓ carrying a constant current I throughout its length. The current density J at some point P at distance r from the dipole, large compared with its length, is then deduced to be the resultant of a current point source at $z = \ell/2$ and a current point sink at $z = -\ell/2$. Then it easily follows that

$$J_r = \frac{I\ell \cos \theta}{2\pi r^3} \tag{6.1a}$$

$$J_\theta = \frac{I\ell \sin \theta}{4\pi r^3} \tag{6.1b}$$

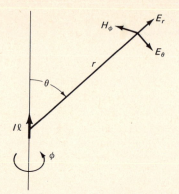

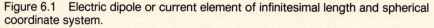

Figure 6.1   Electric dipole or current element of infinitesimal length and spherical coordinate system.

and

$$J_\phi = 0$$

where we have assumed $r \gg \ell$. The development of equation (6.1) is fully analogous to Equations (1.32) and (1.33). See also Chapter 1, Section 1.15.

Using Ampere's law, we now deduce that the corresponding magnetic field H_ϕ is related to J_r via

$$\int_0^{2\pi} H_\phi r \sin \theta \, d\phi = \int_{\phi=0}^{2\pi} \int_0^{\theta} J_r r^2 \sin \theta \, d\theta \, d\phi \tag{6.2}$$

This is a statement that the total current through a ring of radius $r \sin \theta$ is equal to the magnetic field integrated around the ring. It easily follows that

$$H_\phi = \frac{I\ell \sin \theta}{4\pi r^2} \tag{6.3}$$

By symmetry it is evident that

$$H_r = H_\theta = 0$$

An application of Ohm's law to (6.1a) and (6.1b) tells us that

$$E_r = \frac{I\ell \cos \theta}{2\pi\sigma r^3} \tag{6.4a}$$

$$E_\theta = \frac{I\ell \sin \theta}{4\pi\sigma r^3} \tag{6.4b}$$

In fact, (6.3), (6.4a), and (6.4b) are the nonzero field components E_r, E_θ, and H_ϕ of a purely static dipole in the medium of conductivity σ. We can now generalize the results to a time-varying field, with, say, a harmonic time factor $\exp(i\omega t)$, in the following fashion: First of all, we write Ohm's law in its general form,

$$\mathbf{J} = (\sigma + i\epsilon\omega)\mathbf{E} \tag{6.5}$$

so that the electric field near the dipole must behave as

$$E_r = \frac{I\ell\cos\theta}{2\pi(\sigma + i\epsilon\omega)r^3} \tag{6.6}$$

$$E_\theta = \frac{I\ell\sin\theta}{4\pi(\sigma + i\epsilon\omega)r^3} \tag{6.7}$$

Now Maxwell's equations for the present configuration are given by

$$\frac{\partial}{\partial r}(rE_\theta) - \frac{\partial E_r}{\partial\theta} = -i\mu\omega rH_\phi \tag{6.8}$$

$$\frac{\partial}{\partial\theta}(\sin\theta\, H_\phi) = (\sigma + i\epsilon\omega)r\sin\theta\, E_r \tag{6.9}$$

$$\frac{\partial}{\partial r}(rH_\phi) = -(\sigma + i\epsilon\omega)rE_\theta \tag{6.10}$$

By inserting the latter two equations (for E_r and E_θ) into the first, we deduce that

$$\frac{\partial^2(rH_\phi)}{\partial r^2} + \frac{1}{r}\frac{\partial}{\partial\theta}\left[\frac{1}{\sin\theta}\frac{\partial}{\partial\theta}\left(\sin\theta\, H_\phi\right)\right] - \gamma^2 rH_\phi = 0 \tag{6.11}$$

where $\gamma^2 = i\mu\omega(\sigma + i\epsilon\omega)$. An appropriate solution of (6.11) is evidently

$$H_\phi = \frac{I\ell}{4\pi r^2}(1 + \gamma r)e^{-\gamma r}\sin\theta \tag{6.12}$$

This has the appropriate outgoing behavior as $r \to \infty$ and it reduces to (6.3) as $\gamma r \to 0$. Also, as $\omega \to 0$, it tends to the static form (6.3) for any r.

Finally, by using (6.9) and (6.10), we deduce that the electric field components corresponding to (6.12) are

$$E_r = \frac{I\ell\cos\theta}{2\pi(\sigma + i\epsilon\omega)r^3}(1 + \gamma r)e^{-\gamma r} \tag{6.13}$$

$$E_\theta = \frac{I\ell\sin\theta}{4\pi(\sigma + i\epsilon\omega)r^3}(1 + \gamma r + \gamma^2 r^2)e^{-\gamma r} \tag{6.14}$$

These, of course, reduce to the quasi-static forms (6.6) and (6.7) as $|\gamma r| \to 0$, and they also reduce to the purely static forms (6.4a) and (6.4b) as $\omega \to 0$.

Exercise: Show that if we define an auxiliary function A by

$$H_\phi = -\frac{\partial A}{\partial r} \sin \theta \qquad (6.15)$$

A satisfies

$$\frac{\partial}{\partial r}\left(r^2 \frac{\partial A}{\partial r} \right) = \gamma^2 r^2 A \qquad (6.16)$$

and

$$A = \frac{I\ell e^{-\gamma r}}{4\pi r} \qquad (6.17)$$

Exercise: Verify directly that (6.12), (6.13), and (6.14) satisfy Maxwell's equations.

Exercise: Show that when $|\gamma r| \to \infty$,

$$E_\theta \simeq \eta H_\phi \qquad (6.18)$$

where

$$\eta = \left(\frac{i\mu\omega}{\sigma + i\epsilon\omega} \right)^{1/2}$$

and

$$H_\phi \simeq \frac{\gamma I\ell e^{-\gamma r}(\sin \theta)}{4\pi r} \qquad (6.19)$$

For lossless media ($\sigma = 0$) we see that $\gamma = i\beta$, where $\beta = (\epsilon\mu)^{1/2}\omega$

Then we find that

$$H_\phi = \frac{I\ell}{4\pi r^2}(1 + i\beta r)e^{-i\beta r} \sin \theta \qquad (6.20)$$

$$E_r = \frac{I\ell}{2\pi i\epsilon\omega r^3}(1 + i\beta r)e^{-i\beta r} \cos \theta \qquad (6.21)$$

$$E_\theta = \frac{I\ell}{4\pi i\epsilon\omega r^3}(1 + i\beta r - \beta^2 r^2)e^{-i\beta r} \sin \theta \qquad (6.22)$$

If we now think of the current element as two time-varying charges $+q$ and $-q$ separated by a small distance ℓ, we can replace I by dq/dt, or $i\omega q$, and then (6.13) and (6.14) are replaced by

$$E_r = \frac{q\ell}{2\pi\epsilon r^3}(1 + i\beta r)e^{-i\beta r}\cos\theta \tag{6.23}$$

$$E_\theta = \frac{q\ell}{4\pi\epsilon r^3}(1 + i\beta r - \beta^2 r^2)e^{-i\beta r}\sin\theta \tag{6.24}$$

From these we recover the appropriate electrostatic formulas

$$E_r = \frac{q\ell\cos\theta}{2\pi\epsilon r^3} \tag{6.25}$$

$$E_\theta = \frac{q\ell\sin\theta}{4\pi\epsilon r^3} \tag{6.26}$$

in the limit as $\beta r \to 0$.

6.2 THE HERTZ VECTORS

In dealing with the electromagnetic fields produced by current-carrying wires [1], it is convenient to employ the Hertz vector Π. For a homogeneous medium we define it such that its curl leads back to the magnetic field vector $\mathbf{H}$. The constant of proportionality is chosen to agree with accepted convention [2]. Thus we write

$$\mathbf{H} = (\sigma + i\epsilon\omega)\operatorname{curl}\Pi \tag{6.27}$$

Then from Maxwell's equations

$$\mathbf{E} = \operatorname{curl}\operatorname{curl}\Pi \tag{6.28}$$

or

$$\mathbf{E} = -\gamma^2\Pi + \operatorname{grad}\operatorname{div}\Pi \tag{6.29}$$

To avoid conceptual difficulties, we elect to leave Π undefined right at the impressed source(s) of the electromagnetic field. However, the form of the Hertz vector near a source can be determined from prior considerations. An example is given in what follows.

In the case of a current element $I\ell$ located at the origin of a cylindrical coordinate system (ρ, ϕ, z) and oriented in the z direction, we are led to write

$$\Pi = (0, 0, \Pi_z) \tag{6.30}$$

where

$$\Pi_z = \frac{I\ell}{4\pi(\sigma + i\epsilon\omega)r}e^{-\gamma r} \tag{6.31}$$

It can be readily verified that this choice leads to the required form for the fields produced by the current element. For example, we may note that

$$H_\phi = (\sigma + i\epsilon\omega)\mathrm{curl}_\phi\, \Pi = -(\sigma + i\epsilon\omega)\frac{\partial \Pi_z}{\partial \rho}$$

$$= \frac{I\ell}{4\pi r^2}(1 + \gamma r)e^{-\gamma r} \sin\theta \qquad (6.32)$$

and

$$r = (\rho^2 + z^2)^{1/2} \qquad \text{and} \qquad \sin\theta = \frac{\rho}{r}$$

Exercise: Employ (6.28) to obtain explicit expressions for E_ρ and E_z for the current element and show that these are consistent with (6.13) and (6.14).

We argue that any distribution of applied or impressed currents can be represented as a superposition of individual current elements. For example, for an infinitesimal volume bounded by the lines of current flow and two (cross-sectional) surfaces *da* normal to them, the current in the volume element is *J da*, where *J* is the impressed current density. In the case where the vector current density has only a *z* component $\delta\Pi_z$, is given by

$$\delta\Pi_z = \frac{J_z\, da\, \ell}{4\pi(\sigma + i\epsilon\omega)} \frac{e^{-\gamma R}}{R} \qquad (6.33)$$

where ℓ is the length of the current element and R is the distance from the current element to the observation point.

Obviously, now we may replace *da ℓ* by the elemental volume *dv* and integrate over the volume *V* containing all the source elements to yield an expression for the resultant Hertz vector that has only a *z* component:

$$\Pi_z = \frac{1}{4\pi(\sigma + i\epsilon\omega)} \int_V \frac{e^{-\gamma R}}{R} J_z\, dv \qquad (6.34)$$

Similar equations would relate the *x* and *y* components of the Hertz vector to the *x* and *y* components of the impressed currents. Thus in vector form we write

$$\Pi = \frac{1}{4\pi(\sigma + i\epsilon\omega)} \int_V \frac{e^{-\gamma R}}{R} J\, dv \qquad (6.35)$$

where **J** here denotes the impressed current density vector. The corresponding field vectors are then obtained by operating on (6.35) with the use of (6.27) and (6.29). The resulting expressions are valid only for a homogeneous region of infinite extent.

We now return to the basic form of Maxwell's equations that include a source or electric current density **J**. These are

$$\text{curl } \mathbf{E} = -i\mu\omega\mathbf{H} \tag{6.36}$$

$$\text{curl } \mathbf{H} = (\sigma + i\epsilon\omega)\mathbf{E} + \mathbf{J} \tag{6.37}$$

for the usual time factor $\exp(i\omega t)$. Then on using (6.29) and (6.37), we see that

$$\text{curl curl } \Pi - \text{grad div } \Pi + \gamma^2\Pi = \frac{\mathbf{J}}{\sigma + i\epsilon\omega} \tag{6.38}$$

which is applicable for a homogeneous region. This equation can be written in symbolic form

$$(\tilde{\nabla}^2 - \gamma^2)\Pi = -\frac{\mathbf{J}}{\sigma + i\epsilon\omega} \tag{6.39}$$

where

$$\tilde{\nabla}^2 = -\text{curl curl} + \text{grad div}$$

is a vector operator. Actually, in rectangular coordinates we may express it as follows:

$$\tilde{\nabla}^2\Pi = i_x \nabla^2\Pi_x + i_y \nabla^2\Pi_y + i_z \nabla^2\Pi_z \tag{6.40}$$

Using the development of the preceding section, we can now assert that for an unbounded region (6.35) is a solution of the inhomogeneous Helmholtz equation given by (6.39).

The Hertz vector Π that we have introduced is associated with the impressed *electric* current density **J** in the homogeneous region under consideration. Sometimes, it is convenient to employ the magnetic Hertz vector Π°. It is defined in an analogous fashion by saying that for a homogeneous region the electric field may be derived from

$$\mathbf{E} = -i\mu\omega \text{ curl } \Pi^\circ \tag{6.41}$$

Then from Maxwell's equations

$$\mathbf{H} = (-\gamma^2 + \text{grad div})\Pi^\circ \tag{6.42}$$

A good example of a situation where the magnetic Hertz vector is useful is when we wish to compute the field of a small loop of wire carrying a uniform current I. The loop of area dA is located at the origin of a cylindrical coordinate system (ρ, ϕ, z) and oriented in the axial direction. We discussed the static solution for this problem in Chapter 1 where we showed that magnetic fields could be derived from the gradient of a scalar potential. We also introduced the concept of the Hertz vector in that context. The result can be generalized to the dynamic cases by analogy with the above treatment for the electric current element. Thus we are led to write

$$\Pi^\circ = (0, 0, \Pi_z^\circ) \tag{6.43}$$

where

$$\Pi_z^\circ = \frac{I\,dA}{4\pi R}\,e^{-\gamma R} \tag{6.44}$$

where R is the distance from the small loop to the observer. Clearly,

$$(\nabla^2 - \gamma^2)\Pi_z^\circ = 0 \qquad (\text{for } R \neq 0) \tag{6.45}$$

where ∇^2 is the laplacian operator in rectangular or cylindrical coordinates.

Using (6.41) and (6.42), we now see that the nonzero field components of the small loop in spherical coordinates (r, θ, ϕ) are

$$E_\phi = -\frac{i\mu\omega I\,dA}{4\pi r^2}(1 + \gamma r)e^{-\gamma r}\sin\theta \tag{6.46}$$

$$H_r = \frac{I\,dA}{2\pi r^3}(1 + \gamma r)e^{-\gamma r}\cos\theta \tag{6.47}$$

$$H_\theta = \frac{I\,dA}{4\pi r^3}(1 + \gamma r + \gamma^2 r)e^{-\gamma r}\sin\theta \tag{6.48}$$

As $|\gamma r| \to 0$, these expressions reduce to the appropriate static forms. Also, as $|\gamma r| \to \infty$, we see that

$$E_\phi \simeq -\eta H_\theta \simeq \frac{-i\mu\omega I\,dA\,e^{-\gamma r}\sin\theta}{4\pi r} \tag{6.49}$$

To emphasize the duality of the small, linear, electric current element and the small loop carrying a circumferential current, we exploit the magnetic current concept. Thus we set

$$I\,dA = (i\mu\omega)^{-1}M_z\ell \tag{6.50}$$

where M_z is the equivalent z-directed magnetic current and ℓ is the length of this equivalent element. Then (6.44) is rewritten

$$\Pi_z^\circ = \frac{M_z\ell}{4\pi i\mu\omega}\frac{e^{-\gamma R}}{R} \tag{6.51}$$

We now are in the position to generalize the previous discussion by writing Maxwell's equations in the form

$$\text{curl } \mathbf{E} = -i\mu\omega\mathbf{H} - \mathbf{M} \tag{6.52}$$

$$\text{curl } \mathbf{H} = (\sigma + i\epsilon\omega)\mathbf{E} + \mathbf{J} \tag{6.53}$$

where both electric and magnetic source densities $\mathbf{J}$ and $\mathbf{M}$ are allowed. As before, the electric Hertz vector satisfies (6.39), but, in addition, we can assert that the magnetic Hertz vector satisfies

$$(\tilde{\nabla}^2 - \gamma^2)\Pi^\circ = -\frac{\mathbf{M}}{i\mu\omega} \tag{6.54}$$

Employing the previous logic, we can write the solution of (6.54) for an unbounded homogeneous region in the form

$$\Pi^\circ = \frac{1}{4\pi i\mu\omega} \int_V \frac{e^{-\gamma R}}{R} \mathbf{M} \, dv \tag{6.55}$$

where $\mathbf{M}$ is the distribution of the impressed magnetic current density enclosed by the volume V.

In the general case, for a homogeneous region where we have both source electric and source magnetic currents, $\mathbf{J}$ and $\mathbf{M}$, respectively, the fields are obtained from

$$\mathbf{E} = (-\gamma^2 + \text{grad div})\Pi - i\mu\omega \, \text{curl } \Pi^\circ \tag{6.56}$$

$$\mathbf{H} = (-\gamma^2 + \text{grad div})\Pi^\circ + (\sigma + i\epsilon\omega)\text{curl } \Pi \tag{6.57}$$

where Π and Π° satisfy (6.39) and (6.54), respectively.

6.3 SOMMERFELD'S HALF-SPACE PROBLEM

When an antenna is located near a reflecting surface, the resultant electromagnetic fields will be modified. In our earlier examples we considered only the effect of a perfectly conducting ground plane of infinite extent. In such cases we could replace the secondary influence by a geometrical image field. In this way the resultant tangential electric field could be made to vanish at the ground plane.

Unfortunately, the earth's surface is not a perfect conductor, nor is it a plane. This fact greatly complicates any quantitative calculations of the antenna performance. A generic problem that provides useful information was formulated by Arnold Sommerfeld [3] many years ago. Specifically, he considered an oscillating-dipole source to be located over the plane interface separating two homogeneous half-space regions. We will outline his formal solution here and discuss some useful engineering approximations.

The situation is illustrated in Figure 6.2, where we locate an electric current element $I\ell$ at height h over the plane interface at $z = 0$, and it is oriented in the z direction. The upper half space is assumed to be homogeneous with electrical properties σ_0, ϵ_0, and μ_0. The objective is to calculate the field at P, which is any point in the upper half-space.

Because of the symmetry of the problem, we can set $\partial/\partial\phi = 0$. Also, it is obvious that $E_\phi = 0$ even allowing for induced currents in the lower

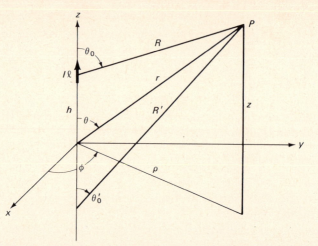

Figure 6.2 Electric dipole or current element (with vertical orientation located at height h over plane interface separating two homogeneous half spaces. Basic cylindrical (ρ, ϕ, z) and spherical (r, θ, ϕ) coordinate systems are shown in addition to spherical coordinate systems (R, θ_0, ϕ) and (R', θ'_0, ϕ) associated with source and image locations, respectively.

half space. We are thus led to express the fields in terms of an electric Hertz vector that has only a z component Π_z. Thus for $z < 0$ we write

$$E_\rho = \frac{\partial^2}{\partial \rho\, \partial z}\, \Pi_z \tag{6.58}$$

$$E_z = \left(-\gamma^2 + \frac{\partial^2}{\partial z^2}\right)\Pi_z \tag{6.59}$$

$$H_\phi = -(\sigma + i\epsilon\omega)\frac{\partial}{\partial \rho}\, \Pi_z \tag{6.60}$$

where

$$\gamma^2 = i\mu_0\omega(\sigma + i\epsilon\omega)$$

For $z > 0$ we add a subscript 0 to γ, σ, and ϵ, while for $z < 0$ we add a subscript 1. This convention will be used in what follows.

Now except at the source, Π_z satisfies

$$(\nabla^2 - \gamma_0^2)\Pi_z = 0 \tag{6.61}$$

while for $z < 0$

$$(\nabla^2 - \gamma_1^2)\Pi_z = 0 \tag{6.62}$$

Here

$$\nabla^2 = \frac{1}{\rho}\frac{\partial}{\partial \rho}\, \rho\, \frac{\partial}{\partial \rho} + \frac{\partial^2}{\partial z^2} \tag{6.63}$$

is the azimuthally symmetric laplacian operator. Solutions of (6.61) are of the form

$$J_0(\lambda\rho)\exp(\pm u_0 z)$$

or

$$Y_0(\lambda\rho)\exp(\pm u_0 z)$$

where $u_0 = (\lambda^2 + \gamma_0^2)^{1/2}$ and λ is a free parameter. The $Y_0(\lambda\rho)$ is singular a $\rho = 0$ for all z, so it is discarded.

We now observe that for $z > 0$

$$\Pi_z = \Pi_z^{\text{prim}} + \Pi_z^{\text{sec}} \tag{6.64}$$

where

$$\Pi_z^{\text{prim}} = \frac{I\ell}{4\pi i \epsilon_0 \omega} \frac{e^{-\gamma_0 R}}{R} \tag{6.65}$$

and where

$$R = [\rho^2 + (z - h)^2]^{1/2}$$

and Π_z^{sec} is, as yet, unknown. Clearly, Π_z^{prim} is the primary influence, while Π_z^{sec} is the secondary influence that results from the presence of the lower half space.

The key step now is to employ the known Sommerfeld integral representation [3] to write

$$\frac{e^{-\gamma_0 R}}{R} = \int_0^\infty \frac{\lambda}{u_0} e^{-u_0|z-h|} J_0(\lambda\rho) \, d\lambda \tag{6.66}$$

This leads us to postulate that for the region $z > 0$

$$\Pi_z = \frac{I\ell}{4\pi i \epsilon_0 \omega} \int_0^\infty \frac{\lambda}{u_0} [e^{-u_0|z-h|} + R_e(\lambda)e^{-u_0(z+h)}] J_0(\lambda\rho) \, d\lambda \tag{6.67}$$

where $R_e(\lambda)$ is a function of λ to be determined. The integration contour here is the real axis of λ running from 0 to ∞. Also, to be definite, we choose the radical such that Re $u_0 > 0$ for all real values of λ. Thus for points $z > h$ the integrand must contain the factor $\exp(-u_0 z)$ to ensure finite fields as $z \to \infty$.

In dealing with region $z < 0$, where there are no sources, the solution must contain the factor $\exp(+u_1 z)$, so we are led to write

$$\Pi_z = \frac{I\ell}{4\pi i \epsilon_0 \omega} \int_0^\infty T_e(\lambda)e^{u_1 z} J_0(\lambda\rho) \, d\lambda \tag{6.68}$$

where $u_1 = (\lambda^2 + \gamma_1^2)^{1/2}$. This representation for Π_z satisfies (6.62), but $T_e(\lambda)$, a function of λ, is not yet known. Again for sake of consistency we choose Re $u_1 > 0$.

The boundary conditions for the problem may now be applied. These require that E_ρ and H_ϕ are continuous at $z = 0$. Such will be the case if both $\partial \Pi_z / \partial z$ and $\gamma^2 \Pi_z$ are continuous. Application to (6.67) and (6.68) leads to the two algebraic equations

$$\lambda[1 - R_e(\lambda)]e^{-u_0 h} = u_1 T_e(\lambda) \tag{6.69}$$

and

$$\frac{\lambda}{u_0} \gamma_0^2 [1 + R_e(\lambda)]e^{-u_0 h} = \gamma_1^2 T_e(\lambda) \tag{6.70}$$

Solving for R_e and T_e, we get

$$R_e(\lambda) = \frac{N^2 u_0 - u_1}{N^2 u_0 + u_1} \tag{6.71}$$

and

$$T_e(\lambda) = \frac{2\lambda}{N^2 u_0 + u_1} e^{-u_0 h} \tag{6.72}$$

where

$$N^2 = \frac{\gamma_1^2}{\gamma_0^2}$$

Inserting $R_e(\lambda)$ and $T_e(\lambda)$ into (6.67) and (6.68), we obtain explicit integral formulas for the Hertz vector that has only a z component. The fact that the boundary conditions are met vindicates our initial assumption that $\Pi_x = \Pi_y = 0$ for this configuration.

When we deal with a horizontal electric dipole or current element I located at $z = h$, the situation is a bit more complicated. But without losing any generality, we may locate $I\ell$ at $x = y = 0$ and orient it in the x direction. Again, the primary fields can be derived from a Hertz vector with an x component Π_x^{prim} parallel to the current element I. It is given by

$$\Pi_x^{\mathrm{prim}} = \frac{I\ell}{4\pi i \epsilon_0 \omega} \frac{e^{-\gamma_0 R}}{R} \tag{6.73}$$

This suggests that we write the following form for the resultant x component of the Hertz vector for $z > 0$:

$$\Pi_x = \frac{I\ell}{4\pi i \epsilon_0 \omega} \int_0^\infty \frac{\lambda}{u_0} [e^{-u_0|z-h|} + R_m(\lambda)e^{-u_0(z+h)}] J_0(\lambda \rho) \, d\lambda \tag{6.74}$$

and for $z < 0$

$$\Pi_x = \frac{I\ell}{4\pi i \epsilon_0 \omega} \int_0^\infty T_m(\lambda)e^{u_1 z} J_0(\lambda \rho) \, d\lambda \tag{6.75}$$

Some experimentation would quickly tell us that the boundary conditions could not be met if Π had only an x component. As Sommerfeld [3] pointed out many years ago, the situation could be remedied if Π was allowed to have x *and* z components. Thus the electromagnetic fields, expressed in rectangular components, would be obtained in the following manner:

$$E_x = -\gamma^2 \Pi_x + \frac{\partial}{\partial x}(\text{div } \Pi) \tag{6.76}$$

$$E_y = \frac{\partial}{\partial y}(\text{div } \Pi) \tag{6.77}$$

$$E_z = -\gamma^2 \Pi_z + \frac{\partial}{\partial z}(\text{div } \Pi) \tag{6.78}$$

$$H_x = (\sigma + i\epsilon\omega)\frac{\partial \Pi_z}{\partial y} \tag{6.79}$$

$$H_y = -(\sigma + i\epsilon\omega)\left(\frac{\partial \Pi_z}{\partial x} - \frac{\partial \Pi_x}{\partial z}\right) \tag{6.80}$$

and

$$H_z = -(\sigma + i\epsilon\omega)\frac{\partial \Pi_x}{\partial y} \tag{6.81}$$

In the present case

$$\text{div } \Pi = \frac{\partial \Pi_x}{\partial x} + \frac{\partial \Pi_z}{\partial z} \tag{6.82}$$

As before, we add a subscript 0 to these quantities when we deal explicitly with the upper half space.

In order to construct the appropriate form for Π_z in the two regions, we first examine the nature of the boundary conditions in the present problem. Specifically, E_x, E_y, H_x, and H_y must be continuous at $z = 0$. Some consideration indicates such is the case if $\gamma^2 \Pi_x$, $\gamma^2 \Pi_z$, $\gamma^2 \, \partial \Pi_x/\partial z$, and div Π are continuous at $z = 0$. We now observe that

$$\frac{\partial}{\partial x} J_0(\lambda\rho) = -\lambda J_1(\lambda\rho)\cos\phi \tag{6.83}$$

and also we require that

$$(\nabla^2 - \gamma^2)\Pi_z = 0 \tag{6.84}$$

where

$$\nabla^2 = \frac{\partial^2}{\partial x^2} + \frac{\partial^2}{\partial y^2} + \frac{\partial^2}{\partial z^2} = \frac{1}{\rho}\frac{\partial}{\partial \rho}\rho\frac{\partial}{\partial \rho} + \frac{1}{\rho^2}\frac{\partial^2}{\partial \phi^2} + \frac{\partial^2}{\partial z^2} \tag{6.85}$$

This suggests that the required forms for $z > 0$ are

$$\Pi_z = \frac{I\ell}{4\pi i \epsilon_0 \omega} \frac{\partial}{\partial x} \int_0^\infty S(\lambda) e^{-u_0(z+h)} J_0(\lambda\rho) \, d\lambda \qquad (6.86)$$

while for $z < 0$

$$\Pi_z = \frac{I\ell}{4\pi i \epsilon_0 \omega} \frac{\partial}{\partial x} \int_0^\infty Q(\lambda) e^{u_1 z} J_0(\lambda\rho) \, d\lambda \qquad (6.87)$$

where S and Q are to be determined. This particular choice for Π_z leads to a convenient form for the divergence. For example, for $z > 0$

$$\text{div } \Pi = \frac{I\ell}{4\pi i \epsilon_0 \omega} \frac{\partial}{\partial x} \int_0^\infty \left\{ \frac{\lambda}{u_0} \left[e^{-u_0|z-h|} + R_m(\lambda) e^{-u_0(z+h)} \right] \right.$$
$$\left. - u_0 S(\lambda) e^{-u_0(z+h)} \right\} \times J_0(\lambda\rho) \, d\lambda \qquad (6.88)$$

In the case where $z < 0$

$$\text{div } \Pi = \frac{I\ell}{4\pi i \epsilon_0 \omega} \frac{\partial}{\partial x} \int_0^\infty [T_m(\lambda) + u_1 Q(\lambda)] e^{u_1 z} J_0(\lambda\rho) \, d\lambda \qquad (6.89)$$

We are now in the position to enforce the continuity of the tangential electric and magnetic fields. The resulting algebraic equations are

$$\gamma_0^2 \frac{\lambda}{u_0} (1 + R_m) e^{-u_0 h} = \gamma_1^2 T_m \qquad (6.90)$$

$$\gamma_0^2 S e^{-u_0 h} = \gamma_1^2 Q \qquad (6.91)$$

$$\gamma_0^2 \lambda (1 - R_m) e^{-u_0 h} = \gamma_1^2 u_1 T_m \qquad (6.92)$$

and

$$\left[\frac{\lambda}{u_0} (1 + R_m) - u_0 S \right] e^{-u_0 h} = T_m + u_1 Q \qquad (6.93)$$

We can immediately solve (6.90) and (6.92) to yield

$$R_m(\lambda) = \frac{u_0 - u_1}{u_0 + u_1} \qquad (6.94)$$

and

$$T_m(\lambda) = \frac{1}{N^2} \frac{2\lambda}{u_0 + u_1} e^{-u_0 h} \qquad (6.95)$$

Then we work with (6.91) and (6.93) to deduce that

$$S(\lambda) = \frac{2\lambda(u_1 - u_0)}{\gamma_0^2(N^2 u_0 + u_1)} = \frac{R_e(\lambda) + R_m(\lambda)}{\lambda} \qquad (6.96)$$

and

$$Q_e = \frac{2\lambda(u_1 - u_0)}{\gamma_1^2(N^2 u_0 + u_1)} e^{-u_0 h} \tag{6.97}$$

This completes the formal integral solution for the electromagnetic fields of the horizontal electric dipole.

Exercise: Consider that a vertical magnetic dipole or small current-carrying loop $I\,dA$ is located at $z = h$ over the horizontal interface $z = 0$ with the same configuration as in Figure 6.2. Using the same notation as before, show that for $z > 0$

$$\mathbf{\Pi} = 0 \qquad \text{and} \qquad \mathbf{\Pi}^{\circ} = (0, 0, \Pi_z^{\circ}) \tag{6.98}$$

where

$$\Pi_z^{\circ} = \frac{I\,dA}{4\pi} \int_0^{\infty} \frac{\lambda}{u_0} \left[e^{-u_0|z-h|} + R_m(\lambda) e^{-u_0(z+h)} \right] J_0(\lambda\rho)\,d\lambda \tag{6.99}$$

where R_m is given by (6.94).

Exercise: Consider that a horizontal magnetic dipole or small current-carrying loop $I\,dA$ is located at $z = h$ and oriented in the x direction (that is, the loop is in the yz plane). In this case show that for $z > 0$

$$\mathbf{\Pi} = 0 \qquad \text{and } \mathbf{\Pi}^{\circ} = (\Pi_x^{\circ}, 0, \Pi_z^{\circ}) \tag{6.100}$$

where

$$\Pi_x^{\circ} = \frac{I_0\,dA}{4\pi} \int_0^{\infty} \frac{\lambda}{u_0} \left[e^{-u_0|z-h|} + R_e(\lambda) e^{-u_0(z+h)} \right] J_0(\lambda\rho)\,d\lambda \tag{6.101}$$

and

$$\Pi_z^{\circ} = \frac{I_0\,dA}{4\pi} \frac{\partial}{\partial x} \int_0^{\infty} S(\lambda) e^{-u_0(z+h)} J_0(\lambda\rho)\,d\lambda \tag{6.102}$$

where R_e and S are given by (6.71) and (6.96), respectively.

Exercise: Show that the preceding expressions for $\mathbf{\Pi}$ and $\mathbf{\Pi}^{\circ}$ for $z < 0$ can be generalized to the case where the permeability is μ_0 for $z > 0$ and is μ_1 for $z < 0$ by redefining the coefficients as follows:

$$R_e(\lambda) = \frac{K_0 - K_1}{K_0 + K_1} \tag{6.103}$$

$$R_m(\lambda) = \frac{N_0 - N_1}{N_0 + N_1} \tag{6.104}$$

and

$$S(\lambda) = [R_e(\lambda) + R_m(\lambda)]\lambda^{-1} \tag{6.105}$$

where

$$K_0 = \frac{u_0}{i\epsilon_0\omega} \qquad K_1 = \frac{u_1}{\sigma_1 + i\epsilon_1\omega}$$

and

$$N_0 = \frac{u_0}{i\mu_0\omega} \qquad N_1 = \frac{u_1}{i\mu_1\omega}$$

Exercise: Consider the further generalization where the lower half space $z < 0$ is composed of a layer of thickness d with properties σ_1, ϵ_1, and μ_1, while the lower semi-infinite region (that is, $z < -d$) has properties σ_2, ϵ_2, and μ_2. The source (electric or magnetic) dipoles are again located at $z = h$ on the z axis. Show that the expression for Π and Π° for $z > 0$ can now be generalized if we set

$$R_e(\lambda) = \frac{K_0 - Z_1}{K_0 + Z_1} \tag{6.109}$$

$$R_m(\lambda) = \frac{N_0 - Y_1}{N_0 + Y_1} \tag{6.107}$$

and

$$S(\lambda) = [R_e(\lambda) + R_m(\lambda)]\lambda^{-1} \tag{6.108}$$

where

$$Z_1 = K_1 \frac{K_2 + K_1 \tanh u_1 d}{K_1 + K_2 \tanh u_1 d} \tag{6.109}$$

and

$$Y_1 = N_1 \frac{N_2 + N_1 \tanh u_1 d}{N_1 + N_2 \tanh u_1 d} \tag{6.110}$$

where

$$K_j = \frac{u_j}{\sigma_j + i\epsilon_j\omega} \qquad j = 0, 1, 2 \tag{6.111}$$

and

$$N_j = \frac{u_j}{i\mu_j\omega} \qquad j = 0, 1, 2 \tag{6.112}$$

The further generalization [4] to an M-layered half space should be evident if the plane wave reflection coefficients from Chapter 4 are employed.

Exercise: Consider the following related problem. A vertical electric dipole or current element is located at height h over a plane interface $z = 0$. The region for $z > 0$ is free space with electrical properties ϵ_0 and μ_0. We now insist that the tangential fields at $z = 0$ satisfy the impedance boundary condition $E_\rho = -H_\phi Z$, where Z is fixed and assumed known. Show that the fields may be derived from an electric Hertz vector that has only a z component given by (6.67) with $R_e(\lambda)$ defined by

$$R_e(\lambda) = \frac{u_0 - i\epsilon_0 \omega Z}{u_0 + i\epsilon_0 \omega Z} \tag{6.113}$$

Exercise: Consider the extension of the previous exercise to the case where the current element is oriented in the x direction. The boundary conditions are now stated in the form

$$E_x = -ZH_y \qquad \text{and} \qquad E_y = ZH_x \qquad \text{at } z = 0$$

Show that for $z > 0$

$$\Pi = (\Pi_x, 0, \Pi_z) \tag{6.114}$$

where Π_x is given by (6.74) with

$$R_m(\lambda) = \frac{u_0 - i\mu_0 \omega Z^{-1}}{u_0 + i\mu_0 \omega Z^{-1}} \tag{6.115}$$

and Π_z is given by (6.86) with

$$S(\lambda) = \frac{2\lambda}{(u_0 + i\epsilon_0 \omega Z)(u_0 + i\mu_0 \omega Z^{-1})} \tag{6.116}$$

Hint: Note that the boundary conditions can be expressed, at $z = 0$, in the form

$$-\gamma_0^2 \Pi_x = i\epsilon_0 \omega Z \frac{\partial \Pi_x}{\partial z} \tag{6.117}$$

and

$$\text{div } \Pi = i\epsilon_0 \omega Z \Pi_z \tag{6.118}$$

6.4 DEBYE POTENTIALS AND THE HALF-SPACE PROBLEM

As we have indicated, there is a systematic procedure for introducing Hertz vectors to deal with dipoles over a conducting half space. But the choice of the type of Hertz vector may not be unique [5]. For example, in the case of a horizontal electric dipole, we can employ both electric *and* magnetic Hertz vectors that are purely z-directed [6]. Then, for example, we would say that

$$\Pi = (0, 0, U) \tag{6.119}$$

and

$$\Pi^\circ = (0, 0, V) \tag{6.120}$$

where U and V are sometimes called the Debye potentials.
For the region $z > 0$ we would write

$$U = U^{\text{prim}} + U^{\text{sec}} \tag{6.121}$$

$$V = V^{\text{prim}} + V^{\text{sec}} \tag{6.122}$$

which is the usual decomposition into primary and secondary influences.
We now need to obtain appropriate forms for the primary Debye potentials. To provide the needed link, we first note that

$$E_z^{\text{prim}} = \left(-\gamma_0^2 + \frac{\partial^2}{\partial z^2} \right) U^{\text{prim}} \tag{6.123}$$

and

$$H_z^{\text{prim}} = \left(-\gamma_0^2 + \frac{\partial^2}{\partial z^2} \right) V^{\text{prim}} \tag{6.124}$$

But we can also write

$$E_z^{\text{prim}} = \frac{\partial^2 \Pi_x^{\text{prim}}}{\partial x \, \partial z} \tag{6.125}$$

and

$$H_z^{\text{prim}} = -i\epsilon_0 \omega \frac{\partial \Pi_x^{\text{prim}}}{\partial y} \tag{6.126}$$

where Π_x^{prim} is given by (6.73). This suggests that the desired forms are

$$U^{\text{prim}} = \frac{\partial}{\partial x} \int_0^\infty f(\lambda) e^{-u_0|z-h|} J_0(\lambda \rho) \, d\lambda \tag{6.127}$$

and

$$V^{\text{prim}} = \frac{\partial}{\partial y} \int_0^\infty f^*(\lambda) e^{-u_0|z-h|} J_0(\lambda \rho) \, d\lambda \tag{6.128}$$

where $f(\lambda)$ and $f^*(\lambda)$ are to be determined by matching the right-hand sides of (6.123) with (6.125), and (6.124) with (6.126).

According to (6.123) and (6.124),

$$E_z^{\text{prim}} = \frac{\partial}{\partial x} \int_0^\infty \lambda^2 f(\lambda) e^{-u_0|z-h|} J_0(\lambda\rho) \, d\lambda \tag{6.129}$$

and

$$H_z^{\text{prim}} = \frac{\partial}{\partial y} \int_0^\infty \lambda^2 f^*(\lambda) e^{-u_0|z-h|} J_0(\lambda\rho) \, d\lambda \tag{6.130}$$

On the other hand, according to (6.125) and (6.126),

$$E_z^{\text{prim}} = \frac{I\ell}{4\pi i\epsilon_0\omega} \frac{\partial}{\partial x} \int_0^\infty (\pm)\lambda e^{\pm u_0(z-h)} J_0(\lambda\rho) \, d\lambda \tag{6.131}$$

where plus is to be used for $z < h$ and minus for $z > h$, and

$$H_z^{\text{prim}} = -\frac{I\ell}{4\pi} \frac{\partial}{\partial y} \int_0^\infty \frac{\lambda}{u_0} e^{\pm u_0(z-h)} J_0(\lambda\rho) \, d\lambda \tag{6.132}$$

with the same convention for the choice of signs.

The right-hand sides of (6.129) and (6.131) are equal if

$$f(\lambda) = (\pm) \frac{I\ell}{4\pi i\epsilon_0\omega\lambda} \tag{6.133}$$

and the right-hand sides of (6.130) and (6.132) are equal if

$$f^*(\lambda) = -\frac{I\ell}{4\pi\lambda u_0} \tag{6.134}$$

When these expressions for $f(\lambda)$ and $f^*(\lambda)$ are inserted into (6.127) and (6.128), we have the desired forms for the primary Debye potentials.

We are now in the position to write the forms for the resultant Debye potentials for the horizontal current element at height h over the interface at $z = 0$. In the case for $z > 0$

$$U = (\pm) \frac{I\ell}{4\pi i\epsilon_0\omega} \frac{\partial}{\partial x} \int_0^\infty \frac{1}{\lambda} \left[e^{-u_0|z-h|} + R_e(\lambda)e^{-u_0(z+h)} \right] J_0(\lambda\rho) \, d\lambda \tag{6.135}$$

and

$$V = -\frac{I\ell}{4\pi} \frac{\partial}{\partial y} \int_0^\infty \frac{1}{\lambda u_0} \left[e^{-u_0|z-h|} + R_m(\lambda)e^{-u_0(z+h)} \right] J_0(\lambda\rho) \, d\lambda \tag{6.136}$$

The corresponding field components are obtained from

$$E_x = \frac{\partial^2 U}{\partial x \, \partial z} - i\mu_0\omega \frac{\partial V}{\partial y} \tag{6.137}$$

$$E_y = \frac{\partial^2 U}{\partial y \, \partial z} + i\mu_0\omega \frac{\partial V}{\partial x} \tag{6.138}$$

$$E_z = \left(-\gamma_0^2 + \frac{\partial^2}{\partial z^2} \right) U \tag{6.139}$$

and

$$H_x = \frac{\partial^2 V}{\partial x\, \partial z} + i\epsilon_0 \omega \frac{\partial U}{\partial y} \tag{6.140}$$

$$H_y = \frac{\partial^2 V}{\partial y\, \partial z} - i\epsilon_0 \omega \frac{\partial U}{\partial x} \tag{6.141}$$

$$H_z = \left(-\gamma_0^2 + \frac{\partial^2}{\partial z^2} \right) V \tag{6.142}$$

Similar forms for U and V hold for the lower half space. Thus for $z < 0$ we would write

$$U = \frac{\partial}{\partial x} \int_0^\infty F(\lambda) e^{u_1 z} J_0(\lambda \rho)\, d\lambda \tag{6.143}$$

and

$$V = \frac{\partial}{\partial y} \int_0^\infty G(\lambda) e^{u_1 z} J_0(\lambda \rho)\, d\lambda \tag{6.144}$$

The corresponding field expressions would be obtained from (6.137)–(6.142) where the following changes are made: $i\mu_0\omega \to i\mu_1\omega$, $i\epsilon_0\omega \to \sigma_1 + i\epsilon_1\omega$, and $\gamma_0 \to \gamma_1$.

We now deduce that the continuity of the tangential fields is guaranteed if the following conditions hold: $\partial U/\partial z$, $\partial V/\partial z$, $(\sigma + i\epsilon\omega)U$, and $i\mu\omega V$ are continuous across the interface at $z = 0$. The algebraic procedure is simplified because the U and V equations separate. Then we easily deduce that $R_e(\lambda)$ and $R_m(\lambda)$ have the forms given by (6.71) and (6.94), respectively.

The procedure outlined can be generalized readily to a layered half space by using the appropriate expressions for $R_e(\lambda)$ and $R_m(\lambda)$. Also, the extension to a horizontal magnetic dipole source is straightforward, and this is left as an exercise for the reader.

The use of z-directed Hertz vectors (of both electric and magnetic types) simplifies the application of the boundary conditions at planar interfaces. The complication is that a representation for the primary field in terms of these Debye potentials is not always self-evident.

Exercise: Show that the horizontal electric dipole problem discussed above can also be formulated in terms of an electric Hertz vector with only x and y components. *Hint:* Express the fields and the source current distribution in terms of a two-dimensional integral of the form

$$\int_{-\infty}^{+\infty} \int_{-\infty}^{+\infty} (\quad) e^{-i\lambda x} e^{-i\beta y} \, d\lambda \, d\beta$$

6.5 RADIATION AND FAR-ZONE FIELDS

The field expressions for dipole sources located over the planar interface all involve integrals of the type

$$P(\alpha, \rho) = \int_0^\infty p(\lambda) \frac{\lambda}{u_0} e^{-u_0\alpha} J_0(\lambda\rho) \, d\lambda \tag{6.145}$$

where α and ρ are linear distances and $p(\lambda)$ is a specified function of λ. We now wish to obtain simplified formulas for field quantities in the far zone. The essential step in the reasoning is that the linear distance $(\alpha^2 + \rho^2)^{1/2}$ is taken to be indefinitely large. What this amounts to in the present context is that we approximate the integral in the following fashion:

$$P(\alpha, \rho) \simeq p(\bar{\lambda}) \int_0^\infty \frac{\lambda}{u_0} e^{-u_0\alpha} J_0(\lambda\rho) \, d\lambda \tag{6.146}$$

$$\simeq \frac{p(\bar{\lambda})e^{-iks}}{s}$$

where $s = (\alpha^2 + \rho^2)^{1/2}$ and $u_0 = (\lambda^2 - k^2)^{1/2}$ and where $k = -i\gamma_0 = (\epsilon_0\mu_0)^{1/2}\omega = \omega/c$. Here we have taken $p(\lambda)$ outside the integral and replaced it by $p(\bar{\lambda})$, where $\bar{\lambda}$ is the predominant value of λ. Of course, the result is exact if $p(\lambda) = 1$, in which case it becomes the Sommerfeld integral.

We can justify this rather cavalier treatment of an infinite integral by an asymptotic analysis (given below), but for the moment we will only indicate its plausibility on physical grounds.

We note that $\alpha = |z - h|$ or $(z + h)$ is the vertical displacement of the observer from the source or geometrical image point. Then, of course, ρ is the horizontal displacement. The geometrical picture in Figure 6.3 may help to visualize the situation.

Now when ρ and α are sufficiently large, the integrand behaves as

$$e^{-u_0\alpha} e^{\pm i\lambda\rho} \simeq \exp[-iks \cos(\delta \mp \bar{\delta})] \tag{6.147}$$

where

$$\alpha = s \cos \bar{\delta} \qquad \rho = s \sin \bar{\delta} \qquad \lambda = k \sin \delta$$

and

$$u_0 = ik \cos \delta$$

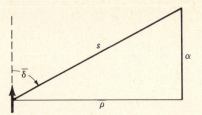

Figure 6.3   Geometrical parameters for far-zone approximation.

The contour, for λ from 0 to ∞, now changes to δ from 0 to $\pi/2$ to $\pi/2 + i\infty$. From the geometry of the situation we see that the important values of δ must be near $\bar{\delta}$, where $\cos \bar{\delta} = \alpha/s$. This suggests that the most significant value of λ, denoted $\bar{\lambda}$, is given by $k \sin \bar{\delta}$. The tacit assumption, of course, is that $p(\lambda)$ is sufficiently slowly varying to be taken outside the integral with $\lambda \simeq \bar{\lambda}$

We now apply the far-zone concept to a few specific cases. For example, consider the formal exact integral representation for Π_z for a vertical electric dipole, as given by (6.67). Clearly, the desired result for the far zone is

$$\Pi_z \simeq \frac{I\ell}{4\pi i\epsilon_0 \omega} \left[\frac{e^{-ikR}}{R} + R_e(\bar{\lambda}) \frac{e^{-ikR'}}{R'} \right] \tag{6.148}$$

where

$$R = [\rho^2 + (z - h)^2]^{1/2} \qquad R' = [\rho^2 + (z + h)^2]^{1/2}$$

and

$$\bar{\lambda} = k \sin \theta_0' = \frac{k\rho}{R'}$$

The geometrical parameters are indicated in Figure 6.3. In the case of a homogeneous lower half space, with $\mu_1 = \mu_0$, we can write

$$R_e(\bar{\lambda}) = \frac{N^2 \cos \theta_0' - (N^2 - \sin^2 \theta_0')^{1/2}}{N^2 \cos \theta_0' + (N^2 - \sin^2 \theta_0')^{1/2}} \tag{6.149}$$

which has the recognizable Fresnel form [2] for a TM plane wave incident at an angle θ_0' onto a half space with complex refractive index $N^2 = \gamma_1^2/\gamma_0^2 = (\sigma_1 + i\epsilon_1\omega)/(i\epsilon_0\omega)$.

To obtain the field components, we need to perform the differentiation operations on Π_z. This calls for some comment when dealing with far-zone fields. For example, for the vertical electric dipole source the magnetic field has only a ϕ component given by

$$H_\phi = -i\epsilon_0\omega \frac{\partial \Pi_z}{\partial \rho} \tag{6.150}$$

Now

$$\frac{\partial}{\partial \rho} \frac{e^{-ikR}}{R} = -ik\left(1 + \frac{1}{ikR}\right)\frac{\rho}{R} \frac{e^{-ikR}}{R}$$

$$\simeq -ik\frac{\rho}{R}\frac{e^{-ikR}}{R} \simeq -ik(\sin\theta_0)\frac{e^{-ikR}}{R} \tag{6.151}$$

where we have discarded terms that vary as $1/R^2$. In a similar fashion, we write

$$\frac{\partial}{\partial \rho} \frac{e^{-ikR'}}{R'} \simeq -ik\sin\theta_0'\frac{e^{-ikR'}}{R'} \tag{6.152}$$

Thus the far-zone approximation for the magnetic field is

$$H_\phi \simeq \frac{ikI\ell}{4\pi}\left[\frac{e^{-ikR}}{R}\sin\theta_0 + R_e(\bar{\lambda})\frac{e^{-ikR'}}{R'}\sin\theta_0'\right] \tag{6.153}$$

If, *in addition*, we say that $R \gg h$, the above may be simplified if we note that

$$R \approx r - h\cos\theta$$

and

$$R' \approx r + h\cos\theta$$

where $r = (\rho^2 + z^2)^{1/2}$ and $\cos\theta = z/r$. As indicated in Figure 6.3, (r, θ) are spherical coordinates about the origin at the earth's surface directly below the source dipole. Then (6.153) becomes

$$H_\phi \simeq \frac{ikI\ell}{4\pi}\frac{e^{-ikr}}{r}[e^{ikh\cos\theta} + R_e(\bar{\lambda})e^{-ikh\cos\theta}]\sin\theta \tag{6.154}$$

where

$$R_e(\bar{\lambda}) \simeq \frac{N^2\cos\theta - (N^2 - \sin^2\theta)^{1/2}}{N^2\cos\theta + (N^2 - \sin^2\theta)^{1/2}} \tag{6.155}$$

It is important to note the distinction between (6.153) and (6.154). In both cases the range (that is, R, R', or r) is assumed to be very large compared with the wavelength $2\pi/k$. But in the latter case the height h of the dipole above the interface is bounded, so that, in effect, the direct and the reflected rays are parallel.

Another interesting feature of the far-zone field expressions is the behavior for grazing angles. For example, if we allow $z = h \to 0$, the angles θ_0, θ_0', θ all tend to $\pi/2$. Then if N is finite (that is, noninfinite), $R_e(\bar{\lambda}) \to -1$, and the field approaches zero. This is simply a statement to the effect that the leading term in the asymptotic expansion varies as $1/r^2$ or some higher inverse power of the range.

But there is still an interesting paradox that remains. If in the general expression for H_ϕ we let $|N|^2 \to \infty$, corresponding to a perfectly conducting ground plane, we see that $R_e \to +1$. This suggests indeed that the leading term in the far-zone expansion is varying as $1/r$. This is an example where we have two limiting processes (that is, N becoming infinite and θ tending to $\pi/2$). The resultant depends on the order in which these limits are taken. Physically, the paradox is connected with the fact that the Brewster angle is nearly $\pi/2$ when $|N|$ is sufficiently large.

As an other example, we consider the vertical magnetic field H_z produced by a horizontal electric dipole or current element. Without approximation, for $z > 0$, we would write

$$H_z = -i\epsilon_0 \omega \frac{\partial \Pi_x}{\partial y} \tag{6.156}$$

where Π_x is given by (6.74). Now in the far zone we may use the form

$$\Pi_x \simeq \frac{I\ell}{4\pi i \epsilon_0 \omega} \left[\frac{e^{-ikR}}{R} + R_m(\bar{\lambda}) \frac{e^{-ikR'}}{R'} \right] \tag{6.157}$$

where for a homogeneous lower half space

$$R_m(\bar{\lambda}) = \frac{\cos \theta_0' - (N^2 - \sin^2 \theta_0')^{1/2}}{\cos \theta_0' + (N^2 - \sin^2 \theta_0')^{1/2}} \tag{6.158}$$

with the same notation as in the previous example. Now we note that

$$\frac{\partial}{\partial y} \frac{e^{-ikR}}{R} = -ik \left(1 + \frac{1}{ikR} \right) \frac{y}{R} \frac{e^{-ikR}}{R}$$

$$\simeq -ik \frac{y}{R} \frac{e^{-ikR}}{R} \simeq -ik \sin \theta_0 \sin \phi \frac{e^{-ikR}}{R} \tag{6.159}$$

Similarly,

$$\frac{\partial}{\partial y} \frac{e^{-ikR'}}{R'} \simeq -ik \left(1 + \frac{1}{ikR'} \right) \frac{e^{-ikR'}}{R'} \left(\frac{y}{R'} \right) \tag{6.160}$$

Thus

$$H_z \simeq \frac{ikI\ell}{4\pi} \left[\frac{e^{-ikR}}{R} \sin \theta_0 + R_m(\bar{\lambda}) \frac{e^{-ikR'}}{R'} \sin \theta_0' \right] \sin \phi \tag{6.161}$$

and if, in addition, $R \gg h$, we find that

$$H_z \simeq \frac{ikI\ell}{4\pi} \frac{e^{-ikr}}{r} \left[e^{-ikh\cos\theta} + R_m(\bar{\lambda}) e^{ikh\cos\theta} \right] \sin \theta \sin \phi \tag{6.162}$$

where

$$R_m(\bar{\lambda}) \simeq \frac{\cos \theta - (N^2 - \sin^2 \theta)^{1/2}}{\cos \theta + (N^2 - \sin^2 \theta)^{1/2}} \tag{6.163}$$

6.6 ASYMPTOTIC EVALUATION OF SOMMERFELD INTEGRALS

In spite of the rather complicated integral forms for the Hertz vectors for dipoles over a plane interface, the resultant radiation fields can be deduced quite readily. A few specific examples were indicated above. Unfortunately, the results become invalid as θ approaches $\pi/2$ or $90°$, corresponding to grazing incidence on the planar interface. In fact, in point-to-point communication from ground-based terminals, the ground-reflected ray may be very near grazing. Thus we really need to examine the integral forms with more finesse. We will outline such an approach here.

The analytical study of what are known as Sommerfeld integrals has had a long history, and the "master" got the ball rolling with his famous 1909 paper [3] that contained an "error" not present in his 1926 paper. The subsequent confusion and resulting misunderstandings by later investigators clouded the subject. Actually, the record shows that Sommerfeld never acknowledged the original error, which amounted to taking the wrong sign for the radical in the argument of the complex error function in the final field expressions.

Benefiting by half a century of hindsight, we outline an asymptotic evaluation of the ground wave field for the vertical electric dipole that yields the correct form [7]. At the same time we reconcile these more complicated field expressions with the simpler forms that we already derived for the radiation fields.

To begin with, we recast (6.67) into the following equivalent form:

$$\Pi_z = \frac{I\ell}{4\pi i\epsilon_0\omega}\left(\frac{e^{-ikR}}{R} + \frac{e^{-ikR'}}{R'} - 2P\right) \tag{6.164}$$

where

$$P = ik\int_0^\infty \frac{\Delta\lambda J_0(\lambda\rho)}{(u_0 + ik\Delta)u_0}\,e^{-u_0(z+h)}\,d\lambda \tag{6.165}$$

where Δ is a dimensionless parameter that plays a vital role. In the case of a homogeneous half space $\eta_0\Delta = u_1/(\sigma_1 + i\epsilon_1\omega)$, where

$$u_1 = (\lambda^2 + \gamma_1^2)^{1/2} \qquad \gamma_1^2 = i\mu_1\omega(\sigma_1 + i\epsilon_1\omega)$$

where σ_1, ϵ_1, and μ_1 are the properties of the lower half space. In arriving at this form for P, we have essentially set

$$R_e = \frac{u_0 - ik\Delta}{u_0 + ik\Delta}$$

so that Δ can be interpreted as a normalized surface impedance.

Exercise: For vertical electric dipole excitation show that at $z = 0$ the tangential electric and magnetic fields can be expressed in the form

$$E_\rho = \int_0^\infty e_\rho(\lambda)\lambda J_1(\lambda\rho) \, d\lambda \tag{6.166}$$

$$H_\phi = \int_0^\infty h_\phi(\lambda)\lambda J_1(\lambda\rho) \, d\lambda \tag{6.167}$$

where

$$e_\rho(\lambda) = -\frac{I\ell\lambda}{4\pi i\epsilon_0\omega}(1 - R_e) \tag{6.168}$$

and

$$h_\phi(\lambda) = \frac{I\ell}{4\pi i\epsilon_0\omega}\frac{\lambda}{u_0}(1 + R_e) \tag{6.169}$$

Then confirm that

$$e_\rho(\lambda) = -\eta_0\Delta(\lambda)h_\phi(\lambda) \tag{6.170}$$

where

$$\Delta(\lambda) = \frac{u_0}{ik}\left(\frac{1 - R_e}{1 + R_e}\right) \tag{6.171}$$

and

$$\eta_0 = (\epsilon_0\omega)^{-1}k = \left(\frac{\mu_0}{\epsilon_0}\right)^{1/2}$$

Now actually, Δ is a function of λ in general, but here we will assume that it is a constant. For example, in the case of a homogeneous half space with $\mu_1 = \mu_0$, we see that

$$\Delta = \frac{ik}{\gamma_1}\left(1 + \frac{\lambda^2}{\gamma_1^2}\right)^{1/2} \simeq \frac{ik}{\gamma_1} \tag{6.172}$$

which will be valid when the ground conductivity is sufficiently large. In any case we will treat Δ as a constant that satisfies the inequality $|\Delta^2| \ll 1$ in what follows below. Relaxing this condition will be considered later.

Our next step is to rewrite the expression for P, given by (6.165), in the form of a complete infinite integral. This is easily accomplished if we note that

$$J_0(\lambda\rho) = \tfrac{1}{2}[H_0^{(1)}(\lambda\rho) + H_0^{(2)}(\lambda\rho)]$$
$$= \tfrac{1}{2}[H_0^{(1)}(\lambda\rho) - H_0^{(1)}(-\lambda\rho)] \tag{6.173}$$

Then clearly,

$$P = \frac{ik\Delta}{2}\int_{-\infty}^\infty \frac{\lambda H_0^{(1)}(\lambda\rho)}{(u_0 + ik\Delta)u_0}e^{-u_0(z+h)} \, d\lambda \tag{6.174}$$

The contour is along the real axis of λ from $-\infty$ to $+\infty$ with a small indentation above the branch point at $\lambda = +k$ and a small indentation below the branch point at $\lambda = -k$. The integral is still not in a convenient form for an asymptotic evaluation.

We introduce a new variable α via the substitution $\lambda = k \cos \alpha$ and at the same time we set $\Delta = \sin \alpha_0$. Now we see that

$$u_0 = (\lambda^2 - k^2)^{1/2} = i(k^2 - \lambda^2)^{1/2}$$
$$= i(k^2 - k^2 \cos^2 \alpha)^{1/2} = ik \sin \alpha$$

Then

$$u_0 + ik\Delta = ik(\sin \alpha + \sin \alpha_0)$$
$$= \frac{ik}{2}\left[\sin\left(\frac{\alpha + \alpha_0}{2}\right) \cos\left(\frac{\alpha - \alpha_0}{2}\right) \right] \tag{6.175}$$

The integral now becomes

$$P = \frac{ik\Delta}{4} \int_{-i\infty}^{\pi + i\infty} \frac{\cos \alpha \, e^{-ik(z+h)\sin \alpha} H_0^{(1)}(k\rho \cos \alpha)}{\sin[(\alpha + \alpha_0)/2]\cos[(\alpha - \alpha_0)/2]} \, d\alpha \tag{6.176}$$

where the integration contour is as shown in Figure 6.4.

Here we restrict attention to large values of $k\rho$ so that the leading term in the asymptotic representation for the Hankel function suffices. That is,

$$H_0^{(1)}(k\rho \cos \alpha) \simeq \left(\frac{2}{\pi k\rho \cos \alpha}\right)^{1/2} e^{ik\rho\cos\alpha} e^{-i\pi/4} \tag{6.177}$$

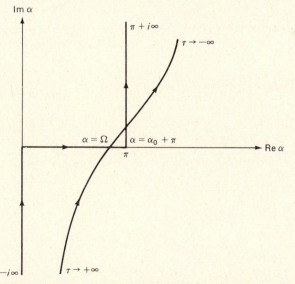

Figure 6.4 Complex α plane showing original contour and deformed or steepest-descent contour through saddle point at $\alpha = \Omega$. Pole location at $\alpha = \alpha_0 + \pi$ is also shown.

Then

$$P \simeq e^{i\pi/4} \frac{\Delta}{2} \left(\frac{k}{2\pi\rho} \right)^{1/2} \int_{-i\infty}^{\pi+i\infty} \frac{(\cos\alpha)^{1/2} e^{-ikR'\cos(\Omega-\alpha)}}{\sin[(\alpha+\alpha_0)/2]\cos[(\alpha-\alpha_0)/2]} \, d\alpha \quad (6.178)$$

where

$$\rho = -R' \cos\Omega \quad \text{and} \quad z + h = R' \sin\Omega$$

We observe that the integrand is a rapidly varying function of α except in a region where $\cos(\Omega - \alpha)$ is near one. In this region there is a major contribution to the integrand. But we must also be concerned about the influence of any pole singularities.

To proceed, we change the integration contour to the steepest-descent path. This is implemented by setting

$$\cos(\alpha - \Omega) = 1 - i\tau^2 \quad (6.179)$$

and letting τ range from $-\infty$ to $+\infty$ through real values. The form of this contour, in the complex α plane, is shown in Figure 6.4. It intersects the real axis of α at $\alpha = \Omega$, where it subtends an angle of $\pi/4$. We note that as α tends to $-i\infty$, τ tends to $+\infty$, and as α tends to $\pi + i\infty$, τ tends to $-\infty$. Also, one should note that

$$\cos(\alpha - \Omega) = 1 - 2 \sin^2 \left(\frac{\alpha - \Omega}{2} \right) \quad (6.180)$$

and thus

$$2^{1/2} \sin \frac{\alpha - \Omega}{2} = e^{i\pi/4} \tau \quad (6.181)$$

Therefore

$$-2^{-1/2} \cos \frac{\alpha - \Omega}{2} \, d\alpha = e^{i\pi/4} \, d\tau \quad (6.182)$$

and

$$d\alpha = \frac{-(2i)^{1/2} \, d\tau}{\{1 - \sin^2[(\alpha - \Omega)/2]\}^{1/2}} = \frac{-(2i)^{1/2} \, d\tau}{[1 - (i\tau^2/2)]^{1/2}} \quad (6.183)$$

Thus we see that an integral of the type

$$\frac{1}{(kR')^{1/2}} \int_{-i\infty}^{\pi+i\infty} G(\cos\alpha) e^{-ikR'\cos(\alpha-\Omega)} \, d\alpha$$

is transformed to

$$-\left(\frac{2i}{kR'} \right)^{1/2} e^{-ikR'} \int_{+\infty}^{-\infty} \frac{G(\cos\alpha)}{[1 - (i\tau^2/2)]^{1/2}} e^{-kR'\tau^2} \, d\tau$$

provided no singularities, such as poles, are crossed in the deformation of the contour. In the present problem

$$G(\cos\alpha) = \frac{(\cos\alpha)^{1/2}e^{-ikR'\cos(\alpha-\Omega)}}{\sin[(\alpha+\alpha_0)/2]\cos[(\alpha-\alpha_0)/2]} \tag{6.184}$$

If $G(\cos\alpha)$ can be regarded as slowly varying in the region $\tau \approx 0$, which is the "saddle point," the factor $G(\cos\alpha)/[1 - (i\tau^2/2)]^{1/2}$ can be expanded in a power series in τ, enabling the integration to be carried out term by term. The leading term of this expansion is the classical saddle-point approximation [8], which is given by

$$(2\pi i)^{1/2}G(\cos\Omega)e^{-ikR'}(kR')^{-1}$$

where succeeding terms contain $(kR')^{-2}$, $(kR')^{-3}$, ... If any pole singularities are crossed in this contour deformation, their residues must be accounted for.

Actually, the direct application of the saddle-point method leads directly to the radiation field expressions. In that case the leading terms vary as $(kR')^{-1}\exp(-ikR')$, where neglected terms contain higher inverse powers of R'.

The important complication that arises is when $\Omega' = (\pi/2) + \theta_0'$ is near π, corresponding to grazing angles (that is, θ_0' near $\pi/2$), as mentioned earlier. In the context of the present development this means that the pole of $G(\cos\alpha)$ at $\alpha = \pi + \alpha_0$ is near the saddle point since α_0 is small and Ω is near π. In other words, the integrand is not slowly varying near the saddle point on account of the factor $\cos[(\alpha-\alpha_0)/2]$ in the denominator. The integral we need to contend with now has the form

$$A = \int_{-\infty}^{+\infty} \frac{e^{-x\tau^2}}{\tau^2 + c}\,d\tau \tag{6.185}$$

where $x = kR'$ and $c = i2[\cos(\Omega-\alpha_0)/2]^2$. The integral for A can be expressed in terms of the complementary error function of argument $(xc)^{1/2}$ in the manner

$$A = \frac{\pi}{c^{1/2}}\,e^{xc}\,\text{erfc}(xc)^{1/2} \tag{6.186}$$

Thus

$$\int_{-i\infty}^{\pi+i\infty} \frac{e^{-ikR'\cos(\Omega-\alpha)}}{\cos[(\alpha-\alpha_0)/2]}\,d\alpha \approx -i2\pi e^{-ikR'}e^{-w}\,\text{erfc}(iw^{1/2}) \tag{6.187}$$

where

$$w = -kR'c = -2ikR'\left[\cos\left(\frac{\Omega-\alpha}{2}\right)\right]^2 \tag{6.188}$$

Using (6.187) and noting that the factor $(\cos\alpha)^{1/2}/\sin[(\alpha+\alpha_0)/2]$ in

(6.184) is replaced by its value at $\alpha = \Omega$, we see that the desired asymptotic form is

$$P \simeq \frac{i(\pi p)^{1/2} e^{-w} \, \text{erfc}(iw^{1/2}) e^{-ikR'}}{R'} \qquad (6.189)$$

where

$$w \simeq p \left(1 + \frac{z+h}{\Delta R'} \right)^2 \qquad (6.190)$$

and

$$p = -\frac{ikR'}{2} \Delta^2 = |p| e^{ib} \qquad (6.191)$$

The complementary error function, as defined in the present context, is

$$\text{erfc}(iw^{1/2}) = \frac{2}{\pi^{1/2}} \int_{iw^{1/2}}^{\infty} e^{-y^2} \, dy \qquad (6.192)$$

where the contour runs from the point $iw^{1/2}$, in the complex y plane, to the right via a line parallel to the real axis.

In the above derivation for the asymptotic form for P, we have really made two key approximations to facilitate the derivations: namely, $|\Delta|^2 \ll 1$ and $kR' \gg 1$. It is just this case where the pole (at $\alpha = \pi + \alpha_0$) and the saddle point ($\alpha = \Omega$) may be in close proximity.

6.7 WAVE OPTICAL INTERPRETATION

Further discussion of the derived form for the integral P is warranted. Clearly, we see that (6.20) is equivalent to

$$P = \frac{(p/w)^{1/2} [1 - F(w)] e^{-ikR'}}{R'} \qquad (6.193)$$

where

$$F(w) = 1 - i(\pi w)^{1/2} e^{-w} \, \text{erfc}(iw^{1/2})$$

The function $F(w)$ is often called the Sommerfeld attenuation function for a complex argument w. The relevant geometry is shown in Figure 6.5.

In the special case for $z = h = 0$ we see that (6.164) reduces to

$$\Pi_z = \frac{I\ell}{2\pi i \epsilon_0 \omega} \frac{e^{-ik\rho}}{\rho} F(p) \qquad (6.194)$$

where

$$p = |p| e^{ib} = -i \frac{k\rho}{2} \Delta^2$$

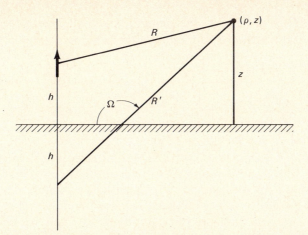

Figure 6.5   Basic geometry for interpreting asymptotic representations.

The corresponding vertical electric field in the air at $z = +0$ is then

$$E_z \simeq -\frac{i\mu_0\omega I\ell}{2\pi\rho}\, e^{-ik\rho}F(p) \tag{6.195}$$

bearing in mind $k\rho \gg 1$.

When $|p| \ll 1$, we may compute values of $F(p)$ from the series expansion

$$F(p) = 1 - i(\pi p)^{1/2} - 2p + i\pi^{1/2}p^{3/2} + \cdots \tag{6.196}$$

As indicated, $F(p) \to 1$ when $|\Delta|^2 \to 0$, corresponding to a perfectly conducting ground plane. In fact, $F(p)$ may often be well-approximated by one even for a finitely conducting ground plane if $|p| \ll 1$.

In the case where $|p| \gg 1$, we may expand the erfc(ip) asymptotically. But here we need to distinguish the two cases depending on whether $\arg p > 0$ or < 0. Thus in this limiting situation, for $b > 0$,

$$F(p) \simeq -2i(\pi p)^{1/2}e^{-p} - \frac{1}{2p} - \frac{1\times 3}{(2p)^2} - \frac{1\times 3\times 5}{(2p)^3} - \cdots \tag{6.197}$$

while for $b < 0$

$$F(p) \simeq -\frac{1}{2p} - \frac{1\times 3}{(2p)^2} - \frac{1\times 3\times 5}{(2p)^3} - \cdots \tag{6.198}$$

In the case of a homogeneous half space

$$\Delta \simeq \frac{ik}{\gamma_1} = \left(\frac{i\epsilon_0\omega}{\sigma_1 + i\epsilon_1\omega}\right)^{1/2} \tag{6.199}$$

will have a phase angle in the range 0 to $\pi/4$, depending on whether $(\epsilon_1\omega/\sigma_1)$ is large or small, respectively. The corresponding range of b is

$-\pi/2$ to 0. Thus in this case the expansion given by (6.198) is appropriate. Then in this situation for sufficiently large values of $|p|$, we see that

$$E_z \simeq \frac{i\mu_0\omega I\ell}{2\pi\rho}\,e^{-ik\rho}\left(\frac{1}{2p}\right)$$

$$\simeq -\frac{i\mu_0\omega I\ell}{2\pi\rho}\,e^{-ik\rho}\left(\frac{1}{ik\rho\Delta^2}\right) \tag{6.200}$$

which is varying as $(1/\rho^2)\exp(-ik\rho)$.

Now when b is positive, the behavior for large $|p|$ is determined from (6.197). In this case the asymptotic behavior for $|p| \to \infty$ is still given by $F(p) \simeq (-1/2p)$, provided $b < \pi/2$. On the other hand, if $b = \pi/2$, the parameter $p = i|p|$ is purely imaginary, corresponding to $\Delta = i|\Delta|$, which means the ground plane is purely inductive. Then we should write (6.197) in the explicit form

$$F(p) \simeq -2e^{i3\pi/4}(\pi|p|)^{1/2}e^{-i|p|} + \frac{i}{2|p|} + \frac{3}{4|p|^2} + \cdots \tag{6.201}$$

where

$$|p| = \frac{k\rho}{2}\frac{X^2}{\eta_0^2} \tag{6.202}$$

in terms of the surface reactance X (that is $Z = iX$, where X is real). Now the vertical electric field varies as

$$\frac{1}{(k\rho)^{1/2}}\exp\left[-ik\left(1 + \frac{X^2}{2\eta_0^2}\right)\rho\right]$$

for sufficiently large values of $k\rho$. This is telling us that a purely reactive boundary will support a trapped surface wave that will certainly dominate the other components of the field.

The parameter p, as defined generally by (6.191), is called the "numerical distance." As envisaged by Sommerfeld [3] and Norton [9], it was introduced to characterize radio wave propagation over plane homogeneous ground. But as we have indicated briefly, the attenuation function $F(p)$ can describe the transmission over the boundary of more complicated situations that could occur, for example, for layered earth models [10]. Thus the parameter p and its appellation "numerical distance" has a more general meaning.

When the source dipole and the observer are both raised above the boundary (that is, $h > 0$ and/or $z > 0$), we need to return to (6.164) to properly describe the resultant fields. It is useful to examine this case in the asymptotic sense such that $|p|$ or $|w| \gg 1$. In this case it is not difficult to show that the following decomposition holds:

$$\Pi_z \simeq (\psi_a + \psi_b + \delta\psi_s)\frac{I\ell}{4\pi i\epsilon_0\omega} \tag{6.203}$$

where

$$\psi_a \simeq \frac{e^{-ikR}}{R} + \frac{C - \Delta}{C + \Delta} \frac{e^{-ikR'}}{R'} \tag{6.204}$$

$$\psi_b \simeq -\left\{ \frac{1}{p[1 + (C/\Delta)]^3} + \frac{1}{2p^2[1 + (C/\Delta)]^5} \right.$$
$$\left. + \frac{1}{4p^3[1 + (C/\Delta)]^7} + \cdots \right\} \frac{e^{-ikR'}}{R'} \tag{6.205}$$

and

$$\psi_s \simeq -\frac{2\Delta}{C + \Delta} [2i(\pi w)^{1/2} e^{-w}] \frac{e^{-ikR'}}{R'} \tag{6.206}$$

where

$$p = -i \frac{kR'}{2} \Delta^2 \quad \text{and} \quad C = \cos \theta' = \frac{h + z}{R'}$$

Also, we have defined $\delta = 0$ for arg $w < 0$ and $\delta = 1$ for arg $w > 0$. The corresponding vertical electric field at (ρ, z) can be found from $E_z = k^2(1 - C^2)\Pi_z$. The geometry of the situation is illustrated in Figure 6.6, where we have chosen to emphasize the vertical scale.

Here we see ψ_a can be identified as the geometrical optical contribution consisting of the primary component $\exp(-ikR)/R$ and a specularly reflected component $\exp(-ikR')/R'$ that is modified by the reflection coefficient $(C - \Delta)/(C + \Delta)$, where $\theta' = \arccos C$ is the local angle of

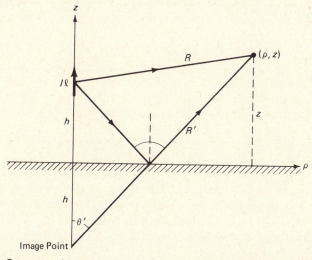

Figure 6.6 Geometry for describing optical form of general asymptotic field representation.

incidence. The specular component can be represented in terms of an equivalent image source located at $z = -h$ with a current moment $I\ell(C - \Delta)/(C + \Delta)$ that obviously varies with the angle of incidence θ'. As $\theta' \to \pi/2$, we see that $\psi_a \to 0$.

The contribution denoted ψ_b is an asymptotic expansion containing terms in $(R')^{-2}$, $(R')^{-3}$, and so on. It is important to note that ψ_b does not vanish as $\theta' \to \pi/2$ corresponding to grazing angles of incidence.

The final contribution denoted $\delta\psi_s$ is only present when arg $w > 0$, and it is only significant if, in addition, w has a small real part (that is, when $b \simeq \pi/2$). The contribution $\delta\psi_s$, of course, is the trapped surface wave mentioned above. For large ranges, where $|p| \gg 1$, it is insignificant because of the exponential damping, except in the rather artificial case where the boundary is purely inductive.

In the asymptotic development of the integral representation for P, we regarded Δ as a small and complex, but fixed, parameter. Now as we have seen, the factor $(C - \Delta)/(C + \Delta)$ emerges as a Fresnel reflection coefficient for a plane wave incident at an angle arccos C. To make this result exact for a homogeneous half-space model of the earth, we should equate

$$\frac{C - \Delta}{C + \Delta} = \frac{K_0 - K_1}{K_0 + K_1} \tag{6.207}$$

where

$$K_0 = \frac{kC}{\epsilon_0 \omega} \qquad K_1 = \frac{\gamma_1}{\sigma_1 + i\epsilon_1 \omega} \left(1 + \frac{k^2 S^2}{\gamma_1^2}\right)^{1/2}$$

and

$$S^2 = 1 - C^2$$

Equivalently, we are requiring that

$$\Delta = \frac{\eta_0}{\eta_1} \left(1 + \frac{k^2 S^2}{\gamma_1^2}\right)^{1/2} \tag{6.208}$$

where

$$\eta_0 = \left(\frac{\mu_0}{\epsilon_0}\right)^{1/2} \qquad \text{and} \qquad \eta_1 = \left(\frac{i\mu_1 \omega}{\sigma_1 + i\epsilon_1 \omega}\right)^{1/2}$$

If Δ is defined in this manner, the form of the geometrical optics contribution ψ_a given by (6.204) holds even if $|\Delta|$ is not small. Of course, if $|kS/\gamma_1|^{-1} \ll 1$, $\Delta \simeq \eta_0/\eta_1$, which is independent of θ'.

Actually, the above redefinition of Δ can be stated in a more general context by saying that we set

$$\eta_0 \Delta = [Z(\lambda)]_{\lambda = ks} = Z(kS) \tag{6.209}$$

where $Z(kS)$ is the surface impedance for a down-coming, TM, polarized plane wave at an angle of incidence arcsin S at the specular point. It is evident that the result applies to stratified earth models [10].

Our asymptotic representation for P is "uniform" in the sense that the correct form of the geometrical optics field emerges in the limit where the numerical distance is sufficiently large.

The whole story has still not been told. The comprehensive generalization of the result to layered earth models should properly account for guided-wave contributions within the layers. These actually emerge as residues that are picked up when the original integration contour is deformed to the steepest-descent path. But the relative importance of these internal guided waves is small if the layers are conductive, in which case the exponential damping is significant.

In the manner that we have formulated the problem, the contour deformation, as indicated in Figure 6.4, has been carried out on the basis that no singularities in the complex α plane were crossed. Now there is a pole at $\alpha = \pi + \alpha_0$. Here $\alpha_0 \simeq \Delta$, where for a homogeneous half space the argument of Δ may lie in the range from 0 to $\pi/4$. Thus in this case the pole is still on the right-hand side of the contour even though it may be close to the contour in the case where $\Omega \to \pi$ (that is, grazing incidence) and when at the same time $|\Delta| \ll 1$. In fact, the relative proximity of the saddle point at $\alpha = \Omega$ and the pole at $\alpha = \alpha_0 \simeq \Delta$ is proportional to the numerical distance.

Actually, the asymptotic result for $F(w)$, as defined by (6.193) is *also* valid in the situation where the argument of Δ is greater than $\pi/4$. In fact, if the boundary at $z = 0$ exhibits a pure inductance, $\Delta \simeq i|\Delta|$, in which case $p \simeq i|p|$. In this limiting situation the surface wave is actually unattenuated, and the amplitude of E_z along the boundary, at $z = 0$, is approximately proportional to $1/\rho^{1/2}$. In such cases the deformed contour has indeed crossed the pole at $\pi + \alpha_0$, and we have accounted for this effect by adopting the definition of the complementary error function given by (6.192).

6.8 INPUT IMPEDANCE OF DIPOLES OVER A HALF SPACE

An important aspect of a radiating system is the power efficiency when the immediate environment has finite conductivity [11]. In the context of the present problem we are interested in the relative amount of power that is absorbed in the lower half space relative to the amount of power that would be radiated by the same dipole located in free space. The simplest approach to this question is to examine the change of the self-impedance Z_0 of the dipole from its value in free space to its value Z in the presence of the finitely conducting half space.

To illustrate our approach, we will deal with the vertical electric dipole or current element Il located at height h over the conducting half space. The impedance change or increment ΔZ is defined by

$$\Delta Z = \bar{Z} - Z_0$$

where

$$Z_0 = \lim_{h \to \infty} \bar{Z}$$

The corresponding resistive and reactive parts are defined by writing $\bar{Z} + \bar{R} + i\bar{X}$, $Z_0 = R_0 + iX_0$, and $\Delta Z = \Delta R + i \Delta X$. It is assumed here that Z_0 is either known or separately accounted for. Then following the prescription of the emf method in antenna theory [1], it follows that

$$Z = \lim_{\substack{z \to h \\ \rho \to 0}} \left(\frac{-E_z^s}{I} \right) \tag{6.210}$$

where E_z^s is the secondary field of the dipole as it approaches the location of the dipole. To be specific, for $z > 0$,

$$E_z^s = E_z - E_z^{\text{prim}}$$

$$= -\frac{I\ell}{4\pi i \epsilon_0 \omega} \int_0^\infty \frac{\lambda^3}{u_0} R_e(\lambda) e^{-u_0(z+h)} J_0(\lambda\rho) \, d\lambda \tag{6.211}$$

where $R_e(\lambda)$ is the reflection function defined, for example, by (6.71) for a homogeneous half space. Thus we see that

$$\Delta Z = -\frac{\ell^2}{4\pi i \epsilon_0 \omega} \int_0^\infty R_e(\lambda) \lambda^3 u_0^{-1} e^{-2u_0 h} \, d\lambda \tag{6.212}$$

and specifically, $\Delta R = \text{Re } \Delta Z$. This is a very simple derivation of a result given by Sommerfeld and Renner [12] for the increment ΔR. They obtained the result using a complicated Poynting vector, power flow formulation.

Exercise: Show that the free-space radiation resistance R_0 of the dipole can be obtained from

$$R_0 = \lim_{\substack{z \to h \\ \rho \to 0}} \text{Re} \, \frac{-E_z^{\text{prim}} \ell}{I} \tag{6.213}$$

where

$$E_z^{\text{prim}} = \left(k^2 + \frac{\partial^2}{\partial z^2} \right) \frac{I\ell}{4\pi i \epsilon_0 \omega} \frac{e^{-ikR}}{R} \tag{6.214}$$

and

$$R = [\rho^2 + (z - h)^2]^{1/2}$$

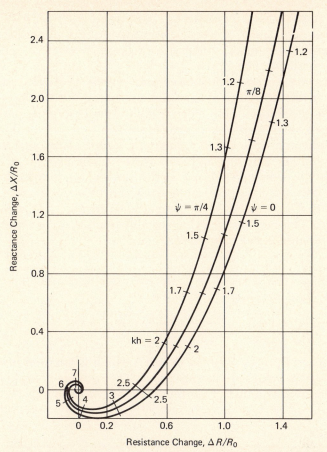

Figure 6.7   Argand plot of impedance change for vertical electric dipole over homogeneous half space of complex refractive index N_1 (in this example $|N_1|^2 = 25$).

To illustrate the behavior of the input impedance, we plot $\Delta R/R_0$ and $\Delta X/R_0$ on the real and imaginary axes, respectively, in Figure 6.7 for several values of the complex refractive index N_1 of the homogeneous half space. Here $N_1^2 = (\sigma_1 + i\epsilon_1\omega)/i\epsilon_0\omega$ is complex, so we write

$$N_1 = |N_1|\exp\left[-i\left(\frac{\pi}{4} - \psi\right)\right] \tag{6.215}$$

Thus

$$|N_1|^2 = \frac{\epsilon_1}{\epsilon_0}\left[1 + \left(\frac{\sigma_1}{\epsilon_1\omega}\right)^2\right]^{1/2} \tag{6.216}$$

and

$$\tan 2\psi = \frac{\epsilon_1\omega}{\sigma_1} \tag{6.217}$$

For the example shown in Figure 6.7, we choose $|N_1|^2 = 25$ while ψ takes the values 0, $\pi/8$, and $\pi/4$, corresponding to the case where the displacement currents are small, moderate, or large, respectively, compared with the conduction currents.

There are various tabulations and graphs of $\Delta R/R_0$ and $\Delta X/R_0$ available in the literature [11], which are based, for the most part, on numerical integration. Both electric and magnetic dipoles with various orientations have been considered [11–15].

As indicated, when kh is sufficiently large, $\Delta R/R_0$ and $\Delta X/R_0$ become negligible compared with one. In fact, even if $kh = \pi$, these ratios are only of the order of 5 percent. In this case the input impedance of the dipole differs only insignificantly from the free-space value. On the other hand, if kh becomes small compared with one, $\Delta R/R_0$ may become quite large. This means that an appreciable fraction of the input power is being dissipated in the lower half space.

Exercise: Show that if $|khN_1| \gg 1$, the following approximate formula for the impedance increment is obtained:

$$\frac{\Delta Z}{R_0} = \frac{\Delta R}{R_0} + i\,\frac{\Delta X}{R_0} \simeq \frac{3i\,\exp(-i2kh)}{(2kh)^3}\left(\frac{N_1 - 1}{N_1 + 1}\right) \tag{6.218}$$

Hint: Replace $R_e(\lambda)$ by $R_e(0)$ in (6.211), which is justified because the most important values of λ are near zero. The result is exact, of course, if $|N_1| = \infty$, corresponding to a perfectly conducting ground.

References

1. J. R. Wait: *Electromagnetic Radiation from Cylindrical Structures*, Pergamon Press, Elmsford, N.Y., 1959. See the Appendix.
2. J. A. Stratton: *Electromagnetic Theory*, McGraw-Hill, New York, 1941.
3. A. N. Sommerfeld: "Ueber die Ausbreitung der Wellen in der Drahtlosen Telegraphie," *Annalen der Physics (4)*, vol. 28, 1909, pp. 665–695, and vol. 81, 1926, pp. 1135–1155.
4. J. R. Wait: *Geo Electromagnetism*, Academic Press, New York, 1982.
5. J. A. Kong: *Theory of Electromagnetic Waves*, Wiley, New York, 1975.
6. J. R. Wait: *Wave Propagation Theory*, Pergamon Press, Elmsford, N.Y., 1981.
7. B. L. van der Waerden: "On the Method of Saddle Points," *Applied Scientific Research*, Sec. B, vol. 2, 1950, pp. 33–45.
8. P. M. Morse and H. Feshbach: *Methods of Theoretical Physics*, McGraw-Hill, New York, 1953.
9. K. A. Norton: "The Propagation of Radio Waves over the Surface of the Earth and in the Upper Atmosphere," *Proceedings of the IRE*, vol. 24, pt. 1, 1936, pp. 1367–1387, and vol. 25, pt. 2, 1937, pp. 1203–1236.
10. J. R. Wait: *Electromagnetic Waves in Stratified Media*, 2d ed., Pergamon Press, Elmsford, N.Y., 1970.

11. J. R. Wait: "Characteristics of Antennas over Lossy Earth," in R. E. Collin and F. J. Zucker (eds.), *Antenna Theory*, McGraw-Hill, New York, 1969, Chap. 23.

12. A. N. Sommerfeld and F. Renner: "Strahlungsenergie und Erdabsorption bei Dipolantennen," *Annalen der Physics*, vol. 41, 1942, pp. 1–36.

13. L. E. Vogler and J. L. Noble: "Curves of Input Impedance Change Due to Ground for Dipole Antennas," *National Bureau of Standards Monograph*, no. 72, 1964.

14. J. R. Wait: "The Electromagnetic Fields of a Horizontal Dipole in the Presence of a Conducting Half-Space," *Canadian Journal of Physics*, vol. 39, 1961, pp. 1017–1028. Gives often-used approximate formulas for different ranges.

15. R. W. P. King, G. S. Smith, M. Owens and T. T. Wu: *Antennas in Matter*, MIT Press, Cambridge, Mass., 1981.

Chapter 7
Propagation and Radiation in Complicated Media

7.1 INTRODUCTION

As we have indicated in prior chapters, the separation-of-variables method works well when we deal with planar, cylindrical, and spherical regions that are at least piecewise homogeneous. Unfortunately, in many practical situations we must cope with the effect of lateral inhomogeneity in the sense that the medium properties change in a nonconvenient fashion. An example is when the radio ground wave passes over a land-sea boundary where there is a distinct change of the electrical properties of the underlying media. Often we may regard such a case as a perturbation from some ideal model. In fact, this is a very fruitful approach in order to derive engineering approximations. At the same time we may examine such mixed-boundary-value problems in terms of integral equations that may be solved even when the perturbations are not small.

The cornerstone for the perturbation formulation is the Rayleigh [1] reciprocity principle that is used extensively in network theory. In fact, many of the concepts such as the compensation theorem have precise analogies in lumped circuit analysis. This is really not surprising because the mathematical and physical bases of circuit formulations are implied by Maxwell's equations. To provide the link with circuit concepts, we will include a brief discussion of relevant topics in Appendix A of this chapter.

This chapter, which certainly is not comprehensive, is intended to illustrate the art of approximation in developing useful formulas for dealing with complicated situations. Such approaches are desirable to complement purely numerical formulations that run into considerable difficulty when the scale of the inhomogeneity is becoming large compared with a wavelength.

7.2 LORENTZ' RECIPROCAL RELATIONS AND RECIPROCITY

Maxwell's equations are written, for the usual time factor $\exp(i\omega t)$,

$$\text{curl } \mathbf{H} = i\omega \mathbf{D} + \mathbf{J} \tag{7.1}$$

$$\text{curl } \mathbf{E} = -i\omega \mathbf{B} - \mathbf{M} \tag{7.2}$$

$$\mathbf{B} = \hat{\mu}\mathbf{H} \tag{7.3}$$

and

$$\mathbf{D} = \hat{\epsilon}\mathbf{E} \tag{7.4}$$

Here $\mathbf{J}$ and $\mathbf{M}$ designate the densities of the impressed electric and magnetic currents; $\hat{\epsilon}$ and $\hat{\mu}$ designate the complex permittivity and permeability of the medium. Thus, for example, we may write $\hat{\epsilon} = \epsilon - i\sigma/\omega$ in terms of the real permittivity and real conductivity.

We now suppose that distributions of impressed currents $\mathbf{J}_1, \mathbf{M}_1$ give rise to field components $\mathbf{E}_1$, $\mathbf{H}_1$, $\mathbf{D}_1$, and $\mathbf{B}_1$. Similarly, source current densities $\mathbf{J}_2$, $\mathbf{M}_2$ give rise to $\mathbf{E}_2$, $\mathbf{H}_2$, $\mathbf{D}_2$, and $\mathbf{B}_2$. The individual fields satisfy (7.1) and (7.2), so the following scalar or dot products are valid:

$$\mathbf{E}_2 \cdot \text{curl } \mathbf{H}_1 = i\omega \mathbf{E}_2 \cdot \mathbf{D}_1 + \mathbf{E}_2 \cdot \mathbf{J}_1 \tag{7.5}$$

and

$$\mathbf{H}_1 \cdot \text{curl } \mathbf{E}_2 = -i\omega \mathbf{H}_1 \cdot \mathbf{B}_2 - \mathbf{H}_1 \cdot \mathbf{M}_2 \tag{7.6}$$

or

$$\mathbf{E}_1 \cdot \text{curl } \mathbf{H}_2 = i\omega \mathbf{E}_1 \cdot \mathbf{D}_2 + \mathbf{E}_1 \cdot \mathbf{J}_2 \tag{7.7}$$

and

$$\mathbf{H}_2 \cdot \text{curl } \mathbf{E}_1 = -i\omega \mathbf{H}_2 \cdot \mathbf{B}_1 - \mathbf{H}_2 \cdot \mathbf{M}_1 \tag{7.8}$$

Now we note that from vector calculus

$$\mathbf{E}_2 \cdot \text{curl } \mathbf{H}_1 - \mathbf{H}_1 \cdot \text{curl } \mathbf{E}_2 = -\text{div}(\mathbf{E}_2 \times \mathbf{H}_1) \tag{7.9}$$

and

$$E_1 \operatorname{curl} H_2 - H_2 \cdot \operatorname{curl} E_1 = -\operatorname{div}(E_1 \times H_2) \tag{7.10}$$

Then it follows that

$$
\begin{aligned}
\operatorname{div}(E_1 \times H_2 - E_2 \times H_1) = &-E_1 \cdot J_2 + E_2 \cdot J_1 + H_1 \cdot M_2 \\
&- H_2 \cdot M_1 - i\omega(E_1 \cdot D_2 - E_2 \cdot D_1) \\
&+ i\omega(H_1 \cdot B_2 - H_2 \cdot B_1)
\end{aligned}
\tag{7.11}
$$

By virtue of (7.3) and (7.4) the terms multiplying $i\omega$ in (7.11) vanish.

We now integrate both sides of (7.11) over a common volume designated by, say, V. Then on using the divergence theorem; we have

$$
\begin{aligned}
\iiint_V \operatorname{div}(E_1 \times H_2 - E_2 \times H_1) \, dv \\[1em]
= \iint_S (E_1 \times H_2 - E_2 \times H_1) \cdot n \, da \\[1em]
= \iiint_V (-E_1 \cdot J_2 + E_2 \cdot J_1 + H_1 \cdot M_2 - H_2 \cdot M_1) \, dv
\end{aligned}
\tag{7.12}
$$

where n is a unit vector directed along the outward normal to S and da is an element of area.

Now clearly, the surface integral in (7.12) is zero if the enclosed volume contains no sources. This important property will be used in what follows.

We now specialize our result and take S to be a closed surface that *encloses all the sources*. This situation is illustrated in Figure 7.1. We also

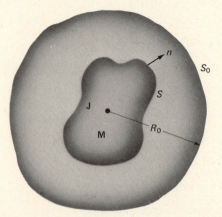

Figure 7.1   Closed surface S that encloses sources J and M and exterior spherical surface of infinitely large radius R_0.

consider another surface designated by S_0 that can be taken as spherical with radius R_0. To facilitate the discussion, we now take R_0 to be sufficiently large that the external region is homogeneous. In fact, R_0 may tend to infinity.

We now invoke the Sommerfeld [2] radiation condition, which stipulates that the field components $\mathbf{E}_1$, $\mathbf{E}_2$, . . . will vanish as $(1/R_0)\exp(-ikR_0)$ as $R_0 \rightarrow \infty$, where $k = (\epsilon_0\mu_0)^{1/2}\omega$ is the wave number in the external homogeneous region. Thus it follows that integrals such as

$$\iint\limits_{S_0} (\mathbf{E}_1 \times \mathbf{H}_2) \cdot \mathbf{n} \, da$$

will remain bounded as $R_0 \rightarrow 0$. But, in addition, we have the important property of radiation fields that

$$\mathbf{E}_1 \times \mathbf{H}_2 \simeq \mathbf{E}_2 \times \mathbf{H}_1 \simeq \frac{1}{\eta_0} \mathbf{E}_1 \cdot \mathbf{E}_2 \mathbf{i}_r \tag{7.13}$$

as $R_0 \rightarrow \infty$, where $\eta_0 = (\mu_0/\epsilon_0)^{1/2}$ and $\mathbf{i}_r$ is a unit vector directed radially outward. This result suggests that

$$\lim_{R_0 \rightarrow \infty} \iint\limits_{S_0} (\mathbf{E}_1 \times \mathbf{H}_2 - \mathbf{E}_2 \times \mathbf{H}_1)\mathbf{n} \cdot da = 0 \tag{7.14}$$

because the integrand tends to zero more rapidly than $1/R_0^2$.

The corresponding surface integral over the volume enclosed by S and S_0 is zero because there are no contained sources! Thus we have demonstrated that

$$\iint\limits_{S} (\mathbf{E}_1 \times \mathbf{H}_2 - \mathbf{E}_2 \times \mathbf{H}_1)\mathbf{n} \cdot da = 0 \tag{7.15}$$

where S, as indicated in Figure 7.1, *encloses all sources.*

7.3 MUTUAL IMPEDANCE FORMULATION

We now consider the very practical situation where we have two antennas located within the medium. The objective is to characterize the coupling between these antennas and, in particular, to determine the mutual impedance between the accessible terminal pairs denoted A and B. The situation is illustrated in Figure 7.2. The surface S is any surface that encloses antenna B but excludes antenna A. The surface S_b, within S, is drawn or chosen small enough so that it just encloses antenna B. Thus

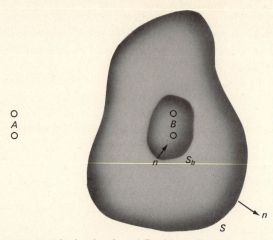

Figure 7.2 Antenna terminal pairs A and B with surface S_b, within S, that just encloses terminal pair B.

in the volume enclosed by S and S_b there are no sources, so we may say that

$$\iint_{S+S_b} (\mathbf{E}_A \times \mathbf{H}_B - \mathbf{E}_B \times \mathbf{H}_A) \cdot \mathbf{n} \, da = 0 \qquad (7.16)$$

Here $\mathbf{E}_A$ and $\mathbf{H}_A$ are the fields of source antenna A when an impressed current I_a is applied to terminals A under the condition that terminals B are open-circuited. Correspondingly $\mathbf{E}_B$ and $\mathbf{H}_B$ are the fields of antenna B when an impressed current I_b is applied to terminals B when terminals A are open-circuited.

We now allow S_b to shrink so it just encloses antenna B. Then over S_b, $\mathbf{E}_a$ can be associated with the open-circuit voltage $-v_b$ at terminals B for a current I_a in terminals A. On the other hand, $\mathbf{H}_B$ is associated with the current I_b between terminals B. Thus it follows that

$$\iint_{S_b} \mathbf{E}_A \times \mathbf{H}_B \cdot \mathbf{n} \, da \rightarrow -v_b I_b \qquad (7.17)$$

We can visualize this transition from wave fields to circuit parameters if we deal explicitly with a concrete antenna model such as a perfectly conducting cylindrical antenna of finite length fed by a circumferential slot or gap of width δ. Then the element of area is $2\pi\rho \, dz$. Also, $\mathbf{E}_A$ has only a significant z component $E_{A,z}$ and the magnetic field H_B has only a significant ϕ component $H_{B,\phi}$. Now $I_b \simeq 2\pi\rho H_{B,\phi}$ and $v_b \simeq -\int_{-\ell}^{\ell} E_{A,z} \, dz$ at the gap, which can be coincident with "terminals" B, as indicated in Figure 7.3.

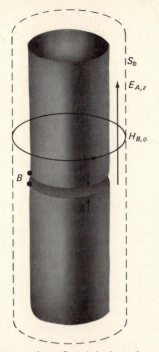

Figure 7.3 Example of how surface S_b might be taken with reference to antenna B with terminal pair at gap.

Now, by definition, $\mathbf{H}_A$ is the magnetic field when terminals B are open-circuited; thus the integral

$$\iint_{S_b} \mathbf{E}_B \times \mathbf{H}_A \cdot \mathbf{n} \, da$$

can be neglected. Then (7.17) reduces to the important result

$$\iint_S (\mathbf{E}_A \times \mathbf{H}_B - \mathbf{E}_B \times \mathbf{H}_A) \cdot \mathbf{n} \, da = I_a I_b Z_{ab} \tag{7.18}$$

where

$$Z_{ab} = \frac{v_b}{I_a} \tag{7.19}$$

is the mutual or transfer impedance between terminals A and B. In fact, (7.18) can be regarded as a general definition of the mutual impedance Z_{ab} between the terminal pairs A and B without having to specify explicitly the form of the antenna structures [3].

It is of interest to note that the procedure described above could be repeated for the case where the integration indicated in (7.16) was taken

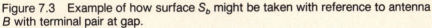

over the surface $S + S_a$, where S_a encloses source antenna A. We then would find that the open-circuit voltage v_a at terminals A, for an impressed current I_b, are related by

$$Z_{ba} = \frac{v_a}{I_b} = Z_{ab} \tag{7.20}$$

where Z_{ab} is given by (7.18) above. This result is a demonstration of the reciprocity theorem in the form that holds for a four-pole passive network.

7.4 MONTEATH'S FIELD-COMPENSATION THEOREM

We now consider two pairs of terminals A and B *outside* a closed surface S, as indicated in Figure 7.4. As before, we denote the mutual impedance between the terminals A and B as Z_{ab} for a specified distribution of complex permittivity and permeability in the medium both inside and outside S. The fields $\mathbf{E}_A$ and $\mathbf{H}_A$ again refer to the fields produced by a current I_A at terminals A for terminals B open-circuited; $\mathbf{E}_B$ and $\mathbf{H}_B$ are defined in similar fashion.

Now we consider that the properties of the region *within S only* are modified. Then for the same terminal currents I_a and I_b the field quantities change to $\mathbf{E}'_A$, $\mathbf{H}'_A$, $\mathbf{E}'_B$, and $\mathbf{H}'_B$, and the mutual impedance between terminals pairs A and B becomes Z'_{ab}. We seek to find an expression for the change $Z'_{ab} - Z_{ab}$.

In accordance with the concept of the compensation theorem [3, 4] in network theory, we regard the changes in the fields outside S to be due to the incremental field changes $\mathbf{E}'_A - \mathbf{E}_A$ and $\mathbf{H}'_A - \mathbf{H}_A$ within S for a current I_a in terminals A. The changes in the open-circuit voltage in terminals B is $I_a(Z'_{ab} - Z_{ab})$. It is important to recognize that we have, in effect, introduced secondary sources in S as a means to represent the modification of the medium.

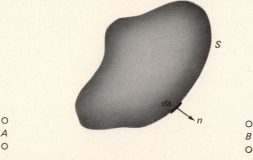

Figure 7.4   Antenna terminal pairs A and B located externally to closed surface S that encloses secondary sources.

A current I_b produces fields $\mathbf{E}_B$ and $\mathbf{H}_B$ in the unmodified region within S, so we are led to write

$$\iint\limits_S [(\mathbf{E}'_A - \mathbf{E}_A) \times \mathbf{H}_B - \mathbf{E}_B \times (H'_A - \mathbf{H}_A)] \cdot \mathbf{n} \, da$$

$$+ I_a I_b (Z'_{ab} - Z_{ab}) = 0 \qquad (7.21)$$

Now

$$\iint\limits_S (\mathbf{E}_A \times \mathbf{H}_B - \mathbf{E}_B \times \mathbf{H}_A) \cdot \mathbf{n} \, da = 0 \qquad (7.22)$$

because terminal pairs A and B are outside S. Thus we obtain the useful result that

$$Z'_{ab} - Z_{ab} = \frac{1}{I_a I_b} \iint\limits_S [(\mathbf{E}_B \times \mathbf{H}'_A) - (\mathbf{E}'_A \times \mathbf{H}_B)] \cdot \mathbf{n} \, da \qquad (7.23)$$

An equivalent form is

$$Z'_{ab} - Z_{ab} = \frac{1}{I_b I_a} \iint\limits_S [(\mathbf{E}_A \times \mathbf{H}'_B) - (\mathbf{E}'_B \times \mathbf{H}_A)] \cdot \mathbf{n} \, da \qquad (7.24)$$

Equations (7.23) and (7.24) are really an explicit statement of Monteath's [3] field equivalent of Ballentine's corollary of the compensation theorem from circuit theory.

7.5 MIXED-PATH GROUND WAVE TRANSMISSION

An excellent example of a problem that can be handled by the compensation theorem is radio wave transmission over an inhomogeneous ground [5–8]. Our illustrative model is a flat earth with two homogeneous portions separated by a straight-line boundary. The situation is illustrated in Figure 7.5, where we have chosen a rectangular coordinate system (x, y, z) where the earth's surface is $z = 0$ and the separation boundary is the y axis.

We assume that the electromagnetic fields satisfy an impedance boundary condition on the earth's surface. Thus for the region $x > 0$

$$\left. \begin{aligned} E_x &= -Z_1 H_y \\ E_y &= +Z_1 H_x \end{aligned} \right]_{z=0} \qquad (7.25)$$

while for the region $x < 0$

$$\left. \begin{aligned} E_x &= -Z H_y \\ E_y &= +Z H_x \end{aligned} \right]_{z=0} \qquad (7.26)$$

and Z_1 and Z are the respective surface impedances.

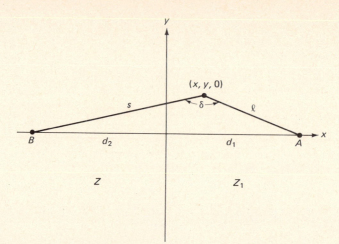

Figure 7.5 Plan view of two-section flat earth. Coordinates of A are $(d_1, 0, 0)$, and coordinates of B are $(d_2, 0, 0)$. Also note that $AB = d = d_1 + d_2$.

We now locate vertical hertzian electric dipoles of effective heights h_a and h_b at A and B, respectively, on the earth's surface, as indicated in Figure 7.5. To simplify the discussion, the line joining A and B intersect the separation boundary at right angles. Thus without further loss of generality we choose A to be located at $(d_1, 0, 0)$ and B to be located at $(-d_2, 0, 0)$.

In the limiting case where the surface impedances of the two regions are the same (that is, $Z_1 = Z$), the solution for the mutual coupling between the dipoles is well known [16]. In that case the mutual impedance Z_{ab} is well-approximated by the expression

$$Z_{ab} \simeq \frac{h_a h_b i \mu_0 \omega}{2\pi d} e^{-ikd} F(d, Z) \tag{7.27}$$

where $k = \omega/c = (\epsilon_0\mu_0)^{1/2}\omega$, $d = d_1 + d_2$, and $F(d, Z)$ is the Sommerfeld attenuation function [2, 9]. The latter is given by

$$F(d, Z) = 1 - i(\pi p)^{1/2}e^{-p} \operatorname{erfc}(ip^{1/2}) \tag{7.28}$$

where $p = -(ikd/2)(Z/\eta_0)^2$ and $\eta_0 = 120\pi$. As discussed elsewhere [9], the above expression for Z_{ab} is valid subject to $kd \gg 1$ and $|Z/\eta_0|^2 \ll 1$.

Now in the case where the path is mixed (that is, $Z_1 \neq Z$), the mutual impedance changes from Z_{ab} to Z'_{ab}. The compensation theorem can then be employed to write

$$Z'_{ab} - Z_{ab} = \frac{1}{I_a I_b} \int_{x=0}^{\infty} \int_{y=-\infty}^{+\infty} [\mathbf{E}_b \times \mathbf{H}'_a - \mathbf{E}'_a \times \mathbf{H}_b]_z \, dx \, dy \tag{7.29}$$

Here $\mathbf{E}'_a$ and $\mathbf{H}'_a$ are the electromagnetic fields of antenna A for an impressed current I_a at its terminals for the modified (that is, inhomogeneous) earth model. On the other hand, $\mathbf{E}_b$ and $\mathbf{H}_b$ are the electromagnetic fields of antenna B for an impressed current I_b at its terminals for the unmodified (that is, homogeneous) earth model. The integration in (7.29) extends over the surface $z = 0$ for the region of the ground that has been modified (that is, $x > 0$).

A crucial point now is to assume that the impedance boundary conditions as stated by (7.25) and (7.26) apply to both primed and unprimed fields. Then it is a simple matter to show that

$$Z'_{ab} - Z_{ab} = \frac{(Z_1 - Z)}{I_a I_b} \int_0^\infty \int_{-\infty}^{+\infty} \mathbf{H}'_{at} \cdot \mathbf{H}_{bt} \, dx \, dy \tag{7.30}$$

where, as indicated, the integrand is the dot product of the two tangential magnetic field vectors at the earth's surface.

To reduce the two-dimensional integral formulation to a useful form, we utilize some convenient normalizations. First of all, we write

$$Z'_{ab} = \frac{h_a h_b i \mu_0 \omega}{2\pi d} e^{-ikd} F'(d, Z, Z_1) \tag{7.31}$$

where F' is the modified attenuation function that we wish to find. Within the hertzian dipole approximation we may also write

$$\mathbf{H}_{bt} = \frac{ikI_b h_b}{2\pi s} e^{-iks} \left(1 + \frac{1}{iks}\right) F(s, Z)(\mathbf{i}_s \times \mathbf{i}_z) \tag{7.32}$$

where $s = [(d_2 + x)^2 + y^2]^{1/2}$, $F(s, Z)$ is the Sommerfeld attenuation function for a homogeneous earth of surface impedance Z, and $\mathbf{i}_s$ and $\mathbf{i}_z$ are unit vectors in the directions of increasing s and z, respectively. In a similar vein, we then also write

$$\mathbf{H}'_{at} \simeq \frac{ikI_a h_a}{2\pi \ell} e^{-ik\ell} \left(1 + \frac{1}{ik\ell}\right) F'(\ell, Z, Z_1)(\mathbf{i}_\ell \times \mathbf{i}_z) \tag{7.33}$$

where $\ell = [(d_1 - x)^2 + y^2]^{1/2}$ and $\mathbf{i}_\ell$ is the unit vector in the direction of increasing ℓ. Also, $F'(\ell, Z, Z_1)$ is the modified attenuation function for a transmission path from dipole A to the point $(x, y, 0)$ over the modified surface.

From (7.30), (7.32), and (7.33) the mutual impedance formula thus becomes

$$Z'_{ab} - Z_{ab} = (Z - Z_1) \frac{k^2 h_a h_b}{4\pi^2} \int_0^\infty \int_{-\infty}^{+\infty} \frac{e^{-ik(s+\ell)}}{\ell s}$$

$$\times \left(1 + \frac{1}{iks}\right)\left(1 + \frac{1}{ik\ell}\right) F(s, Z) F'(\ell, Z, Z_1) \cos \delta \, dx \, dy \tag{7.34}$$

where δ is the angle subtended by $\mathbf{i}_s$ and $\mathbf{i}_\ell$. Actually, this result is a two-dimensional integral equation for the unknown function F' since it appears on both sides of the equation. We will further approximate it in order to get a one-dimensional form.

The key observation is that the major contribution of the integral in the double integral occurs when the phase of the exponential is slowly varying. Cancellation occurs in other cases. Thus we are led to develop the following power series expansion:

$$
\ell + s = d_2 + d_1 + \frac{y^2}{2}\left(\frac{1}{d_2 + x} + \frac{1}{d_1 - x}\right)
$$
$$
+ \frac{y^4}{8}\left[\frac{1}{(d_2 + x)^3} + \frac{1}{(d_1 - x)^3}\right] + \cdots
\tag{7.35}
$$

for $0 < x < d_1$, while

$$
\ell + s = d_2 - d_1 + 2x + \frac{y^2}{2}\left(\frac{1}{d_2 + x} + \frac{1}{x - d_1}\right) + \cdots
\tag{7.36}
$$

for $x > d_1$. A further approximation is to replace both $[1 + (1/iks)]$ and $[1 + (1/ik\ell)]$ by unity, which is justified over the major portion of the integrand when neither A nor B are near the boundary. This approximation is valid even though $\ell \to 0$ as the integration point approaches antenna A (see Appendix B of this chapter).

With the above simplifications we can write

$$
Z'_{ab} - Z_{ab} = \frac{(Z_1 - Z)k^2 h_a h_b}{4\pi^2} e^{-ikd}
$$
$$
\times \int_0^{d_1} \frac{F(d_2 + x, Z)F'(d_1 - x, Z, Z_1)}{(d_2 + x)(d_1 - x)}
$$
$$
\times \left[\int_{-\infty}^{+\infty} e^{-i\alpha^2 y^2}\, dy\right] dx
\tag{7.37}
$$

where

$$
\alpha^2 = \frac{kd}{2(d_2 + x)(d_1 - x)}
$$

The integration extends over x only from 0 to d_1 because the contribution from the integrand for $x > d_1$ is negligible in the presence of the rapidly varying function $\exp(-2ikx)$. Furthermore, we see that $Z'_{ab} \simeq Z_{ab}$ if antenna A is on the left-hand side of the boundary.

The y integration of the problem is now simple:

$$
\int_{-\infty}^{+\infty} e^{-i\alpha^2 y^2}\, dy = \left(\frac{\pi}{i}\right)^{1/2} \alpha
\tag{7.38}
$$

Here we note that the significant values of $|y|$ are of the order of α^{-1} or less. Thus the justification for neglecting the terms in y^4 or higher is that

$$\frac{k}{8}\left[\frac{1}{(d_2+x)^3}+\frac{1}{(d_1-x)^3}\right]\frac{1}{\alpha^4} \ll 1$$

This condition is satisfied if $k(d_1+d_2) \gg 1$, as already assumed, and provided the terminals A and B are not too close to the boundary.

Using (7.31), (7.37), and (7.38), we now find that

$$F'(d, Z, Z_1) \simeq F(d, Z)$$

$$-i\epsilon_0\omega(Z_1-Z)\left(\frac{d}{2\pi ik}\right)^{1/2}$$

$$\times \int_0^{d_1} \frac{F(d_2+x, Z)F'(d_1-x, Z, Z_1)}{(d_2+x)^{1/2}(d_1-x)^{1/2}}\,dx \qquad (7.39)$$

for $d_1 > 0$, while

$$F'(d, Z, Z_1) \simeq F(d, Z) \qquad (7.40)$$

for $d_1 < 0$. However, in view of (7.40) it follows that $F'(d_1-x, Z, Z_1)$, where it occurs in the integrand of (7.39), can be replaced by $F(d_1-x, Z_1)$.

After a simple change of variable the final expression for $F'(d, Z, Z_1)$ can be written in the convenient form

$$F' = F(p_0) - i\left(\frac{p_0}{\pi}\right)^{1/2}(1-K^{1/2})\int_0^{p_1}\frac{F(p)F(p_0-Kp)}{[p(p_0-Kp)]^{1/2}}\,dp \qquad (7.41)$$

where

$$F(p) = 1 - i(\pi p)^{1/2}e^{-p}\,\mathrm{erfc}(ip^{1/2}) \qquad (7.42)$$

$$p_0 = -\left(\frac{ikd}{2}\right)\left(\frac{Z}{\eta_0}\right)^2 \qquad (7.43)$$

$$p_1 = -\left(\frac{ikd_1}{2}\right)\left(\frac{Z_1}{\eta_0}\right)^2 = \frac{p_0V}{K} \qquad (7.44)$$

and

$$K = \left(\frac{Z}{Z_1}\right)^2 \quad\text{and}\quad V = \frac{d_1}{d_1+d_2} = \frac{d_1}{d}$$

It is significant that the resultant transmission formula for a two-section path is given entirely in terms of attenuation functions for transmission over homogeneous sections of surface impedance Z and Z_1. Here we note that p_0 is the numerical distance appropriate for a one-section path from A to B (or B to A) over the plane with surface impedance Z. On the other hand, p_1 is the numerical distance appropriate for transmission over the plane from point B to the boundary (at $x = 0$) over a plane with surface impedance Z_1.

Exercise: Show that an equivalent form of (7.39) is

$$F'(d, Z_1, Z) = F(d, Z_1)$$
$$- i\epsilon_0\omega(Z - Z_1)\left(\frac{d}{2\pi ik}\right)^{1/2} \int_0^{d_2} \frac{F(d_1 + x', Z_1)F(d_2 - x', Z)}{[(d_1 + x')(d_2 - x')]^{1/2}} \, dx' \quad (7.45)$$

for $d_1 > 0$, while

$$F'(d, Z_1, Z) = F(d, Z_1) \quad (7.46)$$

for $d_1 < 0$.

Exercise: Show that $F(d, Z)$ satisfies the integral equation

$$F(d, Z) = 1 - i\epsilon_0\omega Z\left(\frac{d}{2\pi ik}\right)^{1/2} \int_0^d \frac{F(d - x, Z)}{[x(d - x)]^{1/2}} \, dx \quad (7.47)$$

where $F(d, Z)$ is defined by (7.28).

Exercise: Show from (7.41) that if $d_1 \ll d$, we may write

$$\frac{F'}{F(p_0)} \simeq 1 - \frac{2i(1 - K^{1/2})}{\pi^{1/2}}$$
$$\times \left(p_1^{1/2} - \frac{i\pi^{1/2}p_1}{2} - \frac{2p_1^{3/2}}{3} + \frac{i\pi^{1/2}p_1^2}{4} + \frac{4p_1^{5/2}}{15} - \cdots\right) \quad (7.48)$$

Hint: Assume that $F(p)/p^{1/2} \simeq F(p_0)/p_0^{1/2}$ and use the series expansion

$$F(p) \simeq 1 - i(\pi p)^{1/2} - 2p + i\pi^{1/2}p^{3/2} + \cdots$$

Exercise: Show that if the surface impedance $Z(x)$ varies with distance x along the path, the integral equation for the attenuation function $F'(d)$ is

$$F'(d) = 1 - i\epsilon_0\omega\left(\frac{d}{2\pi ik}\right)^{1/2} \int_0^d Z(x)\frac{F'(d - x)}{[x(d - x)]^{1/2}} \, dx \quad (7.49)$$

The reduction of the two-dimensional integral equation to a one-dimensional integral formula for the two-section path was based on a stationary-phase approximation. The validity of this approach was only established if the terminals A or B were not near the boundary. An alternative viewpoint [5] is to examine the two-dimensional representation in the case where the terminal A is near the boundary (that is,

$d_1 \ll d_2$). In this case it can be argued that the important contribution to the surface integration is confined to the region where x and y are also small compared with d_2. As a consequence, it is permissible to employ the following approximations in (7.37):

$$F(s, Z) \simeq F(d_2, Z) \simeq F(d, Z)$$

$$F'(\ell, Z, Z_1) \simeq 1, \qquad \frac{1}{s} \simeq \frac{1}{d_1}$$

$$\cos \delta \simeq \frac{x - d_1}{\ell}, \qquad k(s + \ell) \simeq k(d_2 + x + \ell)$$

The integral formula for the mutual impedance modification now becomes

$$Z'_{ba} - Z_{ba} \simeq (Z - Z_1) \frac{k^2 h_a h_b}{4\pi^2 d_2} e^{-ikd_2} F(d_2, Z)$$

$$\times \int_{x=0}^{\infty} e^{-ikx} \int_{-\infty}^{+\infty} \frac{e^{-ik\ell}}{\ell} \left(\frac{x - d_1}{\ell} \right) \left(1 + \frac{1}{ik\ell} \right) dx \, dy \qquad (7.50)$$

where $\ell = [(x - d_1)^2 + y^2]^{1/2}$.

The integration over y in (7.50) can be carried out immediately if we note that

$$\int_0^\infty \frac{e^{-ik\ell}}{\ell} \, dy = \frac{\pi}{2i} H_0^{(2)}(k|x - d_1|) \qquad (7.51)$$

which is the integral representation for the Hankel function of the second kind of order zero. The resulting integration over x is effected if we note that

$$\int_0^\infty e^{-ikx} \frac{d}{dx} H_0^{(2)}(k|x - d_1|) \, dx$$

$$= \begin{cases} -(1 - ikd_1)H_0^{(2)}(kd_1) + kd_1 H_1^{(2)}(kd_1) & \text{for } d_1 > 0 \qquad (7.52) \\ -(1 - ikd_1)H_0^{(2)}(-kd_1) - kd_1 H_1^{(2)}(-kd_1) & \text{for } d_1 < 0 \end{cases}$$

where $H_1^{(2)}$ is the Hankel function of the second kind of order one.[*]

From the above results it is not difficult to show that for $\alpha = kd_1 > 0$

$$\frac{F'}{F} \simeq 1 + \frac{q}{2} e^{i3\pi/4} e^{i\alpha}[(1 - i\alpha)H_0^{(2)}(\alpha) - \alpha H_1^{(2)}(\alpha)] \qquad (7.53)$$

while for $\alpha = -kd_1 > 0$

---

[*] Equation (7.52) follows from the identity

$$\frac{d}{d\alpha}[\alpha e^{\pm i\alpha}(H_0^{(2)}(\alpha) \mp iH_1^{(2)}(\alpha))] = e^{\pm i\alpha} H_0^{(2)}(\alpha)$$

which is easily verified.

$$\frac{F'}{F} \simeq 1 + \frac{q}{2}\, e^{i3\pi/4} e^{-i\alpha}[(1 + i\alpha)H_0^{(2)}(\alpha) - \alpha H_1^{(2)}(\alpha)] \tag{7.54}$$

where

$$q = \frac{Z - Z_1}{\eta_0}\, e^{-i\pi/4}$$

Equations (7.53) and (7.54) are expected to be valid quite close to and on either side of the boundary of separation. The results, however, are not valid very near and right at the boundary because of the impedance boundary conditions that have been used. However, it is believed that the results are applicable when α is somewhat greater than $|Z/\eta_0|$, which is small compared with unity. Further modification of the theory to remove this restriction would require a more precise description of how the two sections of the path are joined up.

When the distance of A from the boundary is sufficiently large that $\alpha \gg 1$ (but still assuming $|d_1/d_2| \ll 1$), we can write, for $d_1 > 0$,

$$\frac{F'}{F} \simeq 1 + iq\left(\frac{2\alpha}{\pi}\right)^{1/2}\left(1 + i\,\frac{3}{8\alpha} - \frac{5}{128\alpha^2} + \cdots\right) \tag{7.55}$$

while for $d_1 < 0$

$$\frac{F'}{F} \simeq 1 - \frac{q}{2}\left(\frac{1}{2\pi\alpha}\right)^{1/2} e^{-2i\alpha}\left(1 - \frac{i}{8\alpha}\,a \cdots\right) \tag{7.56}$$

It is noted that the right-hand side of (7.55), when the last parenthetical expression is replaced by unity, is in agreement with the first two terms of the series development of (7.48). Any closer agreement is not expected since p_1 has been assumed here to be small. Also, it is to be noted that the departure of the right-hand side of (7.56) from unity is an indication of a *reflected* wave from the boundary that is neglected in the stationary-phase approximation. In most cases it is of minor significance because $|q| \ll 1$.

A number of extensions of the preceding analytical development have been worked out to account for oblique tranverses of the propagation path across the separation boundary and to allow for gentle changes of the elevation profile and nonabruptness of the two segments of the path [5, 7, 10–14].

A major conclusion from the analytical study of these idealized flat-earth models is that the stationary-phase evaluations are adequate to predict the general features of mixed-path transmission. This fact has also been borne out by the careful experimental measurements of R. J. King [7], who employed microwave laboratory models under controlled conditions. But, of course, there may be numerous examples where the two-dimensional aspects of the surface inhomogeneities should be accounted for in an explicit fashion (for example, see [8] and [12]).

7.6 ANTENNA RADIATION OVER INHOMOGENEOUS SURFACE

There is a broad class of antenna problems that can be handled via the compensation theorem. In particular, we often need to know how the immediate environment of the antenna will modify its performance relative to some ideal situation [10, 15–17]. We will give several examples here.

In the first example we consider a vertical electric dipole of moment $I\,ds$ located at the center of a disk of radius a. The disk is characterized by a surface impedance Z_a. The region above the disk is free space (that is, $z > 0$). The disk itself is lying on a flat ground with a surface impedance Z that is a constant everywhere. The objective is to deduce the field at a point P in the upper half space. To simplify the discussion, we allow P to be in the far field such that $kR \gg 1$ and also θ is not too near $\pi/2$.

We can imagine that our receiving antenna at P is a short, hertzian electric dipole oriented in the vertical direction. The situation is illustrated in Figure 7.6. The mutual impedance Z_m between the two dipoles is given by

$$Z_m \simeq \frac{ds\,d\ell\;i\mu_0\omega}{4\pi R_0}(1 + R_0(\theta_0))e^{-ikR_0}\sin^2\theta_0 \tag{7.57}$$

where

$$R_e(\theta_0) = \frac{\cos\theta_0 - (Z/\eta_0)}{\cos\theta_0 + (Z/\eta_0)} \tag{7.58}$$

where $\eta_0 = \mu_0\omega/k = (\mu_0/\epsilon_0)^{1/2}$. Here R_e is the Fresnel reflection coefficient for plane waves incident at an angle θ on the ground plane [16].

Now we are interested in the case where the ground plane is

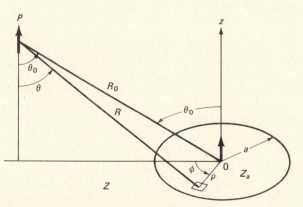

Figure 7.6 Geometry for vertical electric dipole at center of circular region whose surface impedance is Z_a, which differs from value Z.

modified by having the disk present. According to the compensation theorem, the modified mutual impedance Z'_m is given by

$$Z'_m - Z_m = \frac{1}{I_0 I_p} \iint_S (Z_a - Z) \mathbf{H}_{pt} \cdot \mathbf{H}'_{0t} \, dS \tag{7.59}$$

Here $\mathbf{H}_{pt}$ is the tangential magnetic field vector of the dipole at P over the surface S in the unmodified case (that is, $Z_a = Z$), while $\mathbf{H}_{0t}$ is the tangential magnetic field of the central dipole over the surface in the modified case. The corresponding currents at the dipole terminals are I_0 and I_p, respectively.

In this formulation of the problem Z_a, the surface impedance of the disk region in the modified situation, could be a function of the radial coordinate ρ and the azimuthal coordinate ϕ. The element of area is then $\rho \, d\phi \, d\rho$. Thus, in general, we would write

$$Z'_m - Z_m = \frac{1}{I_0 I_p} \int_{\rho=0}^{a} \int_0^{2\pi} (Z_a - Z) \mathbf{H}_{pt} \cdot \mathbf{H}'_{0t} \rho \, d\rho \, d\phi \tag{7.60}$$

To proceed, we now write

$$\mathbf{H}_{pt} = \frac{ik \, d\ell \, I_0}{4\pi R} e^{-ikR} [1 + R_e(\theta)] \sin \theta \, \mathbf{i}_p \tag{7.61}$$

where $\mathbf{i}_p$ is a unit vector in the ground plane and making an angle $\pi/2$ with R. Here $R_e(\theta)$ is the Fresnel reflection coefficient as defined above, but the appropriate angle is θ subtended by R and the vertical direction.

The tangential magnetic vector $\mathbf{H}'_{0t}$ is not known, but here we introduce a useful and very reasonable approximation. We assume that

$$\mathbf{H}'_{0t} \simeq \frac{ik I_0 \, ds}{2\pi \rho} \left(1 + \frac{1}{ik\rho}\right) e^{-ik\rho} \mathbf{i}_\phi \tag{7.62}$$

which would be exact for a dipole located on a perfectly conducting ground plane. This approximation is better for values of ρ that are small, so it would break down when a is large. A more substantive discussion is given elsewhere [15].

While the integral on the right-hand side of (7.60) is now explicit, it is still complicated. To further simplify matters, we let $R_0 \gg a$ so that $\theta \simeq \theta_0$, and as a consequence,

$$\mathbf{i}_p \cdot \mathbf{i}_\phi \simeq \cos \phi \qquad \text{and} \qquad R - R_0 \simeq -\rho \cos \phi \sin \theta_0$$

After some simple algebra we may write

$$Z'_m = Z_m (1 + \Omega) \tag{7.63}$$

where

$$\Omega = -\frac{ik}{2\pi \sin \theta_0} \int_{\rho=0}^{a} \int_{\phi=0}^{2\pi} \frac{Z_a - Z}{\eta_0}$$

$$e^{-ik\rho}\left(1 + \frac{1}{ik\rho}\right)e^{ik\rho\cos\phi\sin\theta_0} \, d\rho \, d\phi \qquad (7.64)$$

This integral is now in manageable form, and since $Z_a = Z_a(\rho, \phi)$ is specified, the dimensional factor $\Omega = \Omega(\theta, \phi)$ is evaluated. We thus have a method to estimate the change of the radiation pattern of the antenna located on the ground that results from the presence of the disk.

An obvious special case is to restrict Z_a to be ρ-dependent only. The ϕ integration is then carried out to give

$$\Omega = \frac{k}{\sin \theta_0} \int_0^a \frac{Z_a - Z}{\eta_0} e^{-ik\rho}\left(1 + \frac{1}{ik\rho}\right)J_1(k\rho \sin \theta) \, d\rho \qquad (7.65)$$

where we have used the integral representation of the Bessel function $J_1(x)$ of argument z.

In the case where the earth is a low-loss dielectric, Z/η_0 is approximately equal to the reciprocal $1/N_1$ of the refractive index of the soil. Then if the disk is metallic, $Z_a \ll Z$, so that (7.65) takes the form

$$\Omega = \frac{-1}{N_1 \sin \theta_0} \int_0^{ka} e^{-ix}\left(1 + \frac{1}{ix}\right)J_1(x \sin \theta_0) \, dx \qquad (7.66)$$

This is a relatively simple integral to evaluate numerically. As an example, we show, in Figure 7.7, a plot of the magnitude $|1 + \Omega|$, expressed in

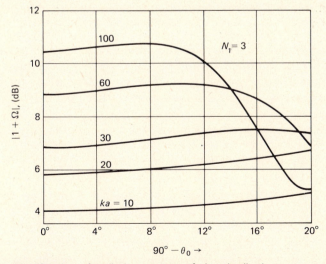

Figure 7.7 Modification of radiation pattern of electric dipole antenna over circular, perfectly conducting disk laid on dielectric ground.

decibels, as a function of the grazing angle $90° - \theta_0$. As indicated by (7.63), the ordinate can be regarded as the modification of the radiation pattern resulting from the presence of the disk. For this numerical example we have chosen $N_1^2 = 9$, and the values of $ka(2\pi \times$ disk radius/wavelength) are as indicated on Figure 7.7. As we see, the enhancement due to the disk is not significant unless the disk is quite large relative to a wavelength.

Many numerical studies [12, 16, 17–21] deal with the influence of extended ground systems on the radiation patterns, including the case of nonazimuthal symmetry (for example, sectors). Essentially, they are based on the formulations discussed above.

7.7 INPUT IMPEDANCE OF ANTENNAS FOR IMPERFECT GROUND PLANES

In our general development for the mutual impedance Z_{ab} between the terminals of antenna A and B, we can include the important special case where the terminals are coincident and the antennas are the same [16]. Then $Z_{ab} \rightarrow Z_{aa}$ is, by definition, the input impedance, which we might write in the form

$$Z_{aa} = Z_0 + \Delta Z \tag{7.67}$$

Here Z_0 is the input impedance of the antenna if the ground plane were everywhere perfectly conducting, and ΔZ is the change of the input impedance that results from the finite conductivity or other modification of the lower half space. Then a direct application of (7.23) tells us that

$$\Delta Z = -\frac{1}{I_0^2} \iint\limits_S (\mathbf{E} \times \mathbf{H}^{(\infty)}) \cdot \mathbf{n} \, da \tag{7.68}$$

where $\mathbf{E}$ is the electric field from the antenna carrying a current I_0 at its terminals for the *modified* ground plane, while $\mathbf{H}^{(\infty)}$ is the magnetic field for the unmodified or perfectly conducting ground plane for the same antenna current I_0. In this case we observe that $(\mathbf{E}^{(\infty)} \times \mathbf{H}) \cdot \mathbf{n}$ is zero over the surface S.

In the case of azimuthal symmetry we see immediately that

$$\Delta Z = -\frac{1}{I_0^2} \int_0^\infty E_\rho(\rho, 0) H_\phi^{(\infty)}(\rho, 0) 2\pi\rho \, d\rho \tag{7.69}$$

where the surface S is the ground plane $z = 0$ and the integration extends from $\rho = 0$ to ∞. This result is applicable, for example, to the case where the antenna is a vertical current-carrying wire. To illustrate how this formulation is used to predict the performance of an antenna with a ground system, we invoke the surface impedance condition:

$$E_\rho(\rho, 0) \simeq -Z_{\text{eff}}(\rho) H_\phi(\rho, 0) \tag{7.70}$$

Here $Z_{eff}(\rho)$ is ρ-dependent but independent of ϕ. To choose a concrete situation, we might assume Z_{eff} is zero out to some radius a. This would be the case when the disk is of perfect conductivity and centrally located beneath the linear antenna. Beyond $\rho = a$, we then set $Z_{eff} = Z_g$, where Z_g is the surface impedance of the ground. Then (7.69) reduces to

$$\Delta Z = \frac{1}{I_0^2} Z_g \int_a^\infty H_\phi(\rho, 0) H_\phi^{(\infty)}(\rho, 0) 2\pi\rho \, d\rho \tag{7.71}$$

For a linear antenna carrying a current $I(z)$ and extending from $z = 0$ to h on the z axis, we know that [16]

$$H_\phi^{(\infty)} = -\frac{1}{2\pi} \frac{\partial}{\partial \rho} \int_0^h \frac{\exp[-ik(z^2 + \rho^2)^{1/2}]}{(z^2 + \rho^2)^{1/2}} I(z) \, dz \tag{7.72}$$

This result is just the same as the broadside field of a linear antenna of length $2h$, with a symmetrical current distribution, located in free space. Once $I(z)$ is specified, $H_\phi^{(\infty)}$ is readily computed or deduced.

Now $H_\phi(\rho, 0)$ is not known, but it can be replaced by $H_\phi^{(\infty)}(\rho\ 0)$ for a first-order approximation, bearing in mind that the important contribution to the integrand in (7.71) occurs for values of ρ that are not too large. Then we have an explicit expression to deduce ΔZ. The corresponding value $I_0^2 \Delta R/2$, where $\Delta R = \mathrm{Re} \, \Delta Z$, is the change of the input power to the antenna that results from changing the ground from a perfectly conducting surface of infinite extent to one that is perfectly conducting only out to a distance a. Of course, from the standpoint of antenna efficiency, one is interested in minimizing ΔR. Related problems that have been examined are ground systems composed of radial wires laid on the earth's surface or lightly buried and wire meshes [10, 16, 18]. In such cases Z_{eff}, the surface impedance, may be a function of ρ within the ground system. Examples of such calculations are available for a wide range of the parameters [16].

References

1. Lord Rayleigh (J. W. Strutt): *Theory of Sound*, 2d ed., Macmillan, London, 1894, p. 155.
2. A. N. Sommerfeld: *Partial Differential Equations in Physics*, Academic Press, New York, 1949.
3. G. D. Monteath: "Application of the Compensation Theorem to Certain Radiation and Propagation Problems," *Proceedings of the IEE* (London), vol. 98, pt. IV, 195, pp. 23–30.
4. H. H. Skilling: *Electrical Engineering Circuits*, 2d ed., Van Nostrand, New York, 1965. See pg. 373.
5. J. R. Wait: "Oblique Propagation of Ground Waves Across a Coastline," *NBS Journal Research*, vol. 67D, 1963, pp. 617–630, and vol. 68D, 1964, pp. 747–752.

6. J. R. Wait: "Propagation of Electromagnetic Waves over a Smooth Multi-section Curved Earth — An Exact Theory," *Journal of Mathematical Physics*, vol. 11, 1970, pp. 2851 – 2860.

7. R. J. King and J. R. Wait: "Electromagnetic Ground Wave Propagation — Theory and Experiment," *Symposia Mathematica* (Academic Press), vol. 18, 1976, pp. 107 – 208.

8. J. R. Wait: "Theories for Radio Ground Wave Transmission over a Multisection Path," *Radio Science*, vol. 15, 1980, pp. 971 – 976.

9. J. R. Wait: *Wave Propagation Theory*, Pergamon Press, Elmsford, N.Y., 1981.

10. G. D. Monteath: *Applications of the Electromagnetic Reciprocity Principle*, Pergamon Press, Elmsford, N.Y., 1973. See p. 48.

11. S. Rotheram: "Ground-Wave Propagation," *Proceedings of the IEE* (London), vol. 128, pt. F, no. 5, 1981, pp. 275 – 295.

12. E. C. Field: "Low Frequency Ground Wave Propagation over Narrow Terrain Features," *IEEE Transactions*, vol. AP–30, 1982, pp. 831 – 836.

13. D. A. Hill and J. R. Wait: "HF Ground Wave Propagation over Mixed Land, Sea and Sea-Ice Paths," *IEEE Transactions*, vol. GE–19, 1981, pp. 210 – 216.

14. R. H. Ott and L. A. Berry: "An Alternative Integral Equation for Propagation over Irregular Terrain," *Radio Science*, vol. 5, 1970, pp. 767 – 771.

15. J. R. Wait: "The Theory of an Antenna over an Inhomogeneous Ground Plane," in E. C. Jordan (ed.), *Electromagnetic Theory and Antennas*, Pergamon Press, Elmsford, N.Y., 1963.

16. J. R. Wait: "Characteristics of Antennas over Lossy Earth," in R. E. Collin and F. J. Zucker (eds.), *Antenna Theory*, McGraw-Hill, New York, 1969, Chap. 23.

17. J. R. Wait: "Low Angle Radiation of an Antenna over an Irregular Ground Plane," *Atti della Foundazione Giorgio Ronchi* (Florence), vol. 35, 1980, pp. 576 – 583.

18. J. R. Wait and K. P. Spies: "Integral Equation Approach to the Radiation from a Vertical Antenna over an Inhomogeneous Ground Plane," *Radio Science*, vol. 5, 1970, pp. 73 – 79.

19. D. A. Hill and J. R. Wait: "HF Radio Wave Transmission over Sea Ice and Remote Sensing Possibilities," *IEEE Transactions*, vol. GE–19, 1981, pp. 201 – 209.

20. C. Teng and R. J. King: "Surface Fields and Radiation Patterns of a Vertical Electric Dipole over a Radial Wire Ground System," *Electromagnetics*, vol. 1, 1981, pp. 101 – 116.

21. K. S. Park, R. J. King, and C. J. Teng: "Radiation Patterns of an HF Vertical Dipole Near a Sloping Beach," *Electromagnetics*, vol. 2, 1982, pp. 129 – 145.

Appendix A. The Compensation Theorem in Network Theory

There is a useful concept that may be borrowed from classical network theory that is very effective in evaluating the resultant influence when an individual element of the circuit is modified. For example, Shea [1] states that "if a network is modified or altered by making a change ΔZ in the impedance Z of a branch, the change in the currents in all other branches is that which would be produced by an e.m.f. $-I\,\Delta Z$ acting in series with the modified branch, where I is the current in Z.

Following Skilling [2], we can illustrate the compensation theorem in a very neat way. In Figure 7.A.1(a) we have a linear network represented by a box but with one branch brought out, which is the one to be modified. Current in this branch is I. Now, as indicated in Figure 7.A.1(b), an incremental impedance ΔZ and a secondary source of electromagnetive force Δe have been inserted in the branch. Now Δe is adjusted until the current is still I. This requires that $\Delta e = I \cdot \Delta Z$. Then, as shown in Figure 7.A.1(c), another secondary source is added, which is equal and opposite; the resulting current is now $I' = I + \Delta I$, as shown in Figure 7.A.1(d). In accordance with the principle of superposition, ΔI is the current in the modified branch by this new source acting alone, as

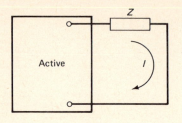

(a)

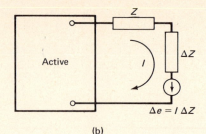

(b)

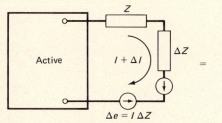

(c)

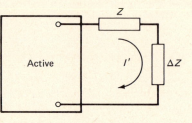

(d)

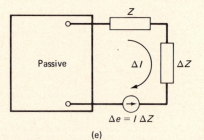

(e)

Figure 7.A.1   Illustrating Skilling's version of network-compensation theorem. See text for description parts (a) to (e).

shown in Figure 7.A.1(e). Then clearly, $\Delta I = -\Delta e/Z'$, where $Z' = Z + \Delta Z$. As we see, ΔZ is positive when ΔI is negative.

There are corresponding increments of current in each branch of the network equal to the currents produced in each branch by the new source alone. But the two secondary sources are equal and opposite, so they may be removed. Then we obtain the original network modified as depicted in Figure 7.A.1(d). Thus we have demonstrated the validity of the compensation theorem as enunciated above.

We now wish to know the effect of a change of various elements of a two-port network upon the mutual impedance of its terminal pairs. Following Monteath [3, 4], we adopt a network of the form shown in Figure 7.A.2 that has terminal pairs A and B. The mutual impedance between A and B is Z_{ab}. We suppose that the impedances $Z_1, Z_2, \ldots ,$

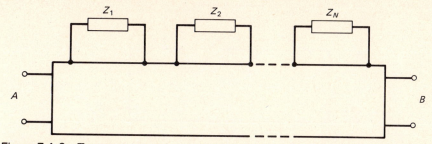

Figure 7.A.2 Two-port network, where indicated impedance elements may be modified.

Z_N of N branches are modified to $Z'_1, Z'_2, \ldots, Z'_N$. The objective is to determine the resulting mutual impedance Z'_{ab} of the modified network.

It is convenient in what follows to deal with a single branch whose impedance Z_n is modified to Z'_n. Then by superposition we may later sum over n from 1 to N.

First of all, we consider the unmodified network. Let I_{an} be the current in Z_n when a current I_a is applied to terminals A with B open-circuited. Similarly, I_{bn} is the current in Z_n when I_b is applied to terminals B with A open-circuited. In the modified network I_{an} becomes I'_{an} and I_{bn} becomes I'_{bn} for the same terminal currents I_a and I_b.

In the unmodified network the voltage v_b at B for a current I_A into B is given by

$$v_b = I_a Z_{ab} \tag{A.1}$$

as indicated in Figure 7.A.3(a). The resulting current in the nth impedance is I_{an}. In accordance with the compensation theorem, the modified $v'_b = v_b + (\Delta v)_n$ can be deduced by inserting a fictitious generator having emf $-I_{an}(Z'_n - Z_n)$ into the branch that is in series with the impedance Z'_n. Then $(\Delta v_b)_n$ is calculated for the case where A and B are open-circuited. To facilitate this latter calculation, we can again invoke the reciprocity theorem. To do this, we now inject the current I_b into terminals B with A open-circuited. The current in the modified branch Z'_n is now $I'_{b,n}$.

Now by reciprocity we have the following equality between current and voltage ratios:

$$\frac{I'_{b,n}}{I_b} = \frac{(\Delta v_b)_n}{I_{a,n}(Z' - Z)} \tag{A.2}$$

Thus

$$\Delta v_b = \sum_{n=1}^{N} \frac{I_{a,n} I'_{b,n}}{I_b}(Z'_n - Z_n) \tag{A.3}$$

is the total change in open-circuit voltage at terminals B, for a current I_a in terminals A, resulting from modifications in N impedance elements. In

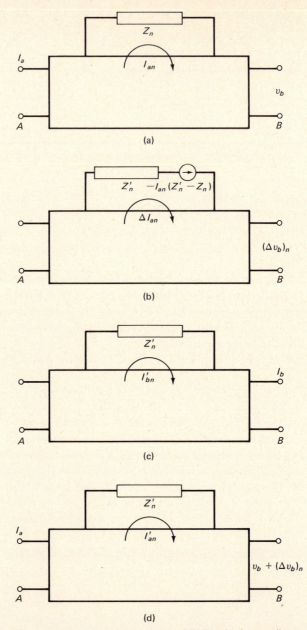

Figure 7.A.3 The progressive development of Ballantine's corollary to compensation theorem illustrated for the nth impedance element of two-port network; Z_n is modified to Z'_n, and we wish to relate this term to change of mutual impedance between terminal pairs A and B.

terms of mutual impedance between terminal pairs A and B, we have

$$Z'_{ab} = Z_{ab} + \frac{1}{I_a I_b} \sum_{n=1}^{N} I_{a,n} I'_{b,n} (Z'_n - Z_n) \qquad (A.4)$$

An equivalent form is obtained by interchanging the role of the terminals A and B. Thus it is found that

$$Z'_{ab} = Z_{ab} + \frac{1}{I_a I_b} \sum_{n=1}^{N} I'_{a,n} I_{b,n} (Z'_n - Z_n) \qquad (A.5)$$

Sometimes (A.4) and (A.5) are described as Ballantine's corollary to the compensation theorem [5]. Within the context of network theory the formulation of the impedance change $Z'_{ab} - Z_{ab}$ is exact, but we still need to deduce the currents $I'_{b,n}$ or $I'_{a,n}$ in the modified network. Actually, in many cases of practical interest, we can replace $I'_{a,n}$ by $I_{a,n}$ in (A.5) if the change $Z'_n - Z_n$ is small compared with Z_n. The choice of which approximation to employ is guided by physical considerations. This approach has been illustrated when dealing with specific field problems involving inhomogeneous media [4, 6–11].

References

1. T. A. Shea: *Transmission Networks and Wave Filters*, Van Nostrand, New York, 1929.
2. H. H. Skilling: *Electrical Engineering Circuits*, 2d ed., Van Nostrand, New York, 1965.
3. G. D. Monteath: "Application of the Compensation Theorem to Certain Radiation and Propagation Problems," *Proceedings of the IEE* (London), vol. 98, pt. IV, 1951, pp. 23–30.
4. G. D. Monteath: *Applications of the Electromagnetic Reciprocity Principle*, Pergamon Press, Elmsford, N.Y., 1973.
5. S. Ballantine: "Reciprocity in Electromagnetic Mechanical, and Acoustical, and Interconnected Systems," *Proceedings of the IRE*, vol. 17, 1929, p. 929.
6. J. R. Wait: "Oblique Propagation of Ground Waves Across a Coastline," *NBS Journal Research*, vol. 67D, 1963, pp. 617–630, and vol. 68D, 1964, pp. 747–752.
7. J. R. Wait: "Propagation of Electromagnetic Waves over a Smooth Multisection Curved Earth—An Exact Theory," *Journal of Mathematical Physics*, vol. 11, 1970, pp. 2851–2860.
8. R. J. King and J. R. Wait: "Electromagnetic Ground Wave Propagation—Theory and Experiment," *Symposia Mathematica* (Academic Press), vol. 18, 1976, pp. 107–208.
9. J. R. Wait: "Theories for Radio Ground Wave Transmission over a Multisection Path," *Radio Science*, vol. 15, 1980, pp. 971–976.
10. J. R. Wait: *Wave Propagation Theory*, Pergamon Press, Elmsford, N.Y., 1981.
11. J. R. Wait: "Nature of the EM Field Reflected From a Coastline," *Electronics Letters*, vol. 1, 1965, pp. 65–66.

Appendix B. Justification for Approximation of Integral Equation

In reducing the two-dimensional integral equation (7.34) to a single line-integral form, we replaced the factor $[1 + (1/ik\ell)]$ by 1. This neglect of the near-field effect of antenna A was claimed to be valid if A was not near the separation boundary. Some justification for this step is needed.

We write

$$Z'_{ab} - Z_{ab} = [Z'_{ab} - Z_{ab}]_\infty + \Delta z_m \tag{B.1}$$

where the term in brackets corresponds to (7.34) in the case where $1 + (ik\ell)^{-1}$ is replaced by 1. Thus we regard Δz_m as the correction that we now wish to estimate. Clearly, it will be most important in the region of the surface integration near A, where $k\ell$ becomes small. In this region we can make the following approximations for the other parameters:

$$\Delta z_m \simeq (Z - Z_1) \frac{k^2 h_a h_b}{4\pi^2} \int_{x=0}^{\infty} \int_{y=-\infty}^{\infty} \frac{e^{-i(k\ell+kx)}}{\ell^2} \left(\frac{x - d_1}{\ell}\right) dx \, dy$$

$$\times \frac{e^{-ikd}}{ikd} F(d, Z) \tag{B.2}$$

where

$$\ell = [(x - d_1)^2 + \ell^2]^{1/2}$$

As we indicated, most of the contribution to the integrand comes in the region where ℓ^2 is small. Thus if d_1 is not too small (that is, A not near the boundary), we can, in effect, replace the lower limit for the x integration by $-\infty$. Then if we change the x, y variables to polar coordinates ℓ, δ via

$$x = d_1 + \ell \cos \delta \qquad \text{and} \qquad dx \, dy = \ell \, d\ell \, d\delta$$

we see that

$$\Delta z_m \simeq (Z - Z_1) \frac{k^2 h_a h_b}{4\pi} \frac{e^{-ikd}}{ikd}$$

$$\int_{\ell=0}^{\infty} \int_0^{2\pi} \frac{e^{-ik\ell}}{\ell} e^{-ik\ell\cos\delta} \cos \delta \, d\delta \, d\ell \qquad (B.3)$$

The integration over δ is effected by noting that

$$J_1(k\ell) = -\frac{1}{2\pi i} \int_0^{2\pi} e^{-ik\ell\cos\delta} \cos \delta \, d\delta \qquad (B.4)$$

which is the integral representation for the Bessel function of order one and argument $k\ell$. The subsequent ℓ integration then follows if we note that

$$\int_0^{\infty} \sin kx \, \frac{J_1(kx)}{x} \, dx = 1 \qquad (B.5)$$

and

$$\int_0^{\infty} \cos kx \, \frac{J_1(kx)}{x} \, dx = 0 \qquad (B.6)$$

Thus the double integral in (B.3) is -2π. This gives our desired estimate:

$$\Delta z_m \simeq \frac{Z - Z_1}{\eta_0} \frac{h_a h_b i \mu_0 \omega}{2\pi d} e^{-ikd} F(d, Z) \qquad (B.7)$$

or more simply

$$\frac{\Delta z_m}{Z_{ab}} \simeq \frac{Z - Z_1}{\eta_0} \qquad (B.8)$$

which has negligible magnitude compared with one.

Appendix A
Some Essential Results from Vector Analysis

A.1 INTRODUCTION

A vector is a generic name for such quantities as velocity, force, electric field, magnetic field, and so on. It has both magnitude and direction; we denote it by **A**, meaning it has magnitude A. But note that A can be a phasor (that is, complex number) when we deal with time-harmonic quantities. Some basic relations follow.

The *scalar product* of two vectors **A** and **B** is defined by

$$\mathbf{A} \cdot \mathbf{B} = AB \cos \psi \tag{A.1}$$

where A and B are the magnitude of **A** and **B**, respectively, and ψ is the angle between them. Note that $\mathbf{A} \cdot \mathbf{B} = 0$ means that the vectors are perpendicular to each other. Scalar multiplication obeys commutative and distributive laws; that is

$$\mathbf{A} \cdot \mathbf{B} = \mathbf{B} \cdot \mathbf{A} \tag{A.2}$$

and

$$(\mathbf{A} + \mathbf{B}) \cdot \mathbf{C} = \mathbf{A} \cdot \mathbf{C} + \mathbf{B} \cdot \mathbf{C} \tag{A.3}$$

respectively, where $\mathbf{C}$ is another vector. Also, it follows that

$$A_x B_x + A_y B_y + A_z B_z = \mathbf{A} \cdot \mathbf{B} \tag{A.4}$$

where A_x, B_x, and so on are the components of $\mathbf{A}$ and $\mathbf{B}$ in rectangular coordinates projected onto the x axis. To be more specific, we consider the vector $\mathbf{A}$ to be drawn from point (x_1, y_1, z_1) to (x_2, y_2, z_2); the direction components are

$$A_x = x_2 - x_1 = \ell \cos \alpha \tag{A.5}$$

$$A_y = y_2 - y_1 = \ell \cos \beta \tag{A.6}$$

$$A_z = z_2 - z_1 = \ell \cos \gamma \tag{A.7}$$

where ℓ is the length or magnitude of the vector and α, β, and γ are the angles the vector makes with the coordinate axes.

The *vector product* $\mathbf{A} \times \mathbf{B}$ is a vector perpendicular to both, pointing in the direction in which a right-handed screw would advance if turned from vector $\mathbf{A}$ into $\mathbf{B}$ through the smaller angle ψ between them. In this context the magnitude of $\mathbf{A} \times \mathbf{B}$ is equal to $ab \sin \psi$, where a and b are the magnitudes of $\mathbf{A}$ and $\mathbf{B}$, respectively. In the case of vector products

$$\mathbf{A} \times \mathbf{B} = -\mathbf{B} \times \mathbf{A} \tag{A.8}$$

and

$$(\mathbf{A} + \mathbf{B}) \times \mathbf{C} = \mathbf{A} \times \mathbf{C} + \mathbf{B} \times \mathbf{C} \tag{A.9}$$

In terms of direction components we have

$$(\mathbf{A} \times \mathbf{B})_x = A_y B_z - A_z B_y \tag{A.10}$$

$$(\mathbf{A} \times \mathbf{B})_y = A_z B_x - A_x B_z \tag{A.11}$$

$$(\mathbf{A} \times \mathbf{B})_z = A_x B_y - A_y B_x \tag{A.12}$$

An equivalent statement, in easy-to-remember form, is

$$\mathbf{A} \times \mathbf{B} = \begin{vmatrix} \mathbf{i}_x & \mathbf{i}_y & \mathbf{i}_z \\ A_x & A_y & A_z \\ B_x & B_y & B_z \end{vmatrix} \tag{A.13}$$

where $\mathbf{i}_x$, $\mathbf{i}_y$, and $\mathbf{i}_z$ are unit vectors in the x, y, and z directions, respectively.

A.2 FUNCTION PROPERTIES

We consider the function $V(x, y, z)$, which could represent a scalar quantity such as pressure, temperature, or potential. Now let ΔV be the change in value of $V(x, y, z_0)$ as we pass from a point (x, y) in the plane $z = z_0$ to a point $(x + \Delta x, y + \Delta y)$ in the same plane $z = z_0$. The average rate of change is $\Delta V / \Delta s$, where Δs is the distance between these points.

As Δs approaches zero, we then have a definition of the directional derivative $\partial V/\partial s$. The partial derivatives $\partial V/\partial x$ and $\partial V/\partial y$ are simply the directional derivatives taken along the x and y coordinate axis, respectively.

Now clearly, we can write

$$\frac{\partial V}{\partial s} = \frac{\partial V}{\partial n} \cos \psi \qquad (A.14)$$

where $\partial V/\partial n$ is the maximum rate of change and ψ is the angle subtended by this direction and the direction we actually take. This is indicated for the two-dimensional (x, y) case in Figure A.1. Here, of course,

$$\frac{\partial V}{\partial n} = \lim_{\Delta n \to 0} \frac{\Delta V}{\Delta n} \qquad (A.15)$$

is the gradient of the function V.

We may extend the argument to three dimensions and write

$$\frac{\partial V}{\partial x} = \frac{\partial V}{\partial n} \cos \alpha \qquad (A.16)$$

$$\frac{\partial V}{\partial y} = \frac{\partial V}{\partial n} \cos \beta \qquad (A.17)$$

$$\frac{\partial V}{\partial z} = \frac{\partial V}{\partial n} \cos \gamma \qquad (A.18)$$

where α, β, and γ are the angles made by the normal to the level surface (that is, $V = $ constant) and the coordinate axes. Then it follows that

$$\frac{\partial V}{\partial n} = \left[\left(\frac{\partial V}{\partial x} \right)^2 + \left(\frac{\partial V}{\partial y} \right)^2 + \left(\frac{\partial V}{\partial z} \right)^2 \right]^{1/2} \qquad (A.19)$$

which is a consequence of the geometrical identity

$$\cos^2 \alpha + \cos^2 \beta + \cos^2 \gamma = 1 \qquad (A.20)$$

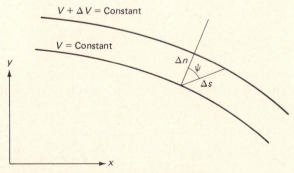

Figure A.1 Equipotential lines in $z = $ constant plane.

In fact, if we multiply (A.16), (A.17), and (A.18) by $\cos \alpha$, $\cos \beta$, and $\cos \gamma$, we find that

$$\frac{\partial V}{\partial n} = \frac{\partial V}{\partial x} \cos \alpha + \frac{\partial V}{\partial y} \cos \beta + \frac{\partial V}{\partial z} \cos \gamma \tag{A.21}$$

An equivalent statement is

$$\frac{\partial V}{\partial n} = \frac{\partial V}{\partial x} \frac{\partial x}{\partial n} + \frac{\partial V}{\partial y} \frac{\partial y}{\partial n} + \frac{\partial V}{\partial z} \frac{\partial z}{\partial n} \tag{A.22}$$

which follows from purely functional considerations.

A.3 NOTE ON COORDINATE SYSTEMS

Three coordinate systems are in very common use. They are as follows:

1. Rectangular or cartesian (x, y, z)
2. Cylindrical (ρ, ϕ, z)
3. Spherical (r, θ, ϕ)

These three systems are illustrated in Figure A.2(a), A.2(b), and A.2(c), respectively. Just from geometry of the figures we can see that the variables are connected by

$$x = \rho \cos \phi = r \sin \theta \cos \phi \tag{A.23}$$

$$y = \rho \sin \phi = r \sin \theta \sin \phi \tag{A.24}$$

$$z = r \cos \phi \tag{A.25}$$

We refer to ρ as the radial coordinate in cylindrical coordinates and r as the radial coordinate in spherical coordinates; the angle ϕ is described as the azimuthal angle; while θ is the polar angle. Of course, z is the axial coordinate in both cartesian and cylindrical systems.

In a general system of coordinates (u, v, w), a point $P(u, v, w)$ is defined as the point of intersection of three surfaces

$$f_1(x, y, z) = u \tag{A.26}$$

$$f_2(x, y, z) = v \tag{A.27}$$

$$f_3(x, y, z) = w \tag{A.28}$$

For convenience we will restrict attention to orthogonal systems. Then the differential distances along the coordinate lines are proportional to the differential of the coordinates. Thus

$$ds_u = h_u \, du \qquad ds_v = h_v \, dv \qquad ds_w = h_w \, dw \tag{A.29}$$

for a general element of length

$$(ds)^2 = (ds_u)^2 + (ds_v)^2 + (ds_w)^2$$
$$= h_u^2 \, du^2 + h_v^2 \, dv^2 + h_w^2 \, dw^2 \tag{A.30}$$

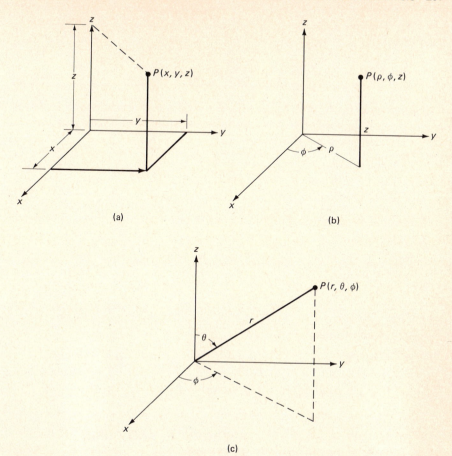

Figure A.2   (a) Cartesian (rectangular) system; (b) cylindrical system; (c) spherical system.

For example, in our three coordinate systems we have the following results. In the cartesian system

$$ds_x = dx \quad ds_y = dy \quad ds_z = dz$$
$$(ds)^2 = dx^2 + dy^2 + dz^2 \tag{A.31}$$

In the cylindrical system

$$ds_\rho = d\rho \quad ds_\phi = \rho \, d\phi \quad ds_z = dz$$
$$(ds)^2 = d\rho^2 + \rho^2 \, d\phi^2 + dz^2 \tag{A.32}$$

In the spherical system

$$ds_r = dr \quad ds_\phi = r \sin \theta \, d\phi \quad ds_\theta = r \, d\theta$$
$$(ds)^2 = dr^2 + r^2 \, d\theta^2 + r^2 \sin^2 \theta \, d\phi^2 \tag{A.33}$$

The volume $d\tau$ of the elementary cell, as depicted in Figure A.2, is given by

$$d\tau = ds_u \, ds_v \, ds_w = h_u h_v h_w \, du \, dv \, dw$$
$$= dx \, dy \, dz = \rho \, d\rho \, d\phi \, dz = r^2 \sin\theta \, dr \, d\theta \, d\phi \qquad \text{(A.34)}$$

The quantities h_u, h_v, and h_w are often called the metrical coefficients for the orthogonal coordinate system under consideration. Note that for our three systems we have the following:

1. For cartesian, $h_x = h_y = h_z = 1$.
2. For cylindrical, $h_\rho = 1$, $h_\phi = \rho$, $h_z = 1$.
3. For spherical, $h_r = 1$, $h_\theta = r$, $h_\phi = r \sin\theta$.

In some situations we need to work with elementary areas. In the (u, v, w) system these are

$$dS_u = ds_v \, ds_w, \quad dS_v = ds_w \, ds_u, \quad \text{and} \quad dS_w = ds_u \, ds_v \qquad \text{(A.35)}$$

or

$$dS_u = h_v h_w \, dv \, dw \qquad dS_v = h_w h_u \, dw \, du \qquad dS_w = h_u h_v \, du \, dv \quad \text{(A.36)}$$

In our three systems,

$$\begin{aligned}
dS_x &= dy \, dz & dS_y &= dz \, dx & dS_z &= dx \, dy \\
dS_\rho &= \rho \, d\phi \, dz & dS_\phi &= d\rho \, dz & dS_z &= \rho \, d\rho \, d\phi \\
dS_r &= r^2 \sin\theta \, d\theta \, d\phi & dS_\theta &= r \sin\theta \, dr \, d\phi & dS_\phi &= r \, dr \, d\phi
\end{aligned}$$

A.4 GRADIENT, LAPLACIAN, DIVERGENCE, AND CURL FOR VARIOUS COORDINATE SYSTEMS

Gradient

We define the gradient of V in the u direction as the directional derivative of V in the direction of u, that is, $\partial V / h_u \, \partial u$ or, what is the same thing, $\partial V / \partial s_u$. Thus the three orthogonal components of grad V are

$$\text{grad}_u \, V = \frac{1}{h_u} \frac{\partial V}{\partial u} \qquad \text{grad}_v \, V = \frac{1}{h_v} \frac{\partial V}{\partial v} \qquad \text{grad}_v \, V = \frac{1}{h_w} \frac{\partial V}{\partial w} \qquad \text{(A.37)}$$

In our three systems,

$$\text{grad}_x \, V = \frac{\partial V}{\partial x} \qquad \text{grad}_y \, V = \frac{\partial V}{\partial y} \qquad \text{grad}_z \, V = \frac{\partial V}{\partial z}$$

$$\text{grad}_\rho \, V = \frac{\partial V}{\partial \rho} \qquad \text{grad}_\phi \, V = \frac{\partial V}{\rho \, \partial \phi} \qquad \text{grad}_z \, V = \frac{\partial V}{\partial z}$$

$$\text{grad}_r \, V = \frac{\partial V}{\partial r} \qquad \text{grad}_\theta \, V = \frac{\partial V}{r \, \partial \theta} \qquad \text{grad}_\phi \, V = \frac{1}{r \sin\theta} \frac{\partial V}{\partial \phi}$$

Of course, we can also write

$$\text{grad } V = \frac{1}{h_u}\frac{\partial V}{\partial u}\,\mathbf{i}_u + \frac{1}{h_v}\frac{\partial V}{\partial v}\,\mathbf{i}_x + \frac{1}{h_w}\frac{\partial V}{\partial w}\,\mathbf{i}_w \tag{A.38}$$

where $\mathbf{i}_u$, $\mathbf{i}_v$, and $\mathbf{i}_w$ are unit vectors in the u, v, and w directions, respectively. Thus the gradient of the scalar function V, denoted grad V or ∇V, is a vector whose orthogonal components are the directional derivatives along the three (orthogonal) coordinate directions.

Divergence

The divergence of a vector $\mathbf{F}$ is defined by

$$\text{div } \mathbf{F} = \lim_{s \to 0} \frac{\iint F_n\, dS}{v} \tag{A.39}$$

where F_n is the normal component of $\mathbf{F}$ over a surface S enclosing the point, and where v is the enclosed elemental volume. We now apply this definition to the cell depicted in Figure A.3. The area in the u direction is dS_u and the flux across this surface is $F_u\, dS_u$. The corresponding flux across the surface at $u + du$ is evidently

$$F_u\, dS_u + \frac{\partial}{\partial u}\,(F_u\, dS_u)\, du$$

In a similar fashion, we can deduce the flux change in the v and the w directions. We then deduce that the residual flux (that is, total flux out minus total flux in) is

$$\frac{\partial}{\partial u}\,(F_u\, dS_u)\, du + \frac{\partial}{\partial v}\,(F_v\, dS_v)\, dv + \frac{\partial}{\partial w}\,(F_w\, dS_w)\, dw$$

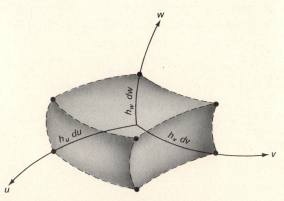

Figure A.3   Basic cell in curvilinear coordinates (u, v, w).

We must then divide this by the volume of the cell (that is, $h_u h_v h_w \, du \, dv \, dw$) to get

$$\text{div } \mathbf{F} = \frac{1}{h_u h_v h_w} \left[\frac{\partial}{\partial u}(h_v h_w F_u) + \frac{\partial}{\partial v}(h_w h_u F_v) + \frac{\partial}{\partial w}(h_u h_v F_w) \right] \qquad \text{(A.40)}$$

In our three coordinate systems we readily deduce that

$$\text{div } \mathbf{F} = \frac{\partial F_x}{\partial x} + \frac{\partial F_y}{\partial y} + \frac{\partial F_z}{\partial z} \qquad \text{(A.41)}$$

$$\text{div } \mathbf{F} = \frac{1}{\rho}\frac{\partial}{\partial \rho}(\rho F_\rho) + \frac{1}{\rho}\frac{\partial F_\phi}{\partial \phi} + \frac{\partial F_z}{\partial z} \qquad \text{(A.42)}$$

$$\text{div } \mathbf{F} = \frac{1}{r^2 \sin \theta}\left[\sin \theta \frac{\partial}{\partial r}\left(r^2 F_r\right) + r\frac{\partial}{\partial \theta}(\sin \theta F_\theta) + r\frac{\partial F_\phi}{\partial \phi} \right] \qquad \text{(A.43)}$$

We note that div $\mathbf{F}$ is often written $\nabla \cdot \mathbf{F}$, which is the dot product of the vector operator

$$\nabla = \mathbf{i}_x \frac{\partial}{\partial x} + \mathbf{i}_y \frac{\partial}{\partial y} + \mathbf{i}_z \frac{\partial}{\partial z} \qquad \text{(A.44)}$$

and the vector

$$\mathbf{F} = \mathbf{i}_x F_x + \mathbf{i}_y F_y + \mathbf{i}_z F_z \qquad \text{(A.45)}$$

when dealing explicitly with cartesian coordinates. As a word of caution, note that the ρ component of $\nabla \cdot \mathbf{F}$ is *not* $\partial F_\rho/\partial \rho$. Similar inconsistencies would arise if one applied the dot product rule in the spherical system. Possibly this is good reason to use the notation div $\mathbf{F}$.

Laplacian

The quantity div grad V, designated by $\nabla^2 V$, written in differential form, follows directly from a combined use of (A.37) and (A.40). Thus we obtain

$$\nabla^2 V = \frac{1}{h_u h_v h_w}\left[\frac{\partial}{\partial u}\left(\frac{h_v h_w}{h_u}\frac{\partial V}{\partial u}\right) + \frac{\partial}{\partial v}\left(\frac{h_w h_u}{h_v}\frac{\partial V}{\partial v}\right) \right.$$
$$\left. + \frac{\partial}{\partial w}\left(\frac{h_u h_v}{h_w}\right)\frac{\partial V}{\partial w} \right]$$

$$\nabla^2 V = \frac{\partial^2 V}{\partial x^2} + \frac{\partial^2 V}{\partial y^2} + \frac{\partial^2 V}{\partial z^2}$$

$$\text{(A.46)}$$

$$\nabla^2 V = \frac{1}{\rho}\frac{\partial}{\partial \rho}\rho\frac{\partial V}{\partial \rho} + \frac{1}{\rho^2}\frac{\partial^2 V}{\partial \phi^2} + \frac{\partial^2 V}{\partial z^2}$$

$$\nabla^2 V = \frac{1}{r^2 \sin \theta} \left[\sin \theta \frac{\partial}{\partial r} \left(r^2 \frac{\partial V}{\partial r} \right) + \frac{\partial}{\partial \theta} \left(\sin \theta \frac{\partial V}{\partial \theta} \right) \right.$$
$$\left. + \frac{1}{\sin \theta} \frac{\partial^2 V}{\partial \phi^2} \right]$$

The operator ∇^2 is called the laplacian. It is also written as $\nabla \cdot \nabla$, or simply as Δ. Again, the dot product of the vector operator ∇ and itself leads to the correct form for $\nabla^2 V$ or $\nabla \cdot \nabla V$ only in cartesian coordinates.

Sometimes, one encounters the operation $\nabla^2 \mathbf{F}$. Unless other information is provided, one should regard this as the vector sum of the laplacian operating on the *cartesian* components of $\mathbf{F}$:

$$\nabla^2 \mathbf{F} = \nabla^2 F_x \mathbf{i}_x + \nabla^2 F_y \mathbf{i}_y + \nabla^2 F_z \mathbf{i}_z$$

Curl

By definition, the u component of the curl $\mathbf{F}$ is the circulation per unit area in the u constant surface at the desired point. If we imagine $\mathbf{F}$ as a mechanical force, $\text{curl}_u \mathbf{F}$ is the work done per unit area by F in moving around the small closed circuit in the u surface. The situation is illustrated in Figure A.4, where we identify the four sides of the small rectangle by 1, 2, 3, and 4. The work done along 1 is $-F_w h_w \, dw$ in the counterclockwise direction. The work done along 2 in the counterclockwise direction is $F_w h_w \, dw + (\partial/\partial v)(F_w h_w \, dw) \, dv$. The work done along 3 in the counterclockwise direction is $F_v h_v \, dv$. The work done along 4 in the counterclockwise direction is $-F_v h_v \, dv - (\partial/\partial w)(F_v h_v \, dv) \, dw$. Thus the total work done or circulation of F, around the circuit is

$$\frac{\partial}{\partial v}(F_w h_w \, dw) \, dv - \frac{\partial}{\partial w}(F_v h_v \, dv) \, dw$$

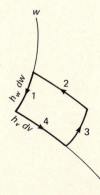

Figure A.4   Small closed circuit in u = constant surface.

This quantity, divided by dS_u or $h_v \, dv \, h_w \, dw$ is, by definition, the u component of the curl. The corresponding results for the v and w component follow immediately. Then, in summary,

$$\mathrm{curl}_u \, \mathbf{F} = \frac{1}{h_v h_w} \left[\frac{\partial}{\partial v} (h_w F_w) - \frac{\partial}{\partial w} (h_v F_v) \right]$$

$$\mathrm{curl}_v \, \mathbf{F} = \frac{1}{h_w h_u} \left[\frac{\partial}{\partial w} (h_u F_u) - \frac{\partial}{\partial u} (h_w F_w) \right]$$

$$\mathrm{curl}_w \, \mathbf{F} = h_u h_v \left[\frac{\partial}{\partial u} (h_v F_v) - \frac{\partial}{\partial v} (h_u F_u) \right]$$

Explicit forms for the three common systems are easily written by using the appropriate valves for the metrical coefficients h_u, h_v, and h_w. For example, in the cartesian system

$$\mathrm{curl}_x \, \mathbf{F} = \frac{\partial F_z}{\partial y} - \frac{\partial F_y}{\partial z} \qquad \mathrm{curl}_y \, \mathbf{F} = \frac{\partial F_x}{\partial z} - \frac{\partial F_z}{\partial x}$$

and

$$\mathrm{curl}_z \, \mathbf{F} = \frac{\partial F_y}{\partial x} - \frac{\partial F_x}{\partial y}$$

In this system it is clear that curl $\mathbf{F} = \nabla \times \mathbf{F}$, or

$$\mathrm{curl} \, \mathbf{F} = \nabla \times \mathbf{F} = \begin{vmatrix} \mathbf{i}_x & \mathbf{i}_y & \mathbf{i}_z \\ \dfrac{\partial}{\partial x} & \dfrac{\partial}{\partial y} & \dfrac{\partial}{\partial z} \\ F_x & F_y & F_z \end{vmatrix}$$

For the cylindrical system we have

$$\mathrm{curl}_\rho \, \mathbf{F} = \frac{1}{\rho} \frac{\partial F_z}{\partial \phi} - \frac{\partial F_\phi}{\partial z}$$

$$\mathrm{curl}_\phi \, \mathbf{F} = \frac{\partial F_\rho}{\partial z} - \frac{\partial F_z}{\partial \rho}$$

$$\mathrm{curl}_z \, \mathbf{F} = \frac{1}{\rho} \frac{\partial}{\partial \rho} (\rho F_\phi) - \frac{1}{\rho} \frac{\partial F_\rho}{\partial \phi}$$

and finally, for the spherical system

$$\mathrm{curl}_r \, \mathbf{F} = \frac{1}{r \sin \theta} \left[\frac{\partial}{\partial \theta} (\sin \theta \, F_\phi) - \frac{\partial F_\theta}{\partial \phi} \right]$$

$$\mathrm{curl}_\theta \, \mathbf{F} = \frac{1}{r \sin \theta} \left[\frac{\partial F_r}{\partial \phi} - \sin \theta \, \frac{\partial (r F_\phi)}{\partial r} \right]$$

$$\mathrm{curl}_\phi \, \mathbf{F} = \frac{1}{r} \left[\frac{\partial}{\partial r} (r F_\theta) - \frac{\partial F_r}{\partial \theta} \right]$$

In fact, in general, for curvilinear coordinates we can write

$$
\text{curl } \mathbf{F} = \frac{1}{h_u h_v h_w}
\begin{vmatrix}
h_u \mathbf{i}_u & h_v \mathbf{i}_v & h_w \mathbf{i}_w \\
\dfrac{\partial}{\partial u} & \dfrac{\partial}{\partial v} & \dfrac{\partial}{\partial w} \\
h_u F_u & h_v F_v & h_w F_w
\end{vmatrix}
$$

where $\mathbf{i}_u$, $\mathbf{i}_v$, and $\mathbf{i}_w$ are unit vectors in the u, v, and w directions, respectively.

Index